A Framework to Guide Selection of CHEMICAL ALTERNATIVES

Committee on the Design and Evaluation of Safer Chemical Substitutions:
A Framework to Inform Government and Industry Decisions

Board on Chemical Sciences and Technology
Board on Environmental Studies and Toxicology

Division on Earth and Life Studies

National Research Council

THE NATIONAL ACADEMIES PRESS
Washington, D.C.
www.nap.edu

THE NATIONAL ACADEMIES PRESS **500 Fifth Street, NW** **Washington, DC 20001**

NOTICE: The project that is the subject of this report was approved by the Governing Board of the National Research Council, whose members are drawn from the councils of the National Academy of Sciences, the National Academy of Engineering, and the Institute of Medicine. The members of the committee responsible for the report were chosen for their special competences and with regard for appropriate balance.

This project was supported by Contract EP-C-09-003, TO# 25 between the National Academy of Sciences and the U.S. Environmental Protection Agency Any opinions, findings, and conclusions, or recommendations expressed in this material are those of the author(s) and do not necessarily reflect the views of the sponsoring agencies or any of their subagencies.

International Standard Book Number-13: 978-0-309-31013-0
International Standard Book Number-10: 0-309-31013-X

Additional copies of this report are available for sale from the National Academies Press, 500 Fifth Street, NW, Keck 360, Washington, DC 20001; (800) 624-6242 or (202) 334-3313; http://www.nap.edu/.

Printed in the United States of America

THE NATIONAL ACADEMIES

Advisers to the Nation on Science, Engineering, and Medicine

The **National Academy of Sciences** is a private, nonprofit, self-perpetuating society of distinguished scholars engaged in scientific and engineering research, dedicated to the furtherance of science and technology and to their use for the general welfare. Upon the authority of the charter granted to it by the Congress in 1863, the Academy has a mandate that requires it to advise the federal government on scientific and technical matters. Dr. Ralph J. Cicerone is president of the National Academy of Sciences.

The **National Academy of Engineering** was established in 1964, under the charter of the National Academy of Sciences, as a parallel organization of outstanding engineers. It is autonomous in its administration and in the selection of its members, sharing with the National Academy of Sciences the responsibility for advising the federal government. The National Academy of Engineering also sponsors engineering programs aimed at meeting national needs, encourages education and research, and recognizes the superior achievements of engineers. Dr. C. D. Mote, Jr., is president of the National Academy of Engineering.

The **Institute of Medicine** was established in 1970 by the National Academy of Sciences to secure the services of eminent members of appropriate professions in the examination of policy matters pertaining to the health of the public. The Institute acts under the responsibility given to the National Academy of Sciences by its congressional charter to be an adviser to the federal government and, upon its own initiative, to identify issues of medical care, research, and education. Dr. Victor J. Dzau is president of the Institute of Medicine.

The **National Research Council** was organized by the National Academy of Sciences in 1916 to associate the broad community of science and technology with the Academy's purposes of furthering knowledge and advising the federal government. Functioning in accordance with general policies determined by the Academy, the Council has become the principal operating agency of both the National Academy of Sciences and the National Academy of Engineering in providing services to the government, the public, and the scientific and engineering communities. The Council is administered jointly by both Academies and the Institute of Medicine. Dr. Ralph J. Cicerone and Dr. C. D. Mote, Jr., are chair and vice chair, respectively, of the National Research Council.

www.national-academies.org

COMMITTEE ON THE DESIGN AND EVALUATION OF SAFER CHEMICAL SUBSTITUTIONS: A FRAMEWORK TO INFORM GOVERNMENT AND INDUSTRY DECISIONS

Members

DAVID C. DORMAN (*Chair*), North Carolina State University
ERIC J. BECKMAN, University of Pittsburgh, PA
PETER BEAK, University of Illinois, Urbana-Champaign
JEROME J. CURA, Woods Hole Group Inc., MA
ANNE FAIRBROTHER, Exponent Inc., WA
NIGEL GREENE, Pfizer, CT
CAROL HENRY, The George Washington University, DC
HELEN HOLDER, Hewlett-Packard Company, CA
JAMES E. HUTCHISON, University of Oregon
GREGORY M. PAOLI, Risk Sciences International, ON
JULIA B. QUINT, California Department of Public Health (retired)
IVAN RUSYN, University of North Carolina, NC
KATHLEEN SHELTON, DuPont, DE
JOEL A. TICKNER, University of Massachusetts, Lowell
ADELINA VOUTCHKOVA, The George Washington University, DC
MARTIN H. WOLF, Seventh Generation Inc., VT

Staff

MARILEE SHELTON-DAVENPORT, Study Director and Senior Program Officer, Board on Life Sciences
KATHRYN HUGHES, Senior Program Officer, Board on Chemical Sciences and Technology
ELLEN K. MANTUS, Senior Program Officer, Board on Environmental Studies and Toxicology
JANET MULLIGAN, Associate Program Officer, Board on Agriculture and Natural Resources
MARILYN FENICHEL, Editor, Cassell & Fenichel Communications
MIRSADA KARALIC-LONCAREVIC, Manager, Technical Information Center
IVORY CLARKE, Senior Program Assistant, Board on Environmental Studies and Toxicology

Sponsor

U.S. ENVIRONMENTAL PROTECTION AGENCY

Board on Chemical Sciences and Technology

BOARD ON ENVIRONMENTAL STUDIES AND TOXICOLOGY

Acknowledgments

This report has been reviewed in draft form by individuals chosen for their diverse perspectives and technical expertise, in accordance with procedures approved by the National Research Council's Report Review Committee. The purpose of this independent review is to provide candid and critical comments that will assist the institution in making its published report as sound as possible and to ensure that the report meets institutional standards for objectivity, evidence, and responsiveness to the study charge. The review comments and draft manuscript remain confidential to protect the integrity of the deliberative process. We wish to thank the following individuals for their review of this report:

Lawrence Barnthouse, LWB Environmental Services
Linda Birnbaum, National Institute of Environmental Health Sciences
Edward Carney, The Dow Chemical Company
Matthew Eckleman, Northeastern University
John Giesy, University of Saskatchewan
Robert Hamilton, Amway/Access Business Group
Philip J. Landrigan, Mount Sinai School of Medicine
Alexander MacLachlan, Retired, E.I du Pont de Nemours & Co.
Jennifer McPartland, Environmental Defense Fund
Warren Muir, Retired, National Academy of Sciences
Kristi Pullen, Natural Resources Defense Council
Mark Ratner, Northwestern University
Patricia Ruiz, Agency for Toxic Substances and Disease Registry
Katherine von Stackelberg, Harvard School of Public Health
James Stevens, The Dow Chemical Company
Alex Stone, State of Washington
Donna Vorhees, Boston University
Margaret Whittaker, ToxServices LLC

Although the reviewers listed above have provided many constructive comments and suggestions, they were not asked to endorse the conclusions or recommendations, nor did they see the final draft of the report before its release. The review of this report was overseen by **Joseph V. Rodricks,** ENVIRON and **Lynn R. Goldman,** The George Washington University. Appointed by the National Research Council, they were responsible for making certain that an independent examination of this report was carried out in accordance with institutional procedures and that all review comments were carefully considered. Responsibility for the final content of this report rests entirely with the authoring committee and the institution.

Contents

BOXES, FIGURES, AND TABLES

BOXES

FIGURES

TABLES

Summary

Informed chemical use in modern society should consider a variety of factors, including performance, costs, potential adverse effects to human health and the environment, and societal impacts. Chemical alternatives assessments are designed to facilitate consideration of these factors by assisting users in identifying alternative chemicals or approaches that are safer and have reduced environmental impact. The Committee on the Design and Evaluation of Safer Chemical Substitutions—A Framework to Inform Government and Industry Decisions was given the task[1] of developing a framework for assessing potentially safer substitute chemicals in terms of human health and ecological risks and demonstrating how the framework could be used. This report presents the committee's consideration of select existing frameworks, the committee's framework, and recommendations for implementation and future research needs.

STATE OF THE ART OF EXISTING FRAMEWORKS FOR ALTERNATIVE ANALYSIS

Alternatives assessment is a process for comparing alternatives, usually to a chemical of concern and identifying those that are safer. It is different from a safety assessment, where the primary goal is to ensure that exposure is below a prescribed standard; different from risk assessment, where risk associated with a given level of exposure is calculated; and different from a sustainability assessment, which considers all aspects of a chemical's life cycle, including energy and material use. The goal of an alternatives assessment is to facilitate an informed consideration of the advantages and disadvantages of alternatives to a chemical of concern, resulting in the identification of safer alternatives.

The development of this committee's framework built upon the work of regulatory agencies, academic institutions, and others who have developed alternatives assessment frameworks. The committee considered ten frameworks and approaches.[2] These frameworks share many common elements, such as assessing human health and ecological hazards, evaluating critical physicochemical properties, performing life cycle analyses, performance, and social assessments. Across these frameworks, assessments of human health hazards evaluate an array of health-related end points, including carcinogenicity, mutagenicity, reproductive and developmental toxicity, endocrine disruption, acute and chronic or repeat dose toxicity, dermal and eye irritation, and dermal and respiratory sensitization. Most frameworks also include some consideration of ecotoxicity, but the focus tends to be primarily on aquatic toxicity. Many existing frameworks compare chemicals of concern and alternatives against a series of mammalian and ecotoxicity metrics. These frameworks often use tools like the United Nations Globally Harmonized System (GHS) for the Classification and Labelling of Chemicals (GHS 2013) and the GreenScreen® for Safer Chemicals hazard assessment tool (Heine and Franjevic 2013) to classify hazards.[3]

The committee identified several elements that were often missing from existing frameworks. For example, despite the known importance of exposure, many frameworks downplay it and focus on inherent hazards of chemicals. This approach assumes that chemical alternatives would result in similar exposure levels to people, animals, and the environment and is in contrast to an approach that addresses both inherent hazard *and* exposure.

---

[1] Official Statement of Task is in Chapter 1.

[2] Frameworks and approaches considered by the committee included BizNGO Alternatives Assessment Protocol, California Safer Consumer Products Regulation, EPA's Design for the Environment (DfE) Program Alternatives Assessments, German Guide on Sustainable Chemicals, Interstate Chemicals Clearinghouse (IC2) Alternatives Assessment Guide, Lowell Center Alternatives Assessment Framework, REACH Guidance on the Preparation of an Application for Authorisation, TURI Alternatives Assessment Process Guidance, UCLA Multi-Criteria Decision Analysis, and UNEP Persistent Organic Pollutants Review Committee General Guidance on Alternatives.

[3] Classification (or benchmarking) tools provide threshold values for toxicological end points of interest, for evaluating data about effects of chemicals. These tools often result in assignment of a score (e.g., low, medium, high) that can be used to compare alternatives.

Many frameworks also do not consider the decision-making process or decision rules used for resolving trade-offs among different categories of toxicity and other factors (e.g., social impact), or the values that underlie such trade-offs. Also absent from several frameworks is the use of novel toxicity data streams, in silico computational models, and methods to estimate physicochemical information. In addition, a lack of consistency is seen in that existing frameworks provide users with a wide range of options on implementation and minimum data sets. Lack of consistency among frameworks is not unexpected given that their development is often motivated by different factors, such as regulatory pressures, industry concerns, and organizational or stakeholder drivers, which understandably affect the variables and elements considered by the author or authoring organization. Because of both gaps in framework elements within existing frameworks and lack of consistency across frameworks, the committee identified no "ideal" framework from the existing set. The existing frameworks that the committee examined, however, helped to inform the development of the framework offered in this study.

THE COMMITTEE'S ALTERNATIVES ASSESSMENT FRAMEWORK

This report provides a description of the committee's 13-step framework (Figure S-1), which is structured to support decision-making about alternatives to chemicals of concern. The framework is flexible enough for an assessor to use a *hybrid approach,* in which certain steps are completed sequentially, in parallel, or iteratively, providing an opportunity for fit-for-purpose decision making. Wherever possible, the committee's detailed guidance on the implementation of its framework is intended to provide users with that flexibility. To that end, some steps or sub-steps are considered optional, as indicated in Figure S-1. Whether or not assessments lacking certain parts of the committee's framework are acceptable will depend on the type of decision made.

Users of the Committee's Framework

The committee identified multiple audiences and users for this report, all of which would benefit from a unified approach to this challenge and a common understanding of the different processes involved in chemical alternatives assessment:

- regulatory agencies at the federal, state, local, and international level;
- industry, including small, medium, and large businesses;
- organizations encouraging the adoption of safer chemicals; and
- developers of chemicals and chemical processes.

The framework is intended to be used by a multidisciplinary team with training and expertise in toxicology (human health and ecotoxicology), chemistry, materials science, exposure assessment, and life cycle assessment. Additional expertise in engineering, social sciences, economics, and cost analysis also might be required. Assessors without such expertise, such as and small- and medium-sized firms, may need user-friendly assessment tools or technical support to carry out parts of the assessment. Examples of such tools are given throughout the report.

Summary of the Committee's Framework

The committee's alternatives assessment framework has the following main activities, including the asterisked optional activities:

- Step 1: Identify Chemical of Concern
- Step 2: Scoping and Problem Formulation
- Step 3: Identify Potential Alternatives
- Step 4: Initial Screening of Identified Alternatives
- Step 5: Assess Physicochemical Properties
- Step 6-1: Assess Human Health Hazards
- Step 6-2: Assess Ecotoxicity
- Step 6-3: Conduct Comparative Exposure Assessment
- Step 7: Integration of Information to Identify Safer Alternatives
- Step 8: Life Cycle Thinking
- Step 9-1: Additional Life Cycle Assessment*
- Step 9-2: Performance Assessment*
- Step 9-3: Economic Assessment*
- Step 10: Integrate Data and Identify Acceptable Alternatives
- Step 11: Compare Alternatives*
- Step 12: Implement Alternatives
- Step 13: Research and Innovation*

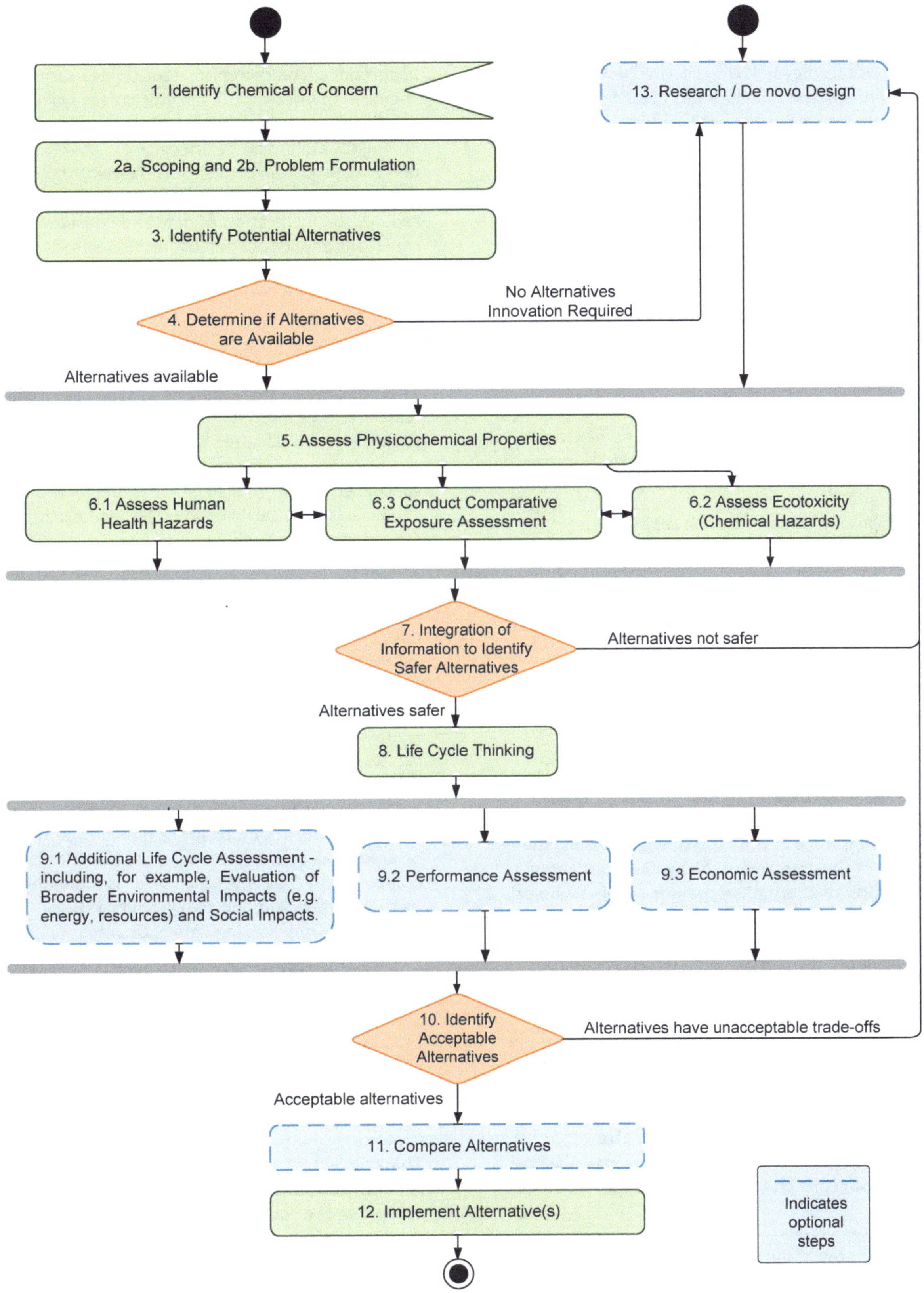

FIGURE S-1 The committee's alternatives assessment framework.

Individuals who currently use other frameworks will quickly recognize familiar elements in the committee's framework. Thus, in many ways, the committee's framework is not a revolutionary new approach; rather, it incorporates ideas from existing approaches into a flexible, inclusive framework.

Additionally, this framework includes **several important unique elements or advancements**, such as:

- a focus on scoping and problem formulation;
- an increased emphasis on comparative exposure assessment;
- increased use of physicochemical properties[4] to assess human health and ecotoxicity hazards;
- a two-tiered approach to evaluating chemical alternatives that includes health and ecotoxicity, followed by a consideration of broader impacts; and
- recognition of the need for research and innovation.

The following sections explain each of these elements in more detail.

A Focus on Scoping and Problem Formulation

An often neglected, but very important, step is that of scoping and problem formulation (Step 2). **This step defines and documents the goals, principles, and decision rules** that will guide all of the following steps in the assessment and thus, the outcome of the assessment. Many decisions about the selection of alternatives are not purely technical, but rather are value-driven or context-dependent. It is important to explicitly articulate and document those assumptions and constraints—which often take the form of decision rules that flow from an organization's goals and principles. The inclusion of a problem formulation and scoping step in the committee's framework is consistent with recent National Research Council (NRC) reports that have recommended similar efforts in other types of assessments (NRC 2014). Specifically, **the preferences of the decision maker need to be made explicit in the form of decision rules or algorithms** to be applied to resolve trade-offs across different attribute domains (e.g., toxicity, material and energy use, and cost) and address uncertainty. The committee anticipates that the chemical of interest and its alternatives will often present different hazards both across domains (e.g., ecological vs. human health hazards) as well as within domains (e.g., neurotoxicity vs. respiratory sensitization). Prioritization of alternatives will require the integration of data and consideration of trade-offs and associated uncertainties. How these trade-offs are resolved is inevitably shaped by applying goals, principles, and decision rules defined in Step 2—aspects that are not scientific judgments. **The user should also describe the decision rules used to identify a "safer" alternative.** This important description of what constitutes "safer" comes into play when considering trade-offs, as described in Chapter 9. When the alternatives assessment is striving to improve the safety of a specific end point (because, for example, the chemical is on a carcinogen list), the alternative will, for pragmatic reasons, need to be an improvement over, or no worse than, the original chemical of concern in the domain that initiated the alternatives assessment. However, a focus on a key end point does not eliminate the need for an assessment of the full range of human health hazard end points and ecotoxicity, or consideration of the life cycle of alternatives. To not include these important elements could lead to the transfer of risks to other parties (burden shifting) and other types of regrettable substitutions. Safer could also be defined in many other ways, including beneficial incremental improvements in one or more domains of interest, or an overall improvement in human health and/or ecotoxicity. What is deemed safer is ultimately context-dependent and also reflects a set of personal, corporate, legislative, or other values.

The problem formulation step (Step 2) also defines the bounds of the assessment, including identifying specific hazards of interest, and establishes the set of steps that will be required to complete the alternatives assessment. At a minimum, the committee recommends consideration of physicochemical properties, comparative exposure, ecotoxicity, human health hazards, and Life Cycle Thinking. Whether or not broader environmental impacts, such as resource use and impact on climate, are within the scope of the assessment should be decided in the problem formulation step. Consideration of economic, performance, and social impact are also optional steps that many assessors will want to consider. Problem formulation also

[4] For the purpose of this report, physicochemical properties are broadly defined as physical properties, solvation properties related to interactions with different media and properties or molecular attributes that define intrinsic chemical reactivity.

defines when and how novel data streams will be used to inform the assessment.

Within problem formulation, the committee found that **characterization of function and performance requirements** are often undervalued parts of alternatives assessment frameworks but are essential for successful prioritization and adoption of alternatives. Characterizing the function of a chemical of concern at the beginning of an alternatives assessment process can help focus the assessment on those functions provided by the chemical of concern. It can also support the identification of a broad range of viable chemical and non-chemical alternatives that meet the functional requirement of that chemical in a particular process or product. A focus on function changes the lens by which chemicals of concern are viewed, from avoidance of such chemicals to identifying the safest, most viable options to meet a particular function in a particular application.

Another crucial item that is embedded in the scoping step is **defining the role of stakeholders**. Stakeholder engagement helps ensure that the assessment will address a broad range of concerns, improves understanding and support of the assessment outcomes, and provides additional review of technical information, analytical methods, and data, improving the overall quality and accuracy of the assessment.

An Increased Emphasis on Comparative Exposure Assessment

The committee recommends an increased **emphasis on comparative exposure assessment** (Step 6.3). The committee found that most of the existing assessment frameworks it studied focus on reducing inherent hazards, with only minor considerations of exposure. The committee believes that consideration of inherent hazard can be a useful initial step for identifying safer alternatives and streamlining assessment. However, an approach that focuses on inherent hazard should only be used when a comparative exposure assessment indicates that the expected routes and amount of exposure are not expected to be substantially different between a chemical of concern and its alternatives. Thus, the committee recommends that the potential for differential exposure (in the absence of exposure-mitigating protection) between the chemical of concern and alternatives be *explicitly* considered rather than *assuming* equivalent exposure.

The committee's increased emphasis on exposure should not be interpreted as a recommendation for more comprehensive risk assessment. The committee concludes that simplified exposure estimates without elaborate exposure modeling can meet the needs of many alternatives assessments. The committee's approach allows for the use of either available exposure models or comparison of critical physicochemical properties as a way to determine the relative exposure potential of alternatives.

Elevating the Role of Physicochemical Property Evaluation

The committee's framework elevates the role of evaluation of *physicochemical properties* (Step 5) in the alternatives assessment process. The committee **broadens the consideration of physicochemical properties** beyond the current practice of evaluating physical hazards such as explosivity and corrosivity. This increased emphasis is consistent with the growing body of literature showing that a number of physicochemical properties are often predictive of ecological and human health hazards and can be used to inform data gaps and guide the chemical design process. Moreover, low-cost and reliable state-of-the-art in silico methods, which are a good source of physicochemical property data, are available to support alternatives assessments. These data also can be obtained experimentally. The physicochemical property data emphasized by the committee's framework can be used to:

- determine the environmental compartment(s) into which the chemicals will partition;
- estimate the potential for bioconcentration and bioavailability;
- estimate the likely route(s) of mammalian exposure and bioavailability; and
- estimate the likelihood for high aquatic toxicity.

A Two-Tiered Approach to Integrating Data

A two-tiered approach to integrating data on chemical alternatives (Steps 7 and 10) is described in the committee's framework. Step 7 primarily focuses on information about comparative exposure, human health, ecotoxicity, and physicochemical properties, with the goal of identifying alternatives that warrant further data gathering and analysis. In most cases, Step 7 is best considered a triage activity rather than a final ranking and selection process because it is followed by further life cycle considerations described in Step 8.

In Step 10, the entire data set for a chemical of concern and its alternatives is considered, including data from optional analyses such as environmental impact, cost, performance and social impact—factors that may require further trade-offs. All this information is added to the mix of data obtained through Step 8. The consideration of trade-offs and uncertainties may impact the identification of suitable alternatives. This process may range from being extremely simple to very challenging. Because of this complexity, as well as the value and context-dependent nature of this process, the committee does not provide a step-by-step algorithm for the completion of Step 10; rather, the committee emphasizes the need to apply the decision rules for resolving trade-offs and uncertainty that were established in Step 2. Similarly, **the committee calls for thorough documentation of the assessment methods, results, and decisions**.

The Need for Research and Innovation

The committee stressed the **need for research and innovation** in its framework (Step 13). Two types of innovation are important: the design of new chemical alternatives and the identification of ways to meet the ultimate needs of industry and the consumer using approaches other than direct chemical substitutions. In cases where no known chemical substitutions are identified, the design of new chemical alternatives by synthetic chemists and other scientists may be part of the solution. While in chemical design, it is current practice to focus on designing new molecules with better performance, **the committee recommends that safety and ecological considerations also be an integral part of early chemical design.** The committee provides specific suggestions for how to do this in Chapter 13.

SCIENTIFIC INFORMATION AND TOOLS REQUIRED TO SUPPORT THE COMMITTEE'S FRAMEWORK

Information that can be used to assess end points of interest (e.g., human health and ecological hazards) includes, but is not limited to, traditional data streams, such as measurement of physicochemical properties, human epidemiologic data, and the results of animal toxicity or ecotoxicity studies. Evaluation of results derived from traditional data streams is often supported by a variety of classification tools (e.g., GreenScreen® and the Globally Harmonized System of Classification and Labelling of Chemicals [GHS]), which categorize the available data into different levels of concern (e.g., low, moderate, high). The committee supports the use of harmonized GHS classification schemes, but **suggests short-term refinements in how they are used, such as supplementing them with additional guidance. The committee recommends more aspirational refinements as well, such as the use of novel in vitro and in silico data**. More information about the scientific information and tools is found throughout the report, as follows:

Human Health

Specifically, in the discussion of human health data (Chapter 8), the committee recommends

- Use of GHS-tied criteria with a few refinements, including using health hazard assessment guidance to classify chemicals for end points where GHS criteria require expert judgment.
- Moving beyond relying solely on traditional types of data associated with GHS or other benchmarking approaches and towards using data from novel high throughput and in silico approaches, for users with adequate scientific resources to do so. The committee specifically emphasizes greater use of available scientific information to fill data gaps when appropriate.
- The eventual development of a well-accepted classification scheme for novel types of data and in silico modeling, analogous to the GHS system, to enhance the use of this information.

Ecotoxicity

In the discussion of ecotoxicity data in Chapter 7, the committee recommends the following refinements:

- Using physicochemical data to determine which environmental compartments a chemical will partition into, and compiling ecotoxicity for these compartments.
- Using relevant high throughput data produced for human health assessment.

Incorporation of High Throughput Data

Developments in toxicity testing have changed dramatically during the past 10 years. Publication of the NRC report entitled *Toxicity Testing in the 21st Century (TT21C): A Vision and a Strategy* (NRC 2007) has spurred new approaches and thinking about chemical hazard assessment. Similarly, advances in chemistry, material sciences, and toxicology will lead to future changes in the conduct of alternatives assessments. **It is critical that the scientific community embrace the challenge and advantages of using novel data streams in the alternatives assessment process.** This report provides the committee's thinking on how these novel in vitro data streams and in silico modeling approaches can be used. In keeping with the spirit of the NRC *TT21C* report, the committee strived not to provide detailed guidance that could restrict future thinking, but rather to demonstrate how these data could be used to support informed decision making. The pharmaceutical industry's experience with integrating novel data and tools early in the product development pipelines serves as an important blueprint for this activity.

Future efforts are needed to develop principles or tools that support the benchmarking and integration of high throughput data on chemical effects, especially in the context of different regulatory requirements. This effort is needed for two types of interrelated activities; first, to address how novel data streams could be used as primary data in human health and ecotoxicity hazard assessments (e.g., the use of in vitro mutagenicity data for DNA reactive chemicals) and second, to address how these data can be used to fill data gaps across a broad range of domains, including health, ecotoxicity, exposure assessment, and physicochemistry.

The committee anticipates that, unlike benchmarking of animal and ecotoxicity data, which have a manageable range of end points and outcomes, the approaches used for novel data streams, especially the broad range of end points provided by high throughput assays, may be less amenable to a formal, endpoint-driven GHS-like classification scheme. Instead, user-defined decision rules and principles will likely guide incorporation of these data into the alternatives assessment process. As a result, the expert, judgment-guided discussions with regulatory bodies may not follow an identical template for all types of chemical alternatives assessments.

Other Considerations

In keeping with the theme of transparency and documentation described earlier, the committee notes the importance of tools to improve communication of assessment methods and information to all stakeholders. Tools that transparently capture how data are considered and integrated into the assessment process, as well as tools to help visualize new types of data, will be critical to facilitating communication of the complex information on chemical alternatives.

The committee's framework is designed to accommodate the advances in tools, including those developed for mixtures and high throughput data, that surely lay ahead, and to allow for the integration of information from a variety of scientific disciplines. The case studies described in Chapter 12 demonstrate how high throughput data and other computational approaches can be used to complete certain steps in the committee's framework. The committee recognizes that the application of these methods may be beyond the scientific capacities of some users, particularly small- and medium-sized companies. Thus, the committee recognizes the importance of developing new tools, education, and technical support networks to assist entities with less capability in implementing novel data streams into the alternatives assessment process.

1

Introduction

There is a rich and growing literature on chemical substitution that dates from the early 1990s, when scientists and regulators in Europe and the United States (U.S.) began to categorize and prioritize chemicals of concern. Chemical alternatives assessment emerged from these regulatory efforts. It refers to a process for identifying, comparing, and selecting safer alternatives to chemicals of concern. The goal of chemical alternatives assessment is facilitating an informed consideration of the advantages and disadvantages of alternatives to a chemical of concern. Over time, government agencies, academic institutions, and professional organizations developed different alternatives assessment frameworks, each with a particular focus. The results from these assessments varied depending on whether the emphasis was on protecting workers, communities surrounding industrial plants, end users of products, or other interests.

RECENT DRIVERS RESULTING IN CHEMICAL ALTERNATIVES ASSESSMENT

Historically, regulations governing chemical use have often focused on the effects of widely used chemicals on human health including their potential to cause cancer and other adverse health effects. As scientific knowledge has expanded, awareness of the mechanisms through which chemicals may exert harmful effects on human health has increased, along with an understanding of their effects on other species and ecosystems. At the same time, many factors, including unprecedented access to information on the internet, have resulted in greater public awareness of potential hazards in the products they use. Along with scientific advances and public awareness, the U.S. Centers for Disease Control and Prevention (CDC) and the U.S. Environmental Protection Agency (EPA) are collecting more information on U.S. citizens' exposure to chemicals. For example, the CDC's Fourth National Report on Human Exposure to Environmental Chemicals published information about the levels of 212 xenobiotic compounds (substances that are not produced by the body) or metabolites in the blood and urine of U.S. study participants (CDC 2009). The report revealed widespread exposure to some commonly used industrial chemicals found in household products, including polybrominated diphenyl ethers (PBDEs), bisphenol A (BPA), and perfluorinated chemicals.

Certain regulatory agencies have identified so-called priority chemicals, those considered to be carcinogenic, mutagenic, reproductive toxicants, and/or fall into the category of PBTs: persistent, bioaccumulative, and toxic chemicals. Many of these chemicals are associated with industrial waste; can contaminate soil, sediment, groundwater, surface water, and air; and are found in plant, animal, and human tissue. In the U.S., examples of priority chemicals may be found on lists developed by some states, including Washington State (Reporting List of Chemicals of High Concern to Children) (WA Department of Ecology 2014) and California (Candidate Chemicals List) (CA DTSC 2010), the EPA's National Waste Minimization program's list of priority chemicals (EPA 2012a), and on lists developed by environmental action groups, retailers, and many manufacturers. The European Union's Candidate List of Substances of Very High Concern for Authorisation (ECHA 2014a) serves a similar purpose abroad. High-priority chemicals are frequent targets for alternatives assessments. Identification of high-priority chemicals and other chemicals of concern has prompted a growing number of state and local governments, as well as major companies, to take steps beyond existing hazardous chemical federal legislation. Between 1990 and 2009, at least 18 states, 6 counties, and 6 city governments enacted laws restricting PBDEs, BPA, lead, chromated copper arsenate, phthalates, dioxin, perchloroethylene, or formaldehyde (Edwards 2009). For example, the Safer Consumer Product Regulations were developed by California's Department of Toxic Substances Control to require manufacturers and other responsible entities to "seek safer alternatives to harmful chemical ingredients in widely used products" (CA DTSC 2013a). Europe's Registration, Evaluation, Authorisation and Restriction of Chemicals (REACH) Substances of Very High Concern list (ECHA 2014b) and Canada's Chemicals Management Plan (Government of Canada 2014) are also driving chemical substitution. In addition, several non-governmental organizations (NGOs) are raising awareness of the need for chemical substitutions, and have developed approaches that have informed alternatives assessments. These efforts include Clean Production Action's GreenScreen® for Safer Chemicals, which is explained later in this report.

In response to these drivers, major companies and retailers (e.g., Bissell, Dell, Hewlett-Packard (HP), Herman Miller, K-Mart, Nike, S.C. Johnson, Sears, Toys R Us, Wal-Mart, Whole Foods, and Volvo) and collective industry efforts (such as the textile industry's Zero Discharge Coalition and the building industry's LEED certification program) have adopted policies to eliminate or phase out particular chemicals. Other manufacturers report they will go beyond regulatory restrictions in selecting the chemicals they will use (Lavoie et al. 2010) as part of their sustainability programs. Other retailers certify that the products they sell exhibit superior environmental performance. Collectively, these activities represent a trend toward more market- and product-based considerations of chemical safety.

Interest in approaches and policies that ensure that any new substances substituted for chemicals of concern are assessed as carefully and thoroughly as possible has also burgeoned (Hogue 2013). The overarching goal of these approaches is to avoid regrettable substitutions. Regrettable substitutions occur when a toxic chemical is replaced by another chemical that later proved unsuitable because it, too, turned out to be a PBT, or because of other concerns. One example of a regrettable substitution occurred in the 1990s and involved the replacement of methylene chloride with *n*-hexane in automotive brake cleaners. Although *n*-hexane performed well as a brake cleaner, some auto mechanics exposed to *n*-hexane developed peripheral neuropathy (Wilson et al. 2007). Similarly, recent research has raised concerns about the toxicity and estrogenic activity of plastic materials that served as a replacement for BPA (Kuruto-Niwa et al. 2005; Viñas and Watson 2013).

GOVERNMENTAL EFFORTS TO DRIVE ADOPTION OF SAFER CHEMICALS

The U.S. government initiated efforts to drive adoption of safer chemicals as early as the 1950s (Lofstedt 2014). Over the years, both regulatory and non-regulatory policies have been enacted that require, conduct, or support the development of chemical alternatives assessments. U.S. government efforts include EPA's Significant New Alternatives Policy (SNAP) program (EPA 2014a), which requires companies to seek approval for substitution of ozone-depleting substances. Also, the Pollution Prevention Act of 1990 includes "reformulation or redesign of products [and] substitution of raw materials" as an approach to reduce sources of pollution. The EPA's Chemical Management program (EPA 2013a) is also affecting chemical substitution. Since December 2009, EPA has published action plans for ten chemicals or chemical classes, which include various recommendations for rule making under the 1976 Toxic Substances Control Act (TSCA) and recommendations for conducting alternatives assessments under EPA's Design for the Environment program, which is specifically for BPA, PBDEs, hexabromocyclododecane, and phthalates (EPA 2012b). With a work plan for 83 chemicals, EPA now has a guide that will be used to focus its activities over the next several years.

At the state and local level, many jurisdictions have enacted requirements that government suppliers report on chemicals of concern and make substitutions. This practice enables government agencies to "lead by example" by using the least toxic alternatives for a particular chemical or product class. Examples of policies that establish requirements for use of safer alternatives in procurement include New York Executive Order No. 4, Establishing a State Green Procurement and Agency Sustainability Program. This policy "directs state agencies, public authorities and public benefit corporations to green their procurements and to implement sustainability initiatives" and establishes processes for agencies to follow in identifying preferred products, such as cleaning products. It also includes a list of chemicals to avoid when making purchasing decisions.

Also notable is the establishment of new organizational structures for government agencies that enable them to collaborate and share information on chemicals and alternatives and develop consistent approaches. For example, the Interstate Chemicals Clearinghouse (IC2) is an association of state, local, and tribal governments that shares information on chemical hazards and priorities, chemical use in products, and safer alternatives. One of the organization's goals is to develop consistent frameworks for assessing chemical alternatives.

In addition to U.S. efforts, other countries have developed regulations that include the substitution principle and require industry to transition to safer alternatives if they are available. In addition to the European REACH program and Canada's Chemicals

Management Plan mentioned above, the Swedish Non-Toxic Environment program (KEMI 2014) is another example. These policies and programs stipulate that replacements should be made, even in the absence of quantitative risk estimates, if changing a chemical substance or its design can reduce risks to the environment and human health (Hansson and Ruden 2007).

GROWTH IN EVALUATIVE APPROACHES

Over the past decade, the number of approaches for evaluating chemical toxicity has grown substantially in response to many factors, ranging from advances in molecular biology to public pressure. Alternatives assessment policies have also evolved as governments grapple with developing procedures to avoid regrettable substitutions. Earlier alternatives assessment policies did not always address the issue of which alternatives should be allowed to replace a chemical of concern or how alternatives should be evaluated.

TSCA Reform and the EPA's Development of Tools

Running in parallel with other efforts to drive safer chemical adoption are attempts to reform TSCA. This law "authorizes the EPA to regulate chemicals that pose an unreasonable risk to human health or the environment" (GAO 2005). However, the agency has had difficulty demonstrating that specific chemicals pose an unreasonable risk, leading to questions about whether TSCA provides the agency with enough regulatory force to protect people and the environment against chemical hazards.

In recent years, the EPA has begun implementing new ideas for managing toxic chemicals under its existing TSCA authority, drawing on more than 20 years of scientific effort to develop tools to predict toxicity of chemicals. The EPA, in collaboration with other federal entities (Collins et al. 2008), is also conducting research and developing toxicity testing and in silico[5] approaches to characterize, predict, and communicate the potential of existing and new chemicals to pose human health and ecological risks. These data, methods, and tools are resulting in an increased ability to conduct chemical alternatives assessments. Because the universe of untested chemicals is vast, even if TSCA is eventually reformed, the approaches being developed are likely to be used by the EPA and other stakeholders, including industry, to ensure chemical safety in the short and intermediate term. By making more chemical data, including information about exposure, hazard, and dose-response relationships, more easily accessible through a variety of databases and dashboards, the EPA is improving the ability of interested parties to evaluate chemical substitutes. Having more institutions and companies complete chemical substitution assessments helps enhance the EPA's objective of ensuring safer chemistry.

Green Chemistry

Another influence on how alternative chemicals are considered is the growing "green chemistry" movement, recognized through the EPA's Presidential Green Chemistry Challenge Awards. These awards recognize the use of green chemistry, defined by the EPA as the "design of chemical products and processes that reduce or eliminate the use or generation of hazardous substances" (EPA 2014b). One goal of green chemists is to design new chemicals that are inherently safer. This involves a consideration of safer chemical synthesis approaches, the environmental and biological fate of chemicals, and how and where a chemical is transported. According to Paul Anastas, one of the green chemistry movement's advocates, chemists who follow these principles can simultaneously "bring about environmental improvement benefiting human health and economics and profitability" (Harris 2012).

THE COMMITTEE'S TASK

Members of the Committee on the Design and Evaluation of Safer Chemical Substitutions—A Framework to Inform Government and Industry Decisions were selected for their expertise in chemistry, chemical engineering, computational modeling, toxicology, ecotoxicology, risk assessment, and public health. The committee was specifically asked to accomplish the following task:

> *An ad hoc committee shall develop and demonstrate a decision framework for evaluating potentially safer substitute chemicals as determined by human health and ecological risks. The committee shall*

---

5 The term *in silico* is used in this report to describe prediction and modeling (typically computational modeling) of effects based on information about a chemical's structure or physicochemical characteristics, including but not limited to structural alerts and structure-activity relationship analysis.

identify the scientific information and tools required by regulatory agencies and industry to improve and increase consideration of potential health and environmental impacts early in the chemical design process. The decision framework shall be capable of integrating multiple and diverse data streams to support early consideration of potential health and environmental impacts as a part of fit-for-purpose decision making.

The framework shall discuss how risk (hazard and potential for human exposure and toxicity) and environmental impact (ecological risks) can be characterized for chemical substitutions within the context of the full range of benefits and shortcomings of substitutes, and how tradeoffs between these risks and factors such as product functionality, product efficacy, process safety and resource use can be quantified.

In its report, the committee shall describe the framework and provide at least two examples that demonstrate how different users in contrasting decision contexts with diverse priorities can apply the framework. These examples shall include demonstration of how high throughput and high content data streams could inform assessment of potentially safer substitutes early in the chemical development process.

Approach to the Study

Two recent National Research Council (NRC) reports that explored new approaches to assessing chemical safety influenced the committee's development of its alternatives assessment framework. The NRC's 2007 report, *Toxicity Testing in the 21st Century: A Vision and a Strategy*, provides a synopsis of how advances in systems biology, in vitro testing in cells and tissues, and related fields could fundamentally change chemical hazard assessment. This new approach to toxicity testing shifts the focus from animal studies to the use of human cells or cellular components (i.e., in vitro testing) to study chemicals' effects on biological processes. While this approach is not without its critics, the report (NRC 2007) and its advocates state that it has the potential to provide information about toxicity much more quickly than conventional animal-based testing.

The committee also considered the NRC's 2009 report, *Science and Decisions: Advancing Risk Assessment*, which concluded that the risk assessment process used by the EPA to estimate the effects of exposure to chemicals was often hindered by disconnects between available scientific data and the information needs of regulators. The report recommended that the EPA streamline the risk assessment process to allow for the appropriate use of available scientific data and ensure that assessments are tailored to meet the specific needs of the problem. To do so, the report recommended that the EPA adopt a three-phase framework that begins with enhanced problem formulation and scoping, a step that identifies the types of technical analyses needed to evaluate and discriminate among the available risk management options (NRC 2009).

In evaluating the literature, the committee found that various definitions have been applied to the terms *alternatives assessment* and *alternatives analysis*. For this report, the committee has used these terms interchangeably to describe the framework for safer chemical substitutions as a structured approach for considering human health and environmental hazards associated with different chemicals or chemical-dependent processes. Safer chemical substitutions can involve two chemical-based approaches: (1) substituting a chemical with another existing one or (2) synthesizing a new chemical to meet the original chemical's functional role. The second approach illustrates how the principles of green chemistry have become an integral component of alternatives assessment. The committee's framework incorporates elements of this philosophy.[6]

Many assessments focus on the intended use or functionality of the chemical (e.g., surfactant, solvent, anti-oxidant). In these cases, manufacturers and other parties select chemical alternatives to obtain the same or similar functionality. The ultimate goal of this process is to lessen the risk by reducing the inherent hazard associated with a chemical or chemical-based process. In some cases, manufacturing or synthetic methods can be redesigned in order to remove the need for a hazardous chemical or process. Therefore, the committee also sought to develop a framework that

---

[6] Although changes to materials or designs might also provide alternatives to chemicals of concern, the Statement of Task specifically directs that the framework should address safer substitute chemicals, and thus the committee's framework is focused on the case of chemical substitution. Finding a non-chemical approach to achieve the desired function was not the committee's focus but is touched on in Chapter 13.

could consider the intended use of a chemical in a manufacturing process or end product.

Box 1-1 is a more detailed description of the committee's definition of alternatives assessments and chemical substitution.

The committee also developed working definitions for the following terms that are used throughout the report:

- Framework: As used by the committee, a *framework* is a high-level organizational approach to rigorously compare chemical alternatives to determine which substitute(s) are safer. A framework conventionally involves a sequential series of steps or a process flowchart. Both the decision points and the order that the steps must be carried out are generally fixed. Frameworks for chemical substitution include steps and may prescribe which tools are used. Some frameworks disclose their underlying logic.
- Step: A *step* is a series of task(s) that need(s) to be completed in a given step or box in the analysis framework. A step is often an established method or approach that also can be used—and is usually valid—as a stand-alone analysis. Examples of steps include performance assessment, hazard assessment, analyses of cost and availability, analysis of life cycle impacts, and assessments of social impacts.
- Tools: The technical methods, approaches, software, or databases used to execute each step in the committee's Safer Chemical Substitution framework are considered *tools*. Which tools can be used to complete a given step may or may not be defined by the framework. Examples of applicable tools include the freely available GreenScreen® for Safer Chemicals, which can be used for hazard screening, and SimaPro, which can be used for evaluating life cycle impacts.
- Transparency: The committee adapted the EPA's description of transparency in risk assessment to alternatives assessment: Transparency is "fully and explicitly disclos[ing] the assessment methods, default assumptions, logic, rationale, extrapolations, uncertainties, and overall strength of each step in the alternative assessment" (EPA 2012c).

Transparency promotes broad participation by stakeholders in the alternatives assessment. The committee recognizes that while transparency is a goal to strive for, it cannot always be expected from private entities. In any case, the committee calls for internal documentation of the assessment methods, default assumptions, logic, rationale, extrapolations, uncertainties, and overall strength of each step in the alternatives

BOX 1-1
CHEMICAL ALTERNATIVES ASSESSMENT

What is a Chemical Alternatives Assessment?

The committee defined alternatives assessment as a process for identifying, comparing, and selecting safer alternatives to chemicals of concern on the basis of their hazards, comparative exposure, performance, and economic viability.[a] A chemical of concern can be a chemical in any material, process, or technology. A safer alternative represents an option that is less hazardous to humans and the environment than the existing chemical or chemical process. A safer alternative to a particular chemical of concern may include a chemical substitute or a change in materials or design that eliminates the need for a chemical alternative.

The Differences between Alternatives Assessment and Other Approaches

To further clarify its task, the committee noted the differences between an alternatives assessment and other approaches. The definitions below explain three other assessments used. Typically, alternatives assessments do *not* include these factors.

- A *safety assessment* is when the primary goal is to ensure that exposure to a particular substance is below some prescribed standard.
- A *risk assessment* is a calculation of the risk associated with a given level of exposure.
- A *sustainability assessment* examines all aspects of the life cycle of a chemical and alternatives, including energy and material use. Ideally, in an alternatives assessment, it is important to at least consider all life cycle segments that would be affected by chemical substitutions to get the most comprehensive view of potential impacts and trade-offs. However, such a detailed assessment is rarely attainable given the limits in current life cycle assessment tools and could potentially lead to inaction.

[a] This definition, with the addition of comparative exposure, builds upon but significantly modifies the definition from the meeting, Building a Chemical Commons: Data Sharing, Alternatives Assessment and Communities of Practice (BizNGO 2013).

assessment even if the documentation is not publicly disclosed.

The committee also considered existing alternatives assessment frameworks and tools. Rather than conduct a systematic review of the literature, the committee took advantage of several recently published reviews. For example, several frameworks and tools were identified in the Organisation for Economic Development (OECD) report, *Current Landscape of Alternatives Assessment Practice: A Meta-Review* (OECD 2013a). In this report, the OECD's Ad Hoc Group on Substitution of Harmful Chemicals compiled extensive information on frameworks, methods, and tools that can be used for assessing alternatives to chemicals of concern (OECD 2013a). Another recent literature review examined more than 20 alternatives assessment frameworks (Edwards et al. 2011). Based on the committee's analysis of these reviews, a subset of existing frameworks were identified for more detailed consideration; this selection was based on: (1) availability in the public domain, (2) consideration of one or more elements (e.g., human toxicity, ecotoxicity) deemed important to the committee, and (3) use by one or more regulatory body.

Frameworks considered by the committee included:

- BizNGO Alternatives Assessment Protocol (Rossi et al. 2012)
- California Safer Consumer Products Regulation (CA DTSC 2013a)
- Design for the Environment Chemical Alternatives Assessments (EPA 2014c)
- German Guide on Sustainable Chemicals (Reihlen et al. 2011)
- Interstate Chemicals Clearinghouse Alternatives Assessment Guide (IC2 2013)
- Lowell Center Alternatives Assessment Framework (Rossi et al. 2006)
- REACH Guidance on the Preparation of An Application for Authorisation (ECHA 2011)
- TURI Alternatives Assessment Process Guidance (TURI 2006a)
- UNEP Persistent Organic Pollutants Review Committee General Guidance on Alternatives (UNEP 2009)

In addition to these frameworks, the committee considered two tools. The committee looked at the GreenScreen® for Safer Chemicals tool in detail because it is integral or related to several of the frameworks and is specifically intended for comparative chemical hazard assessment (Clean Production Action 2014). The committee also considered the University of California at Los Angeles (UCLA) multi-criteria decision analysis tool (Malloy et al. 2011).

The structure of each framework was evaluated and helped guide the development of the committee's framework. Throughout the process, several key decisions, listed below, were made, which also helped determine the framework's structure.

- The framework is to be used by a multidisciplinary team of individuals with training and expertise in toxicology (human health and ecotoxicology), exposure assessment, chemistry, and life cycle assessment. Additional expertise in engineering, epidemiology, social sciences, economics, and cost analysis may also be required. Assessors without such expertise, such as small- and medium-sized firms, may need user-friendly assessment tools or technical support to carry out parts of the committee's framework.
- The framework should provide maximum flexibility to the user while identifying critical steps that should be retained in all alternatives assessments.
- The framework should not be overly prescriptive by specifying all steps or tools needed to conduct an alternatives assessment. This approach provides the greatest flexibility to end users, allowing them to incorporate different steps and tools into the framework.
- The committee decided to focus its attention on technical aspects of the framework rather than offer opinions on policy decisions that are inherent in alternatives assessment.
- Certain activities, while important to alternatives assessment, were deemed to be beyond the scope of the current project or were not well suited to the committee's scientific expertise. The committee provides sufficient information for the reader to understand the general approach needed, but is directed to more detailed references for additional information on topics such as:
    - The criteria and processes used for identifying chemicals of concern.

- A complete discussion of life cycle analysis (LCA) practice.
- A detailed guidance on conducting economic or social impact assessments.

Organization of Report

The report is organized into 13 chapters and four appendices. Chapter 2 provides an overview of the frameworks that were considered by the committee. Chapter 3 introduces the overall structure of the committee's framework. Chapter 4 details the initial steps (scoping, problem formulation, and initial screening) of an alternatives assessment, after a chemical of concern has been identified. Chapter 5 addresses physicochemical properties that should be considered during an alternatives assessment. Chapter 6 presents the concept of comparative exposure, a key part of the committee's framework that differentiates it from other approaches. Chapters 7 and 8 address hazard assessment for ecotoxicity and human health, respectively. Chapter 9 discusses how to integrate the information about the chemical and its potential alternatives to make informed decisions. This is followed by Chapter 10, which presents an overview of contextual information that the committee did not comment on in great detail, including how to consider the impact of alternatives at various stages of the life cycle and impacts that are broader than human and ecological hazard. Chapter 11 describes the final steps in the framework: identifying acceptable alternatives, selecting final or preferred ones from the options, and implementing the selected alternatives. In Chapter 12, two examples of how to implement the committee's thinking are presented in an alternatives analysis of glitazone and decabromodiphenyl ether. Finally, Chapter 13 describes innovation in process and chemical design, including specifics on how to consider properties up-front when developing new chemical entities. Appendix A provides biographic information on the committee. Appendix B accompanies Chapter 6 and provides an overview of how other frameworks considered ecotoxicity. Appendix C describes the visualization tool ToxPi. Appendix D is a supplement to Chapter 8, providing additional information on the United Nations Economic Commission for Europe's Globally Harmonized System of Classification and Labelling of Chemicals (GHS).

2

Existing Frameworks and Approaches

The literature base for alternatives assessment and chemical substitution describes how methods, substitution guidance, and case studies have developed, leading to the availability of different approaches. These approaches tend to have varying sets of criteria with different weighting systems for evaluating current chemicals and possible substitutes. For example, some manufacturers may seek substitutes for priority or controversial chemicals and precursor materials that appear on a list developed by one or more regulatory agencies ("list-based" alternatives assessment). Other manufacturers report that they go beyond regulatory restrictions in selecting the chemicals they will use (Lavoie et al. 2010) as part of their sustainability programs. Retailers may seek to certify that the products they sell exhibit superior environmental performance. Different assessment frameworks can yield different results depending on the focus of the framework.

SPECIFIC FRAMEWORKS CONSIDERED BY THE COMMITTEE

As discussed in Chapter 1, for a more detailed consideration, the committee identified a subset of publicly available frameworks and approaches used to conduct alternatives assessments. Several frameworks were identified in the Organisation for Economic Development (OECD) report, *Current Landscape of Alternatives Assessment Practice: A Meta-Review* (OECD 2013a). In this report, the OECD's Ad Hoc Group on Substitution of Harmful Chemicals compiled extensive information on frameworks, methods, and tools that can be used for assessing alternatives to chemicals of concern. The primary attributes of these frameworks and approaches are presented in Table 2-1 and described here.

BizNGO Alternatives Assessment Protocol: The Business-Nongovernmental Organization Working Group's BizNGO Chemical Alternatives Assessment Protocol (BizNGO CAAP[7]) became publicly available in 2011. The Business-NGO Working Group, a project of the non-profit Clean Production Action organization, designed the protocol to codify practices that have been shown to work well for businesses that are "downstream users" of chemicals. According to the Business-NGO Working Group, such businesses are not invested in the use of any particular chemical but rather tend to focus on the function the chemicals provide to achieve product performance (Rossi et al. 2012).

The BizNGO CAA protocol (Rossi et al. 2012) is based around a 7-step decision tree. The BizNGO CAA protocol recommends ordered steps for carrying out an alternatives assessment without prescribing how to carry out each step (OECD 2013a). For example, the protocol includes life cycle assessment and risk assessment as two separate steps, noting that they are not always necessary or appropriate for selecting an alternative (OECD 2013a). The protocol calls for applying *Life Cycle Thinking* to identify concerns related to potential substitutes' life cycle and exposure.

California's Alternatives Analysis program joins alternatives assessment to a decision process for selecting a course of action intended to decrease toxic threats (Kuczenski et al. 2010). California's 2008 Safer Consumer Product laws[8] require manufacturers or other responsible entities to seek

---

[7] The BizNGO CAAP (Rossi et al. 2012) builds upon many existing frameworks, including: the Lowell Center for Sustainable Production's Alternatives Assessment Framework (Rossi et al. 2006); the U.S. Environmental Protection Agency Design for the Environment program's Alternatives Assessment framework (EPA 2014c); the United Nations Environment Program Persistent Organic Pollutants Review Committee's "General guidance on considerations related to alternatives and substitutes for listed persistent organic pollutants and candidate chemicals" (UNEP 2009); the methodology derived from the University of Massachusetts at Lowell's Toxic Use Reduction Institute's (TURI) 2006 Five Chemicals Alternatives Assessment study (TURI 2006a); and the Interstate Chemicals Clearinghouse's (IC2) Safer Alternatives Assessment Wiki (IC2 2014).

[8] California's 2008 Safer Consumer Product Regulations (California's Assembly Bill 1879, or CAB 1879). The regulations took effect on October 1, 2013. Article 5 (California 1879 article 5) codifies the state's approach to safer chemical substitutions.

safer alternatives to harmful chemical ingredients in widely used products (CADTSC 2013a). The overarching goal of the regulations is to create a predictable and systematic process for reducing toxic ingredients in consumer products (Kuczenski et al. 2010). The law prescribes which elements need to be included in identifying and evaluating safer chemical substitutions and engineering design alternatives, including analyses of use-based exposure and risk, cost and availability, life cycle impacts, and social impacts. California's Alternatives Analyses includes two required phases. The first phase is a screening process focusing on identifying what alternatives will be considered and asking whether the chemical itself or a replacement chemical or design is necessary to achieve the function of the chemical of concern.

The second phase takes a much more in-depth look at the alternatives. Several evaluation modules with methods for examining exposure pathways and life cycle phases are included in this second step. The state mandates that a large number of different criteria be evaluated in its chemical alternatives analyses, using methods that are transparent and well documented. To support implementation of the process, the state is producing guidance for alternatives analysis.

Design for the Environment Chemical Alternatives Assessments: The U.S. Environmental Protection Agency (EPA) Office of Chemical Safety and Pollution Prevention (OCSPP) created a Design for the Environment (DfE) Program Alternatives Assessment framework in 2011 (EPA 2014). This 7-step framework was developed with input from the agency's Toxic Substances Control Act (TSCA) New Chemicals Program and DfE's Cleaner Technology Substitutes Assessments.

EPA's DfE's alternatives assessment process includes specific guidelines for evaluating chemicals for carcinogenicity, mutagenicity, reproductive and developmental toxicity, acute and repeat dose toxicity, toxicity to aquatic organisms, and environmental fate (Whittaker and Heine 2013). DfE has also developed specific Criteria for Hazard Evaluation (EPA 2011a), which define low, moderate, and high hazard designations for alternatives assessments. Both experimental and modeled data can be used in assigning these hazard designations. In the absence of experimental data, measured data from a suitable analog are preferred over estimated data (Whittaker and Heine 2013).

EPA has applied its DfE alternatives assessment methodology to nonylphenol ethoxylates (NPEs), surfactants, flame retardants in furniture and printed circuit boards, and decabromodiphenyl ether (decaBDE) in building materials, textiles, wiring insulation, and plastics. The agency is currently assessing alternative chemicals that can be used in place of certain phthalates, BPA in thermal paper, and hexabromocyclododecane in expandable foam for insulation.

German Guide on Sustainable Chemicals: The German Federal Environmental Agency's Guide on Sustainable Chemicals (German Guide) (Reihlen et al. 2011) is intended to help business enterprises systematically implement sustainable chemistry in their daily practice. Published in 2011, it includes specific guidelines for evaluating intrinsic chemical hazards and analyzing social and life cycle impacts (OECD 2013a).

Interstate Chemicals Clearinghouse (IC2) Alternatives Assessment Guide: The Interstate Chemicals Clearinghouse's (IC2) developed the Alternatives Assessment Guide (IC2 2013) based on input from experts from California, Connecticut, Michigan, Minnesota, New York, and Oregon and is funded by a grant from the EPA to Washington State.

The IC2 Alternatives Assessment Guide focuses on "reducing risk by reducing hazard" (OECD 2013a). The guidance includes a set of principles for alternatives assessments (OECD 2013a), and three decision-making framework options: sequential, simultaneous, and hybrid (IC2 2013). IC2's frameworks stand out for including flexibility as a principle and mentioning the role of green chemistry as an approach for designing safer chemicals (OECD 2013a). The IC2 framework includes seven modules, each evaluating a different consideration for assessing potential alternatives, which users can choose among to conduct an assessment. It also outlines the minimum set of modules that are recommended for a good alternatives assessment. IC2 has also created a Safer Alternatives Assessment Wiki (IC2 2014) to share resources and approaches.

Lowell Center Alternatives Assessment Framework: The University of Massachusetts at Lowell's Center for Sustainable Production's Alternatives Assessment Framework (Lowell AAF) (Rossi et al. 2006) grew out of a 2004 workshop (Lowell 2005) and builds on a methodology developed at the Center's sister organization, the University of Massachusetts at Lowell's Toxics Use Reduction Institute (TURI) (Rossi et al. 2006).

Like the Biz-NGO framework, the Lowell AAF lays out a series of steps and modules to evaluate alternatives, but does not specify methods or tools for completing analyses. The framework is intended to facilitate the relatively quick assessment of "safer and more socially just alternatives to chemicals, materials, and products of concern" (Rossi et al. 2006). It was created to be an open-source approach to foster collaborative development, sharing, and growth of methods, tools, and databases that facilitate decision-making.

REACH Guidance on the Preparation of an Application for Authorisation: The European Chemicals Agency's Chemical Safety Assessment protocol's Guidance on the Preparation of an Application for Authorisation (ECHA 2011) is intended to support the implementation of European Union (EU) Registration, Evaluation, Authorisation, and Restriction of Chemicals (REACH) regulations (ECHA 2014b). REACH has been a major governmental driver for chemical substitution (Tickner et al. 2013). It requires that manufacturers, importers, and downstream users seeking authorization to use identified "chemicals of concern" conduct an assessment of the chemical alternatives. Where the analysis demonstrates that suitable substitutes exist, the applicant must develop a timetable for proposed actions. Based on the suitability of the alternatives, EU government authorities determine whether or not they will continue to authorize applicants to use the substance(s) of concern.

The guidance details how to prepare chemical safety authorization applications, including alternatives assessments. Once possible alternatives have been identified, it specifies that the analysis should involve assessing the alternatives for: technical feasibility; potential risks to the environment and human health; economic feasibility; suitability and availability; as well as identifying relevant research and development. The recommendations for how to assess alternative chemicals' costs, performance, and socioeconomic impact are particularly detailed (OECD 2013a).

TURI Alternatives Assessment Process Guidance: The University of Massachusetts at Lowell's TURI was established as part of a 1989 Massachusetts law requiring manufacturing firms to undertake toxics use reduction planning. In 2005, the Massachusetts state legislature requested that TURI evaluate alternatives to five chemicals of concern. TURI's Alternatives Assessment Process Guidance (TURI 2006a) is an outgrowth of the resulting Five Chemicals Alternatives Assessment Study (TURI 2006b). The objective of the guidance document was to define a consistent process for setting priorities, studying and evaluating the alternatives for the five chemicals (lead, formaldehyde, perchloroethylene, hexavalent chromium, and di (2-ethylhexyl) phthalate [DEHP]) (TURI 2006a). The document recommends steps for carrying out an alternatives assessment without prescribing how to carry out each step (OECD 2013a).

UNEP Persistent Organic Pollutants Review Committee General Guidance on Alternatives: The United Nations Environment Programs (UNEP) Persistent Organic Pollutants Review Committee's "General guidance on considerations related to alternatives and substitutes for listed persistent organic pollutants and candidate chemicals" (UNEP General Guidance on Alternatives) (UNEP 2009) was adopted in 2009. Similar to BizNGO and Lowell, the UNEP guidance suggests a series of steps that can be used to assess potential alternatives to persistent organic pollutants (POPs), and provides narrative guidance on how each step might be executed. It also provides examples of ways to present results from the assessment, but does not give guidance on weighting of factors or resolving trade-offs between different domains, except to require the screening out of other POPs.

SPECIFIC TOOLS CONSIDERED BY THE COMMITTEE

GreenScreen® for Safer Chemicals: In addition to the frameworks above, the committee considered the GreenScreen® for Safer Chemicals tool because it is integral or related to several of the frameworks and is specifically intended for comparative chemical hazard assessment (Clean Production Action 2014; GreenScreen® hazard assessment tool; Heine and Franjevic 2013). GreenScreen® was developed by Clean Production Action, an organization developing tools and strategies in the green chemical space. GreenScreen® is a tool for "benchmarking[9]" the data on chemicals' ecotoxicity and human health hazard data. Benchmark 1 is "Avoid chemicals of high concern." Benchmark 2 is "Use but search for safer substitutes." Benchmark 3 is "Use but still opportunity for improvement." Finally, Benchmark 4 is "Safe chemical." Specific hazard and assessment criteria are defined for each of these benchmarks, as described in Chapter 8.

[9] Classification (or benchmarking) tools provide threshold values for toxicological end points of interest, for evaluating data about effects of chemicals. These tools often result in assignment of a score (e.g., low, medium, high) that can be used to compare alternatives.

UCLA Multi-Criteria Decision Analysis: MCDA methods are a decision analytic tool designed to provide a clear, formal approach to allow decision-makers to evaluate alternatives (Malloy et al. 2011). They present a comparative evaluation of the alternatives based upon provided criteria, taking into account the relative importance of those criteria (Kuczenski et al. 2010). More specifically, the application of Multi-Criteria Decision Analysis tools to alternatives assessment has been most notably explored by The University of California at Los Angeles (UCLA) Sustainable Technology and Policy Programand is sometimes referred to as the UCLA MCDA framework, as listed in Table 2-1. This framework or application of MCDA tools is the outgrowth of a pilot project to develop and evaluate an alternatives analysis methodology that is consistent with California's Safer Consumer Product Regulations (Kuczenski et al. 2010). The project involved using two different MCDA approaches and supporting decision-analysis software. According to Malloy (Malloy et al. 2011), the results demonstrate that the models can produce a transparent evaluation that ranks alternatives and explains how the alternatives' performance on various criteria affected their ordering. The models also allow the methods' assumptions to be adjusted (Malloy et al. 2011).

OVERVIEW OF CHARACTERISTICS FOUND IN EXISTING FRAMEWORKS FOR CHEMICAL ALTERNATIVES ASSESSMENTS

Most of the Chemical Alternatives Assessment frameworks evaluated by the committee characterize hazard, environmental fate, ecotoxicity, human health, and physicochemical properties, although each framework varies in how those attributes are assessed (OECD 2013a). A comparison of several attributes that vary amongst the alternatives assessment frameworks is presented in Table 2-1.

Hazard, human health and physicochemical properties are each assessed by all of the frameworks evaluated by the committee: BizNGO, CA SCP, EPA DfE, German Guide, IC2, Lowell, REACH, TURI, UCLA MCDA, and UNEP. Eight of the ten frameworks examined include environmental fate (BizNGO, EPA DfE, German Guide, IC2, REACH, TURI, UCLA MCDA, UNEP) and ecotoxicity (BizNGO, CA SCP, EPA DfE, IC2, REACH, TURI, UCLA MCDA, UNEP) in their analyses. Some, but not all, frameworks consider life cycle analysis (or Life Cycle Thinking depending upon the framework; see Chapter 10 for a description of Life Cycle Thinking) and the chemical's functional use or application. Life Cycle Thinking identifies hazards from chemical manufacture through product manufacture, use, and disposal and can also help identify important consumer, worker, and environmental exposure pathways. This can be especially important for consumer products. The types of end points (e.g., mammalian toxicity, ecotoxicity), range of outcomes (e.g., toxicological thresholds), and categories used to categorize hazards within a framework can vary somewhat between alternatives assessment frameworks and are considered in greater detail in subsequent chapters of this report.

Many groups have developed more specific "principles" to inform the assessment process. For example, the OECD report identified several sets of principles that are intended to guide the evaluation of safer chemical substitutes, including principles from the Interstate Chemicals Clearinghouse (IC2), the Commons Principles For Alternatives Assessment (BizNGO 2013) and the EPA's Design for Environment Program (OECD 2013a). These "principles" have a number of commonalities, and while some are not necessarily scientific principles, they are meant to guide an informed and thoughtful scientific review process for evaluating alternatives to chemicals of concern. These commonalities among the reviewed frameworks can be summarized as:

- Reduce hazard,
- Reduce exposure,
- Use the best available information,
- Ensure transparency in methods, criteria, and data used,
- Identify and mitigate trade-offs, and
- Take a flexible, iterative approach.

TABLE 2-1 Comparison of Selected Attributes Found in Selected Frameworks

Framework	Exposure at the Use Phase	Cost & Availability	Other Life-Cycle Impacts	Social Impacts	Includes Comparison of Materials and/or Processes
BizNGO (Rossi et al. 2012)	As needed	Yes	As needed	Not mentioned	Yes
CA SCP (CA DTSC 2013a)	Yes	Yes	Yes	Yes	Yes
DfE (EPA 2014c)	As needed	As needed	As needed	As needed	Can be added
German Guide (Reihlen et al. 2011)	Yes	Yes	Yes	Yes	No
IC2 (IC2 2013)	Yes	Yes	As needed	As needed	As needed
Lowell Center (Rossi et al. 2006)	Not mentioned	Yes	Not mentioned	Yes	Yes
UNEP (UNEP 2009)	Yes	Yes	As needed	Yes	As needed
REACH (EC 2011)	Yes	Yes	As needed	Yes (but in the Socio-Economic Analysis)	Yes
TURI (TURI 2006a)	Yes	Yes	Yes	Yes	Yes
UCLA MCDA (Malloy et al. 2011)	Yes	Yes	Yes	Not mentioned[a]	Can be added

[a] MCDA tools should be able to accommodate this impact, even if not mentioned in the UCLA application of them.
SOURCE: Adapted from OECD, 2013a.

3
The Committee's Framework

To develop its framework, the committee assessed the frameworks and tools [Lowell (Rossi et al. 2006); TURI 2006a; UNEP 2009; REACH (ECHA 2011); UCLA MCDA (Malloy et al. 2011); German Guide (Reihlen et al. 2011); BizNGO (Rossi et al. 2012); CA SCP (CA DTSC 2013a); IC2 2013; DfE (EPA 2014c)] identified in Chapter 2 to determine whether they included the elements identified in the committee's statement of task. Most of the frameworks included some, but not all, elements in the task statement. Thus, the committee viewed its role as developing a framework that captures common elements of the frameworks, which reflect more than 20 years of experience in this field, while ensuring that its framework included all the elements identified in the task statement. On the basis of its assessment, the committee made several decisions that influenced the development of its final framework. These decisions are summarized below.

- The statement of task specifically states that the framework should address safer chemical substitution. Therefore, the committee's alternatives assessment framework represents a structured approach for comparing human health and environmental hazards associated with different chemicals or chemical-dependent processes. Although changes to materials or designs might also provide alternatives to chemicals of concern, the framework does not focus on this option.
- The framework is intended to be used by a multidisciplinary team that has training and expertise in toxicology (human health and ecotoxicology), chemistry, materials science, exposure assessment, and life cycle assessment. Additional expertise in engineering, social sciences, economics, and cost analysis might also be required. Assessors without such expertise, such as small- and medium-sized firms, may need user-friendly assessment tools or technical support to carry out parts of the assessment.
- The framework should identify critical elements to be included in all chemical alternatives assessments but also provide flexibility to adopt different steps and tools, when appropriate. The committee emphasizes that the framework outlines the core considerations that should be included in a thorough alternatives assessment. In many cases, an assessor will not have the resources to conduct the most comprehensive assessment options as outlined in this report. However, the framework is meant to be sufficiently flexible so a particular user can at least thoughtfully consider each step of the process and undertake the assessment as information, time, and resources allow. The case study of decabromodiphenyl ether in Chapter 12 demonstrates how the framework might be applied by a user with limited resources.
- The framework is focused on the technical aspects of evaluating alternatives rather than establishing values that inform decisions and policies. For example, the framework does not select the factors to be used to determine whether an alternative is safer than the chemical of concern because this decision is context-dependent and based on value judgments. Those decisions are left to the discretion of the entity conducting the assessment.
- Certain activities, although important to evaluating alternatives, were deemed to be beyond the scope of the current project. The committee provides sufficient information for an understanding of the general approach, but if more information is needed, the references supplied should be used. Those topics that may warrant more information include criteria and approaches for identifying and prioritizing chemicals of concern, a full discussion of life cycle analysis (LCA) practice, and detailed guidance on conducting performance, economic, or social impact assessments.

In addition to the frameworks, the committee also considered principles intended to inform the assessment process (Chapter 2) and other relevant references. Although some of the principles are not necessarily scientific ones, they are meant as a guide to a thoughtful, scientific review process for evaluating alternatives to chemicals of concern.

Therefore, the committee adopted some of them and applied them when constructing its framework. Those principles that fall in this category include the following:

- The goal of chemical alternatives assessments conducted using the committee's framework is to identify safer alternatives that can be used to replace chemicals of concern in products or processes, thereby protecting and enhancing human health and the environment.[10] It is understood that the safer alternatives would also meet other requirements, such as cost and performance. An approach for replacing chemicals of concern with safer chemicals or non-chemical alternatives is what the EPA refers to as "informed substitution" (EPA 2014c). As EPA notes, practicing informed substitution is meant to "minimize the likelihood of unintended consequences, which can result from a precautionary switch away from a chemical of concern without fully understanding the profile of potential alternatives, and to enable a course of action based on the best information—on the environment and human health—that is available or can be estimated" (EPA 2012d). Although no approach can completely eliminate the possibility of unintended consequences of chemical substitutions, the committee's framework is intended to provide a structured, thoughtful evaluation of the advantages and disadvantages of alternatives, helping to support informed transition to safer chemicals.

- To be considered safer, an alternative will, for pragmatic reasons, need to be an improvement over, or no worse than, the original chemical of concern in the domain that prompted the alternatives assessment. However, a focus on a key end point does not eliminate the need for an assessment of the full range of human health hazard end points and ecotoxicity, or consideration of the life cycle of alternatives, and the alternative should also have a lower overall negative impact on worker and public health and the environment than the chemical of concern.[11] Addressing the original areas of concern can be achieved by the direct improvement or elimination of the hazardous attributes of the chemical of concern. It could also include reducing exposure potential, such as by replacing an aquatic toxicant with another chemical that has some aquatic toxicity, but breaks down quickly or has low solubility. The definition of "lower overall negative impact to human health and the environment" is context-dependent and based on value judgments; therefore, the selection of hazard end points for comparison and their relative importance are left to the discretion of the entity conducting the assessment.

- Expected exposures should be understood to help assessors determine the relevance of certain hazards, identify areas of potential concern, identify cases in which an alternative could end up in the environment or vulnerable populations, and identify the need for and appropriate type of monitoring that would be required after implementation of an alternative.

- It is important to integrate knowledge from multiple sources and disciplines to support informed substitution and to document assumptions, data, and methods clearly.

- Even safer alternatives might present some risk to human health or the environment, so chemical alternatives assessments should identify relevant trade-offs and mitigation options or continuous improvement goals that would minimize the potential for unintended consequences.

- Chemical alternatives assessments should be an iterative and flexible process so that they can be adapted to different decision contexts, goals, and conditions.

- Stakeholder engagement should occur throughout the chemical alternatives assessment.

- The chemical alternatives assessment framework should encourage innovation in chemical and process design to meet a particular chemical function for situations in which no alternatives are available, the currently

---

[10] This objective is different from that of a safety assessment, where the primary goal is to ensure that the exposure to a particular substance is below a prescribed safety standard.

[11] Requiring alternatives to offer improvements that address the original areas of concern as part of the definition of safer might sometimes result in excluding potential alternatives that offer substantial improvements in other impact areas while only offering marginal improvements in the original areas. This approach might limit the adoption of incrementally better alternatives that could act as interim solutions while better solutions are developed. These situations could be handled on a case-by-case basis as long as the acceptance of such an interim solution is consistent with the entity's values.

available options do not perform as well as the chemical of concern, or alternatives present their own significant hazards.

- The framework should encourage the direct initiation of chemical alternatives assessments for innovative green chemistry alternatives and sustainable designs, instead of only conducting chemical alternatives assessments when there is a chemical of interest to replace.

This concept is consistent with the NRC report, *Science for Environmental Protection: The Road Ahead,* which states: "the focus on problem identification sometimes occurs at the expense of efforts to use scientific tools to develop safer technologies and solutions. Defining problems without a comparable effort to find solutions can diminish the value of applied research efforts" (NRC 2012, p.7).

Considering these frameworks, decisions, and principles, the committee developed its own framework, as shown in Figure 3-1. The committee's framework identifies critical elements as steps, and places them at key points in the assessment process. At the same time, however, the framework allows flexibility in that other elements may be included in a less rigid order. Indeed, in some cases, those elements might not be needed.

Thus, the proposed framework can be reconfigured, rearranging the simultaneous steps into an order chosen by the user. Figure 3-1 is a diagram of the framework. The discussion that follows provides an overview of each of the framework's steps. For each step, the goal, inputs, outputs, and other frameworks that contain a similar step are described. In subsequent chapters, each step is described in more detail.

STEP 1: IDENTIFY CHEMICAL OF CONCERN

Although four frameworks (DfE, BizNGO, Lowell, and German Guide) address the identification or prioritization of chemicals of concern, this topic was outside of the scope of the committee's task. Therefore, Step 1 is merely the entry point for a chemical of concern into the alternatives assessment process. A chemical might enter the framework because concerns have been raised about it, resulting in a regulatory requirement, obligation, market, or policy incentive to substitute or evaluate alternatives for it. The framework might also be used to help design or evaluate new chemicals that could be potential alternatives for chemicals of concern.

STEP 2: SCOPING AND PROBLEM FORMULATION

Goal: Establish scope of assessment and plan for assessment. This step should determine appropriate stakeholder engagement; identify goals, principles, and decision rules that will guide the assessment; gather information on the chemical of concern; and determine assessment methods that will be used.

Input: Identity of the chemical[12] of concern.

Outputs: Information and parameters needed for the assessment, including goals, principles, and decisions rules for the assessment; stakeholder-engagement plan; information on the chemical of concern; methods and tools for each assessment step; and procedures on how data gaps and uncertainty will be handled.

Frameworks: BizNGO, CA SCP, DfE, German Guide, IC2, Lowell, REACH, TURI, UNEP, and UCLA MCDA

All the frameworks include some preparatory work before beginning the technical portion of an assessment. The 2009 NRC report, *Science and Decisions: Advancing Risk Assessment,* also recommends scoping and problem formulation. Scoping is a discussion between decision makers and stakeholders in which assessors have a supporting role, and problem formulation is a discussion between decision makers and assessors (and technically-oriented stakeholders) to develop a detailed technical plan for the assessment that reflects the broad conceptual design developed in the scoping stage. The committee incorporates scoping and problem formulation into its framework as Steps 2a and 2b.

Step 2a: Scoping—Determine Appropriate Stakeholder Engagement and Describe Goals, Principles, and Decision Rules

Seven frameworks (IC2, DfE, Lowell, UCLA MCDA, TURI, UNEP, and German) advise consulting stakeholders as part of an assessment. The committee included this activity within Step 2a because stakeholder engagement helps ensure that the assessment will address a broad range of concerns, improve stakeholder understanding and

[12] Chemical of concern could be a chemical that is used in a manufacturing process or a chemical in an end product.

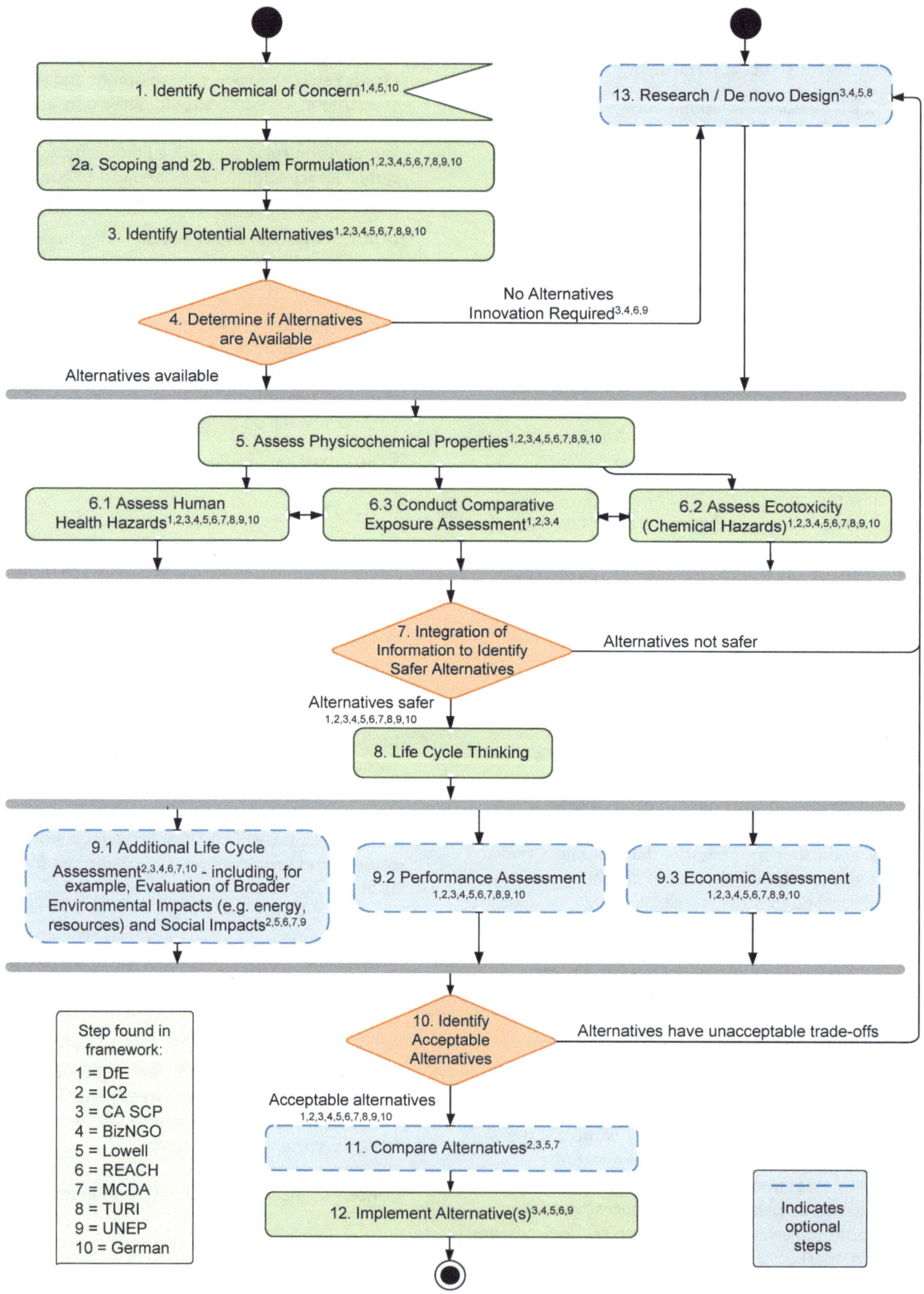

FIGURE 3-1 Framework developed by the committee.

support of the outcomes, and provide additional review for technical information, analytical methods, and other data. These benefits will improve the overall quality and accuracy of the assessment.

Formal external stakeholder engagement might not be necessary when the assessment is conducted within a single business. In that case, it can still be useful to consult employees with different roles within the company, such as product design and procurement, to capture different perspectives and priorities with respect to the assessment. It might be advisable to engage a broader range of stakeholders when a state or federal agency intends to use a chemical alternatives assessment to inform regulations or policy. A more complete discussion of stakeholder engagement can be found in Chapter 4.

Five frameworks (IC2, DfE, BizNGO, Lowell, and German Guide) provide principles that are intended to influence the assessment process. Although some regulatory frameworks (CA SCP, REACH, and UNEP) do not explicitly articulate principles, their approaches and requirements reflect implicit values of the regulators. Framework developers have attempted to embed organizational or corporate values into the frameworks because different entities can have different opinions of what would be considered a good outcome, and in many instances, developers would like to influence the outcome so that it aligns with their own values. Therefore, the committee has included within the assessment process the activity of describing or establishing goals, principles, and decision rules expected to affect basic assumptions or constraints. The reason this activity has been included is that many aspects of substitution decisions are not purely technical, but rather are value-driven or context-dependent. It is important to explicitly articulate and document those assumptions and constraints because they can strongly influence the conclusions and recommendations of an assessment, especially with respect to trade-off resolution. Also, thorough documentation allows for more effective critical evaluation of chemical alternatives assessment results and comparability across assessments.

Assessors themselves typically will not establish the goals, principles, and decision rules. The agency, organization, or corporation usually determines them, but assessors will need to document them. A more complete discussion of goals, principles, and decision rules and their impacts on alternatives assessments can be found in Chapters 4 and 9.

Step 2b: Problem Formulation—Gather Information on the Chemical of Concern and Determine Assessment Methods

All the frameworks include the collection of preliminary information about the original chemical to facilitate the assessment. This activity also has been included in the committee's framework. Information to be collected includes the following:

- The identity of the chemical of concern (and any relevant structurally-related chemicals) must be clearly established because the scope of the assessment and the range of potential alternatives can be affected by this determination. For example, if the flame-retardant pentabromodiphenyl ether was the chemical of concern, decabromodiphenyl ether could be considered as a potential alternative, but it would not be considered a viable alternative if all polybrominated diphenyl ethers were defined as the chemicals of concern.
- The function that a chemical serves or the properties that it gives to a product or process must be defined because viable alternatives must produce acceptable functional results (Lavoie et al. 2010). Clearly defining the chemical's functional and performance requirements can lead to the identification of options for achieving the desired result through non-chemical means, such as material substitutions or design changes. And finally, if the chemical of concern does not perform a necessary function, simple elimination of the chemical might be considered as an alternative, and a formal chemical alternatives assessment would not be necessary. Use scenarios need to be defined to evaluate comparative exposure.
- To determine human health and ecological effects, use scenarios, exposure pathways, and life cycle segments that warrant particular attention in light of socioeconomic, environmental, or other impacts. As described in Chapter 4, any issues about the chemical of concern should be documented before starting an assessment. Defining those elements provide a baseline for comparisons of potential alternatives. Clearly articulating the negative effects of the original substance also helps in establishing human health and ecological goals for the alternatives. A more complete discussion of the information that should be gathered in this step, and the benefits of doing so can be found in Chapter 4.

Formal assessment planning is included in four frameworks (IC2, CA SCP, Lowell, and UCLA MCDA) and has been included in the committee's framework. Specific tools and steps will need to be selected for the assessment, and decisions will need to be made about how to handle data gaps and uncertainty. A more complete discussion of the planning activities associated with this step can be found in Chapter 4.

STEP 3: IDENTIFY POTENTIAL ALTERNATIVES

Goal: Identify chemical, material, and design alternatives on the basis of the requirements established in Step 2. If needed, conduct initial screening to identify alternatives that are clearly not viable substitutes to narrow the number of alternatives to evaluate.

Inputs: Scope established in Step 2 and results of research and consultation with stakeholders.

Output: List of potential alternatives to be evaluated.

Frameworks: BizNGO, CA SCP, DfE, German Guide, IC2, Lowell, REACH, TURI, UNEP, and UCLA MCDA

All the frameworks include a process for identifying potential chemical, material, and design alternatives on the basis of the established requirements in Step 2. Alternatives identification is critical in any framework to establish the alternatives to be assessed relative to the chemical of concern. Therefore, this step has been included in the committee's framework.

BizNGO, CA SCP, DfE, IC2, Lowell, and TURI also include some level of initial screening (i.e., prescreening) of certain factors, such as predicted performance or presence on restricted chemical lists. Reducing the list of potential alternatives for assessment might be needed when resources for conducting assessments are limited, when the list of potential alternatives is too large, or when certain selection criteria can be used to exclude obviously nonviable alternatives. Initial screening also might involve some data gathering on alternatives, but would not normally be considered a complete assessment of any domain. When initial screening is used in an assessment, care must be taken to ensure that overly conservative predictions of alternatives' performance do not lead to the elimination of potentially viable alternatives that could be further developed to meet technical and economic goals. Likewise, promising alternatives should not be disqualified because of data gaps that could be filled later. Alternatives eliminated from consideration at this step should be documented both for transparency purposes and in case it is determined that they should be re-examined at later stages of the assessment. Chapter 4 includes a more complete discussion on identifying and screening potential alternatives.

STEP 4: DETERMINE IF ALTERNATIVES ARE AVAILABLE; REFER CASES WITH LIMITED OR NO ALTERNATIVES TO RESEARCH AND DEVELOPMENT

Goal: Initiate research to develop new alternatives or improve existing ones when no (or limited) alternatives are available.

Inputs: List of potential alternatives from Step 3 and results of initial screening, if available.

Output: Information on how each alternative failed to meet the requirements established in Step 2, which should help research and development efforts.

Frameworks: BizNGO, CA SCP, REACH, and UNEP

This step is an early decision point to determine if alternatives to evaluate further are available. Four frameworks allow or encourage the development of new or improved alternatives when alternatives are not available or those available could be improved. Similarly, this early step has been included in the committee's framework to address those situations so that the process for developing safer substitutes (Step 13) can be initiated earlier. Chapter 13 has a more complete discussion on innovation and the design of safer chemical substitutes.

STEP 5: ASSESS PHYSICOCHEMICAL PROPERTIES

Goal: Gather information on physicochemical properties to facilitate steps that evaluate hazard and exposure.

Inputs: List of potential alternatives from Step 4.

Outputs: Physicochemical properties for each alternative (and for the chemical of concern, if not already determined in Step 3).

Frameworks: BizNGO, CA SCP, DfE, German Guide, IC2, Lowell, REACH, TURI, UNEP, and UCLA MCDA

All the frameworks include a step to gather information about the physicochemical properties of alternatives. These properties contribute to the inherent hazards of a chemical, including its ability to interfere with normal biological processes. Physicochemical properties also define a chemical's physical hazards and influence its environmental fate, such as degradation and persistence. The committee's framework includes a step to determine the physicochemical properties of alternatives and those of the chemical of concern, if not already established in Step 3. Determining physicochemical properties is done early in the assessment because these data can be obtained quickly and inexpensively in the initial stages, and they can potentially be used to screen out chemicals likely to exhibit particular physical and toxicological hazards. Those characteristics are likely to be similar among structurally related chemicals, so such information can help focus later hazard and exposure evaluations on end points and pathways of greatest concern. Chapter 5 has a complete discussion about determining the physicochemical properties of alternatives.

STEP 6: ASSESS HUMAN HEALTH, ECOTOXICITY, AND COMPARATIVE EXPOSURE

This step includes the following three parts:

- Step 6.1: An assessment of hazards to human health
- Step 6.2: An assessment of ecotoxicity hazards
- Step 6.3: An assessment of comparative exposure

Steps 6.1, 6.2, and 6.3 could be completed concurrently because the findings are interrelated, and assessments or conclusions from one step may affect the conclusions from other steps.

Goal: Evaluate human health and ecological hazards and assess comparative exposures.

Inputs: List of potential alternatives and preliminary data on each alternative from Step 3 and physicochemical properties from Step 5. The magnitude of Step 6.3 may also be influenced by results of Life Cycle Thinking performed in Step 8.

Output: Human health and ecological hazards, exposures, and data gaps for each alternative.

Frameworks: BizNGO, CA SCP, DfE, German Guide, IC2, Lowell, REACH, TURI, UNEP, and UCLA MCDA

Every framework includes a step that evaluates human health and ecological hazards associated with the chemical of concern and identified alternatives. Four frameworks (BizNGO, CA SCP, IC2, REACH) include exposure assessment as a part of their chemical alternatives assessment. In line with the committee's belief that understanding exposure is important to understanding the relevance of hazards, a comparative exposure assessment step has been included (Step 6.3). This step includes further evaluation of the exposure potential and impacts of hazards through qualitative or quantitative exposure assessment methods.

The committee's task statement also requires evaluation of "potentially safer substitute chemicals as determined by human health and ecological risks." Therefore, the committee's framework includes steps to examine the human health and ecological hazards and exposures.

Step 6.1: Assess Human Health Hazards

This step identifies the types of adverse effects on human health that are potentially caused by exposure to the chemical of concern and its alternatives and characterizes the quality and relevance of the supporting evidence. Chapter 8 includes a complete discussion of assessing the human health hazards of alternatives.

Step 6.2: Assess Ecotoxicity

This step assesses ecological hazards associated with alternatives and compares them across alternatives. Depending on where the chemical might partition in the environment, this step can include the determination of toxicity to aquatic, sediment, or terrestrial organisms. If not completed in Step 5, this step might also include an evaluation of the persistence of chemicals in the environment and their potential to bioaccumulate in the food chain. Chapter 7 has a more complete discussion about assessing the ecological hazards of alternatives.

Step 6.3: Conduct Comparative Exposure Assessment

This step assesses whether the expected exposures from the chemical of concern and the alternatives would be substantially equivalent. If the expected exposures are not substantially equivalent, then a more detailed exposure assessment might be needed. Understanding the expected exposure is useful when interpreting the relevance of hazards identified in Steps 6.2 and 6.1. Chapter 6 has a more complete discussion of exposure assessment within the context of chemical alternatives assessment.

STEP 7: INTEGRATION OF INFORMATION TO IDENTIFY SAFER ALTERNATIVES

Goals: Identify safer alternatives on the basis of information compiled in previous steps. If no alternatives are considered safer than the chemical of concern, initiate research to develop new alternatives or improve existing alternatives.

Inputs: Results of evaluations of each alternative from Steps 5, 6.1, 6.2, and 6.3.

Outputs: List of safer alternatives and supporting documentation for each, including actions needed to offset trade-offs or detect unintended consequences. List of unacceptable alternatives, including information on how each alternative failed to meet the requirements established in Step 2 or the trade-offs that made the alternatives unacceptable. This information can inform additional research and development efforts.

Frameworks: BizNGO, CA SCP, DfE, German Guide, IC2, Lowell, REACH, TURI, UNEP, and UCLA MCDA

Every framework explicitly or implicitly integrates the findings from human health and ecological assessments to provide decision makers with the potential impacts of the alternatives. The committee's framework also includes a step to integrate human health and ecological information from Step 6. Step 7 acts as a decision point, meaning that if there are no safer alternatives for further assessment, additional research can be initiated to develop new alternatives or improve existing ones. The research will be informed by information on how each alternative failed to meet the requirements established in Step 2 or on the trade-offs that made the alternatives unacceptable. Chapter 9 explains how to integrate information from Steps 6.1, 6.2, and 6.3 to identify safer alternatives, including strategies for making decisions when there is uncertainty in the data and trade-offs to resolve.

STEP 8: LIFE CYCLE THINKING

Goal: Determine whether risks to human health, the environment, or society exist at a place or time beyond the point of use or application, and if those risks are expected to differ between the chemical of concern and proposed alternatives, to determine if additional analysis is needed to inform a substitution decision.

Inputs: List of alternatives from Step 7.

Outputs: Decision about whether further life cycle assessment is needed to inform a substitution decision and areas of concern identified.

Frameworks: BizNGO, CA SCP, German Guide, IC2, REACH, and UCLA MCDA

Step 8 addresses the portion of the statement of task related to whether resource use is considered a potential issue. This step aligns the committee's framework with several other frameworks with regard to this concern. In addition, this step is intended to determine whether human health, environmental, and social equity impacts might occur at a place or time other than the point of use of the chemical of concern. This consideration will serve to determine whether additional assessments are required to compare alternatives. IC2, BizNGO, and the German Guide evaluate whether life cycle concerns indicate a need for a more formal life cycle assessment. Additionally, three other frameworks (CA SCP, REACH, and MCDA) suggest or consider factors, such as greenhouse gas emissions, that would normally be addressed through a life cycle assessment. The committee's framework uses Life Cycle Thinking to complete this analysis.

Life Cycle Thinking is also used to determine whether a more detailed evaluation of social impact is needed to inform a substitution decision. It does so by considering whether there are worker issues (such as child labor or forced labor), consumer issues (such as end-of-life responsibility), local issues (such as respect of indigenous rights), and society-wide issues (such as preventing and mitigating armed conflicts and reducing corruption) that are not addressed by other steps and whether the differences between alternatives are expected to be significant.[13] Five frameworks and tools (IC2, Lowell, REACH, UNEP, and UCLA MCDA) support an option to consider such social impacts beyond those already addressed in other steps. Despite the fact that these impacts are not being routinely included in many assessments currently being performed, this consideration was included in the committee's framework in recognition of growing interest in

---

[13] The UNEP/SETAC Guidelines for Social Life Cycle Assessment of Products (UNEP/SETAC 2009) contain a list of stakeholder groups and impact categories that might be useful to consider.

environmental justice issues and social life cycle assessments.

Many social impacts, such as worker health and safety, will also be addressed by other steps in the framework. However, it might be necessary to consider whether there are worker impacts, local community impacts, or societal issues that have not been addressed by other steps. Chapter 10 presents a more complete discussion of this step.

STEP 9: OPTIONAL ASSESSMENTS

At a minimum, Steps 1-8 of the framework shown in Figure 3-1 should be considered in each assessment. At this stage, the committee's framework includes several optional assessments, identified in the bullets listed above. Whether or not a particular assessment is within the scope and capability is determined during the scoping and problem formulation stage and is also influenced by the outcome of preceding steps.

Step 9 includes the following three optional parts:

- Step 9.1: Additional Life Cycle Assessment
- Step 9.2: Performance Assessment
- Step 9.3: Economic Assessment

Step 9.1: Additional Life Cycle Assessments, Including Evaluation of Broader Environmental and Social Impacts

Goal: Use additional life cycle assessment methods to estimate energy consumed and materials emitted and consumed by a product. This can be done by incorporating different alternatives over part or all of a product's life cycle and estimating the broader environmental impacts associated with these flows. Use life cycle assessment methods to assess potential social and socioeconomic impacts of each alternative over its life cycle.

Inputs: List of alternatives from Step 7 and result of Life Cycle Thinking (Step 8).

Outputs: Assessment of the relative life cycle impacts of alternatives.

Frameworks: BizNGO and IC2

Broader environmental impacts of alternatives can be informed by comparing the life cycles of the alternatives and their implications for how alternatives differ in resource consumption and materials emitted. Two frameworks (BizNGO and IC2) support conducting full life cycle analyses within an alternatives assessment. Also, CA SCP requires the consideration of factors, such as greenhouse gas emissions, that could be addressed through a life cycle analysis. A life cycle assessment step has been included in the committee's framework to support conducting such analyses when needed (as determined in Step 8) and to meet the objective in the task statement, which states that the framework should be able to balance other relevant considerations, such as resource use, with human health and ecological hazards. This step is also consistent with other frameworks.

It should be noted that the goal of this step is to assess the *relative* life cycle impacts of alternatives to uncover trade-offs that might need to be considered and resolved in later decision steps (Step 10). Therefore, the scope of additional life cycle assessment might be adjusted on the basis of topics of concern identified in Step 8.

Potential social and socioeconomic impacts of each alternative over its life cycle may also be assessed, but providing detailed guidance on conducting social impact assessments is outside the scope of the committee. If a social impact assessment is needed, two of the reviewed frameworks (IC2 and REACH) provide specific guidance; however, the most current literature at the time of the assessment should be consulted for the latest in methodological guidance and best practices. Once relevant social issues are identified for alternatives, either in this step or in Step 8 (Life Cycle Thinking), a qualitative assessment might be sufficient to inform substitution decisions. Chapter 10 has a more complete discussion of this step.

Step 9.2: Performance Assessment

Goal: Assess the performance of alternatives against the requirements set in Step 2.

Inputs: List of alternatives from Step 7 and performance requirements from Step 2.

Outputs: Assessment of the performance of each alternative.

Frameworks: BizNGO, CA SCP, IC2, REACH, TURI, and UNEP

Given the critical importance of performance to the viability of an alternative, all the frameworks include some level of performance analysis. Six frameworks include it as a key step, and the other frameworks allow for it elsewhere. The task statement specifically instructs the committee to consider product function and efficacy, so a performance assessment is also included as a possible step in the committee's framework. A performance assessment can range from a simple verification that an alternative can meet the requirements determined in Step 2 to a full characterization of each alternative's performance. If detailed performance requirements have not been established in Step 2, they should be established in this step. Chapter 10 has a more complete discussion of this step. The committee notes that there will be situations in which alternatives' performance cannot be evaluated, such as when a regulator, consortium, or public-private partnership performs the chemical alternatives assessment.

Step 9.3: Economic Assessment

Goal: Assess economic impacts associated with each alternative if an economic analysis is within the scope/formulation (Step 2), is needed to inform a substitution decision, and if there is sufficient information available to complete an economic assessment.

Inputs: List of alternatives from Step 7.

Outputs: Assessment of the economic impacts of each alternative.

Frameworks: BizNGO, CA SCP, IC2, REACH, TURI, and UNEP

Although the task statement does not require the committee to address economic factors, understanding the potential financial impacts of alternatives is important in most substitution decisions. Frameworks considered by the committee include an economic analysis, and this step has been included in the committee's framework.

In cases when regulators require an economic assessment, as with CA SCP or REACH, this step must be completed. However, there will be situations in which financial analyses are not necessary (for example, when alternatives are already in the market or simple calculations show an economic benefit) or cannot be completed (for example, when there is insufficient financial information for a thorough economic evaluation, such as when a regulator, consortium, or public-private partnership conducts the alternatives assessment). In those cases, economic analyses can be deferred to later stages of the assessment or delegated to users of the final report. Providing detailed guidance on conducting economic assessments is outside the scope of the committee, but a more complete discussion of this step can be found in Chapter 10.

STEP 10: IDENTIFY ACCEPTABLE ALTERNATIVES AND REFER CASES WITH NO ALTERNATIVES TO RESEARCH AND DEVELOPMENT

Goals: Identify acceptable alternatives on the basis of information compiled in previous steps, and document findings. Address situations where no alternatives are currently viable by initiating research and development to develop new alternatives or improve existing ones.

Inputs: Results of evaluations of each alternative.

Outputs: List of acceptable alternatives and supporting documentation for each, including actions needed to offset trade-offs or detect unintended consequences. If no alternatives are acceptable, document the information describing why each alternative failed to meet the requirements. That information is used to inform additional research to develop alternatives.

Frameworks: BizNGO, CA SCP, DfE, German Guide, IC2, Lowell, REACH, TURI, UNEP, and UCLA MCDA

Each framework that was considered by the committee includes a step for integrating information across different domains to identify acceptable alternatives. In fact, one framework (UCLA MCDA) is a tailored form of decision analysis, which is a logical procedure for balancing factors from different domains to make decisions (Belton and Stewart 2002). A step to integrate information across different domains to enable identification of acceptable alternatives has also been included in the committee's framework. Inclusion of this step is not only consistent with other frameworks, but also the task statement, which states that the framework should be able to consider the full range of benefits and shortcomings of substitutes, including balancing such factors as product functionality, product efficacy, process safety, and resource use.

Another important aspect of this step is that it is a critical point for documenting the findings of all the analyses performed throughout the assessment. As noted at the beginning of this chapter, thorough documentation of findings allows for a more effective critical evaluation of alternatives assessment results and comparability across assessments. This step also acts as a decision point, meaning if there are no acceptable alternatives, additional research can be initiated to develop new alternatives or improve existing ones that is informed by information on how each alternative failed to meet the requirements established in Step 2. Chapter 11 has a more complete discussion of this step.

STEP 11: COMPARE OR RANK ALTERNATIVES

Goal: Select a single alternative for implementation or differentiate between acceptable alternatives by applying the preferred comparison method.

Input: List of acceptable alternatives from Step 10.

Output: A selected alternative or a ranked or categorized list of alternatives.

Frameworks: CA SCP, IC2, Lowell, and UCLA MCDA

Several frameworks include ranking or categorizing alternatives to select the best ones for the specific application (CA SCP, IC2, Lowell, and UCLA MCDA). For example, CA SCP requires the comparison of the original priority product to each of the alternatives under consideration. Although some frameworks do not explicitly require a ranking step, several imply that a ranking or categorization step will be completed as the assessment process concludes and implementation begins. A ranking step has been included as an option in the committee's framework because it might be necessary to differentiate between potential alternatives to a greater extent than is accomplished in Step 10 to make a substitution decision. Chapter 11 includes a more complete discussion of this step.

STEP 12: IMPLEMENT ALTERNATIVES

Goal: Plan and execute the transition to alternatives, including mitigating trade-offs and monitoring for unintended consequences, as needed.

Input: List of acceptable alternatives and their associated mitigation and monitoring requirements.

Output: Implementation plan created and executed.

Frameworks: BizNGO, CA SCP, Lowell, REACH, and UNEP

Several frameworks either include a step to create a substitution plan after successfully identifying safer alternatives (CA SCP, Lowell, REACH, and UNEP) or stress that assessments should result in the implementation of the identified safer alternatives (BizNGO). California's Safer Consumer Product Regulation not only requires an implementation plan but also requires confirmation that the plan has been executed. An implementation step has been included in the committee's framework to ensure that safer substitutes are implemented (when supported by the findings of the assessment), that those implementations are successful (even when unanticipated challenges are encountered during the transition), and that any unintended consequences are quickly identified once a substitution has been fully implemented. In cases where alternatives have been assessed through consortia or public-private partnerships rather than through the entity that will ultimately implement the change, this step can be adjusted to include other actions that would support implementation, such as creating industry-wide voluntary phase-out dates for the original chemical of concern, market-based incentives for phase-out (such as labeling or approved ingredient lists), or even potential recommendations for regulatory action. Chapter 11 includes a more complete discussion of this step.

STEP 13: RESEARCH / DE NOVO DESIGN

Goal: Create new designs and safer solutions to support replacing chemicals of concern and improving the overall safety of chemical products.

Inputs: Design objectives or list of potential alternatives from Step 3 and information on how each failed to meet the requirements from Step 2.

Output: New chemicals, materials, or designs for assessment.

Frameworks: BizNGO, CA SCP, Lowell, and TURI

Four frameworks (BizNGO, CA SCP, Lowell, and TURI) encourage the development of new or improved alternatives. In addition, new chemicals, materials, or designs under development might need to be evaluated for their potential health and ecological impacts early in the chemical design process. The committee anticipates that situations will arise where replacements for a chemical of concern do not exist, or existing alternatives are not viable in their current form. To address those situations, a step involving research and de novo design has been included in its framework. There are two paths to Step 13: (a) research might be initiated when no alternatives are available at the end of Step 4, 7, or 10, or (b) a new chemical might be in development. Chapter 13 has a more complete discussion on de novo design.

4

Scoping, Problem Formulation, and Identifying Alternatives

Early in the chemical alternatives assessment process, the assessor needs to determine the level of stakeholder engagement and delineate the goals, principles, and decision rules that will provide the context for and guide the assessment. This step is called scoping. The assessor will also need to determine the assessment boundaries and methods—problem formulation—and identify the alternatives that will be considered.

These early steps are often overlooked in existing alternatives assessment frameworks. However, they are important and can improve efficiency by focusing limited resources on a reasonable range of viable alternatives, increase transparency of the assessment, and support informed substitution processes and minimize regrettable substitutions. The goal of the chemical alternatives assessment is to identify safer alternatives that can be used to replace chemicals of concern in products or processes, so it is important that the steps outlined in this chapter support efficient, scientifically informed alternatives assessment processes and do not lead to over-analysis, or so-called "paralysis by analysis."

This chapter describes the elements of scoping and problem formulation (see Box 4-1, Step 2 in the committee's framework) and discusses the process for identifying alternatives (see Box 4-1, Step 3 in the committee's framework).

SCOPING, PROBLEM FORMULATION, AND IDENTIFYING ALTERNATIVES IN OTHER FRAMEWORKS

Most of the frameworks examined by the committee have some reference to scoping, problem formulation, and alternatives identification, although the steps typically are not well developed. Only a few provide explicit guidance on scoping that notes stakeholder engagement or decision rules to guide the assessment. For example, the Lowell Center framework includes an initial element called "Alternatives Assessment Foundation," in which goals, principles, and decision rules are established; examples are provided (Rossi et al. 2006). The IC2 Guide includes a "Stakeholder" module in which decisions are made concerning which stakeholders should be involved in the assessment. The IC2 Guide describes specific decision rules and principles that guide its framework (IC2 2013).

The majority of frameworks contain some type of problem formulation element, but most do not include an extensive characterization of the chemical of concern. For example, the Biz-NGO framework includes a "Characterize End Use and Function" step (Rossi et al. 2012). The TURI framework includes a "functional use prioritization" step (TURI 2006a). CA SCP specifies that regulated entities identify life

BOX 4-1

ELEMENTS OF STEPS 2 AND 3 IN THE COMMITTEE'S FRAMEWORK

Step 2a: Scoping at a Glance

1. Identify the relevant stakeholders and determine their role in the assessment process.
2. Describe the goals, principles, and decision rules that will be used in the assessment.

Step 2b: Problem Formulation at a Glance

1. Gather information on the chemical of concern.
2. Determine the assessment methods that will be used.

Step 3: Identifying Potential Alternatives at a Glance

1. Identify alternatives from expert and stakeholder input and literature review.
2. Gather preliminary data on potential alternatives.
3. Conduct initial screen, if indicated.

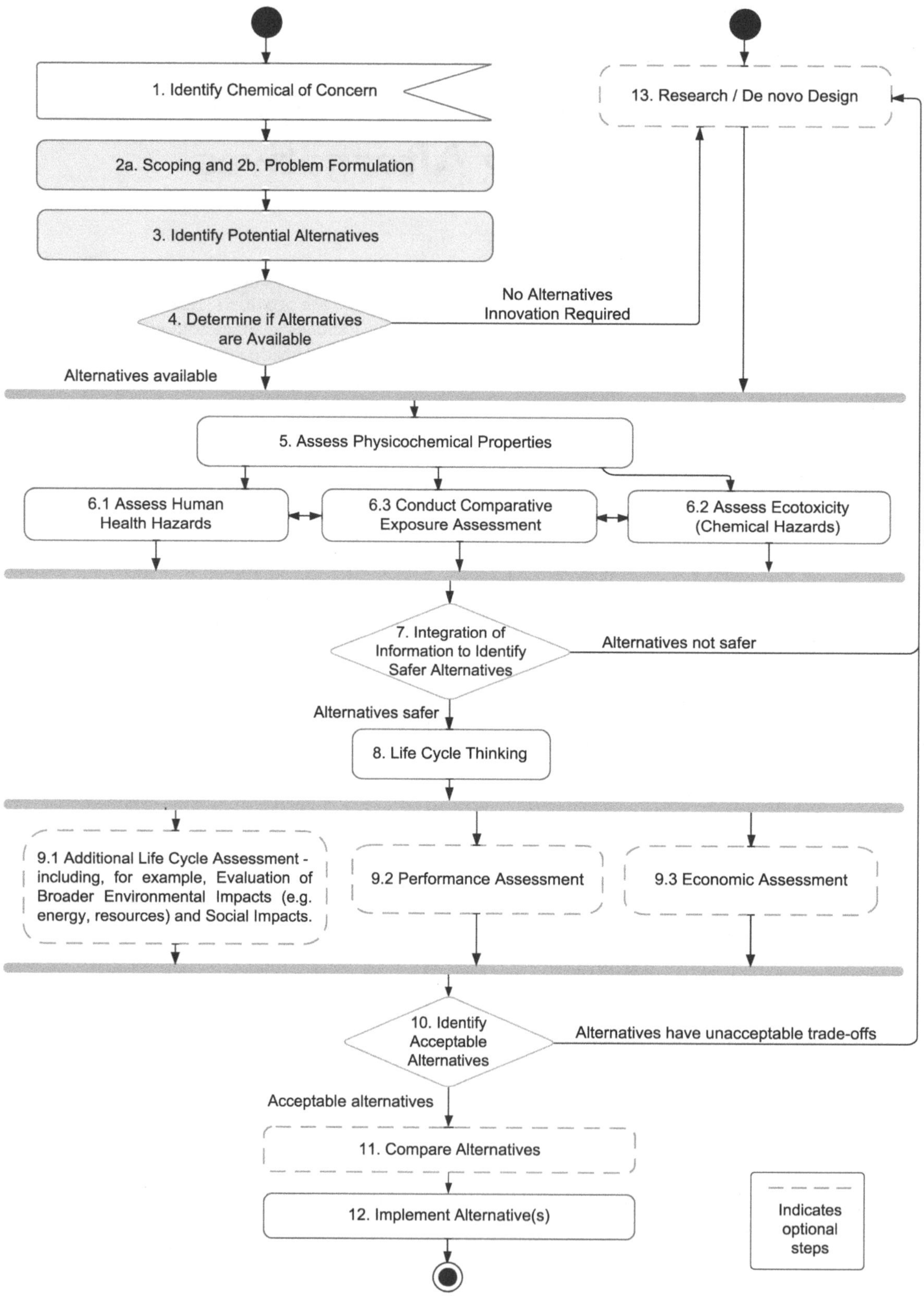

FIGURE 4-1 Committee's framework highlighting scoping, problem formulation, and identification of potential alternatives.

cycle segments and exposures that are most likely to be of concern for alternatives in bounding their alternatives assessments (CA DTSC 2013a). The IC2 framework contains a "Framework" module in which the decision framework and assessment modules to be included in the alternatives assessment are chosen, and an initial screening of attributes of concern for alternatives is conducted (IC2 2013). All the frameworks include some type of alternatives identification step, although only some, such as the TURI framework, contain an initial screening element.

SCOPING IN THE COMMITTEE'S FRAMEWORK

In the committee's framework, scoping is the initial process in an alternatives assessment, in which the level of stakeholder engagement is determined, and the goals, principles, and decision rules are described (see Box 4-2 for the committee's definitions of these terms). Scoping decisions are generally driven by particular policy mandates or organizational or corporate values. That is, how stakeholders will be engaged in the assessment and what goals, principles, and decision rules will guide the process are determined at the corporate level. Thus, the individual assessor or team will not make those decisions per se but will describe for each assessment the level of stakeholder engagement and the goals, principles, and decision rules that will be applied in the assessment process. The committee notes that for broad participation in the alternatives assessment process, transparency will be necessary. The following sections detail the elements of the scoping step.

BOX 4-2

DEFINITIONS OF GOAL, PRINCIPLES, AND DECISION RULES

Goal: Desired outcome of an agency, organization, or corporation. For example, a goal could be "to support the informed transition to functional, cost-effective, and safer alternatives."

Principle: A value or tenet of an agency, organization, or corporation. For example, a principle could be "to protect children's health."

Decision rule: A specific action that helps to implement or enact the principles. For example, a decision rule could be "do not accept reproductive and development hazards as viable alternatives."

Stakeholder Engagement in the Chemical Alternatives Assessment Process

Alternatives assessments are interdisciplinary by nature because they draw on organizational expertise in chemistry, engineering, toxicology, exposure assessment, cost analysis, and other disciplines. Expert advisors who are able to provide critical information and advice to inform the assessment process might also be needed. The multidisciplinary teams, however, might not fully understand the options, hazards, trade-offs, and barriers to adoption of an alternative; thus, it is important to involve stakeholders in the alternatives assessment process. The term *stakeholder* is broadly defined by the committee and includes internal and external members of an organization. Stakeholders are not necessarily expert advisors because they tend to be identified by the fact that they might be positively or negatively affected by the particular decision and are not usually an integral part of the decision-making team.

Stakeholder engagement in the committee's alternatives assessment process spans the length of the assessment, from scoping, problem formulation, and identification of alternatives through to the ultimate adoption of an alternative. The committee's use of stakeholders is consistent with best practices in assessment processes that advocate stakeholder input from problem definition through ultimate decision-making (NRC 1996, 2009). Stakeholder engagement is also included in several other existing alternatives assessment frameworks (Edwards et al. 2011; OECD 2013a). As noted, the extent of stakeholder engagement is generally defined by legal mandates or organizational values and probably will depend on who is conducting the assessment.

Roles for Stakeholders in the Alternatives Assessment Process

The committee identified several critical reasons for stakeholder engagement in the alternatives assessment process. First, stakeholder engagement can help identify alternatives for a chemical of concern that might not be identified by an organization's chemists or process engineers. For example, one group of stakeholders includes workers, who use a chemical in a product or process and might have ideas about alternatives that are not readily apparent to engineers or designers.

These stakeholders might be able to provide critical information on performance requirements that might lead to favoring one alternative over another. Suppliers, another group of stakeholders, might have information on alternatives that a manufacturer might not consider.

Second, stakeholders might have access to data on chemical hazards that are not readily accessible. These data could include important information on chemical use and potential exposures that should be considered in the alternatives assessment process. Such stakeholder engagement helps avoid potential unintended consequences of substitution processes. Third, if the assessment methods and assumptions are made known to relevant stakeholders, they might be able to provide useful input and help identify or solve possible problems or major data gaps.

Fourth, stakeholder engagement is critical to the adoption of alternatives. An alternative will not be viable if the end user rejects it. Adoption of alternatives might require changes in process conditions or work habits. Although such changes do not provide a rationale to avoid substitution, it is important to engage affected stakeholders so that they understand specific changes and can develop training and work practices needed to support the effective adoption of an alternative.

Fifth, some laws require stakeholder engagement in the alternatives assessment process. CA SCP specifically requires stakeholder consultation in reviewing the lists of chemicals of concern, the product or chemical combinations for which alternatives assessments will be required, and the alternatives assessment results (CA DTSC 2013a). Likewise, the Massachusetts Toxics Use Reduction Act mandates that workers be involved in the alternatives assessment process (MGL Chap. 211). Specific third- party certification processes, such as Green Seal, have requirements for stakeholder engagement in defining criteria for safer products and in their specific review (Green Seal 2009).

Level of Stakeholder Engagement in the Alternatives Assessments Process

The extent of stakeholder engagement depends on the context of the alternatives assessment process, which includes legal mandates, organizational values, and potential implications of a substitution. At this stage, it is particularly important to identify those stakeholders who might be able to provide important input in the identification, evaluation, or adoption of alternatives and to determine the degree to which they will participate in the process. Depending on the alternatives assessment, those stakeholders might include workers, trade organizations, regulators, community members, suppliers, and customers.

The following three levels of stakeholder engagement were described in the IC2 framework and should be considered when using the committee's alternatives assessment framework.

1. A corporate or organizational exercise that identifies potential stakeholders, their concerns, and how their concerns might be addressed in the alternatives assessment.
2. A process that identifies potential external stakeholders and actively seeks their input in a formal and structured process.
3. A process in which stakeholders are invited to participate in all aspects of the alternatives assessment, from scoping to adoption of an alternative. Stakeholders could also serve on the assessment team and review the final assessment product.

The different levels of stakeholder engagement have increasing resource and process requirements. As such, it is important to identify stakeholder engagement needs at the earliest point in the assessment to gain the most benefit from stakeholder involvement. It is also important to avoid overextending such engagement, causing the assessment process to become too cumbersome or paralyzed by stakeholder input. Additional guidance concerning stakeholder engagement can be found in the EPA's DfE (EPA 2014c) and the IC2 (IC2 2013). The result of this step should be a clearly documented plan for stakeholder engagement that outlines processes, roles, and responsibilities.

Goals, Principles, and Decision Rules

Assessment goals are most often set by the organization or entity responsible for performing the assessment. Thus, the goals and principles that guide an alternatives assessment process often reflect whether the assessment ultimately will be used to support regulatory, corporate, or other decision-making processes. As in most scientific assessment processes, a number of implicit or explicit values underlie the decisions. Previous NRC reports note that given the underlying science policy and context-dependent nature of risk-assessment processes, transparency in values and assumptions is critical (NRC 1996, 2009). To that end, the committee recommends that an important scoping step is documentation of goals, principles, and decision rules guiding the assessment. Once they are established, the appropriate methods and tools for completing the assessment become clearer.

Goals and Principles

The overall goal of the assessment should be explicitly stated. As noted in Chapters 1 and 3, an overarching goal of alternatives assessments conducted using the committee's framework is to identify and support the informed transition to functional, cost-effective, and safer alternatives. This broad goal is consistent with other frameworks and many organizational goals. For example, SC Johnson produced the Green List evaluation process with the goal of moving toward the safest chemical ingredients for particular applications (SC Johnson 2014). Additionally, California's SCP program has a goal "to reduce toxic chemicals in consumer products, create new business opportunities in the emerging safer consumer products economy, and reduce the burden on consumers and businesses struggling to identify what's in the products they buy for their families and customers" (CA DTSC 2014). The SCP regulations "aim to create safer substitutes for hazardous ingredients in consumer products sold in California" (CA DTSC 2014). Government agencies, such as EPA's Office of Pollution Prevention (EPA 2012d), also have overarching programmatic goals that promote pollution prevention and the use of safer chemicals.

The *principles* that represent desirable outcomes and help guide the actions of an organization should also be explicitly stated. As noted in Chapter 2, various frameworks have identified principles for alternatives assessment. For example, the EPA's DfE program has adopted a set of principles to ensure the value and utility of its analyses, such as alternatives must be commercially available, technologically feasible, and have an improved health and environmental profile (EPA 2012e). Chapter 3 describes the committee's thinking underlying the development of its alternatives assessment framework. The principles and thinking described in Chapters 1 and 3 can provide the basis for each organization to develop the goals and principles underlying its assessments.

Decision Rules

In addition to goals and principles, organizations need to develop *decision rules* to guide the assessment process. They are typically derived from the goals and principles of the assessment, implemented during the evaluation steps, and can help facilitate the assessment when resources are limited. They can be helpful in reducing the number of alternatives to be evaluated in detail; for example, by eliminating from consideration specific alternatives on the basis of early performance, toxicity, or regulatory concern indicators.

Examples of some decision rules might include (Rossi et al. 2006):

- Avoid specific types of chemicals, such as persistent, bioaccumulative, and toxic (PBT) chemicals or carcinogens, regardless of exposure potential.
- Avoid chemicals that might affect critical populations, such as children.
- Evaluate only alternatives made in manufacturing facilities that have strong human rights records.

As described in detail in Chapter 9, there also can be decision rules on, for example, how missing data might be addressed, how to consider trade-offs between domains (for example, between human health and ecotoxicity) or how to weight end points within a domain. In some cases, decision rules might be dictated by policies, such as regulations in the CA SCP, which require examination of hazards and potential exposures throughout the chemical or product life cycle.

Collectively, the goals, principles, and decision rules help guide the assessment process used for choosing the best alternatives and can help resolve trade-offs that might result from integrating results across different attribute domains, such as toxicity, material and energy use, and cost. For example, the California Safer Consumer Products regulations require that alternatives be better than the original chemical in the areas of concern (CA DTSC 2013a). The Biz-NGO framework specifically focuses on hazard reduction as a key goal for alternatives (Rossi et al. 2012), and the GreenScreen® tool lays out specific criteria for lower hazard chemicals (Clean Production Action 2014). Some alternatives assessment frameworks, such as UCLA MCDA, include specific steps aimed at understanding stakeholder values that can guide choices in resolving complex trade-offs (Malloy et al. 2011). As emphasized earlier, the goals, principles, and decision rules should be clearly documented and communicated to assessors completing later steps of the framework.

PROBLEM FORMULATION IN THE COMMITTEE'S FRAMEWORK

The goal of problem formulation is to establish a baseline and boundaries for the assessment that will help focus resources and outline a plan for the assessment. This step could be termed the "planning" phase of the assessment because it involves determining what health effects, exposure pathways, life cycle segments, and performance attributes will need to be considered. At the conclusion of this exercise, the assessor might be able to anticipate where trade-offs will occur in the substitution process.

Gathering Information on the Chemical of Concern

As noted in Chapter 3, to assess alternatives successfully, it is important to characterize the chemical of concern, including its chemical identity, functions, applications, performance requirements, toxicity, and potential exposure pathways. Understanding those characteristics and properties will help focus the assessment on functions or applications of greatest concern and provide a baseline for comparing and identifying potentially viable alternatives. The following discussion outlines the information that is needed for problem formulation.

Chemical Identity

Defining the chemical of concern clearly is the first part of information-gathering process. For example, is it an individual chemical, a chemical mixture, an entire chemical class, or an unintended by-product, or breakdown product of a specific chemical? How the chemical of concern is defined (for example, all polybrominated diphenyl ethers) can be driven by public policy or by the principles and decision rules of an organization. Identification of the chemical entity (or process) will serve to define chemical functions and limit the number of alternatives that need to be considered.

Function and Application

Before determining the chemical requirements and identifying potential alternatives, the assessor must first understand the functions, applications, and processes associated with the chemical of concern. The committee makes a distinction between function and application. A *function* is the service that the chemical broadly provides, such as solvent, adhesive, or coating. An *application* is the more specific use of the chemical, such as a solvent in a cleaning formulation, an adhesive in a specific electronic device, or a coating in food containers. These distinctions help identify appropriate alternatives (see Box 4-3). The committee's framework focuses primarily on assessment of chemical substitutions, although substitutions could involve process or product redesign.

To evaluate function, the assessor should consider the following questions:

- *What is the particular function of the chemical, and how is it used in a particular application?* At a company level, this characterization will be narrow and might be focused on one function and application. At a government or purchaser (such as a hospital) level, there might be several functions and applications to consider for a chemical of concern.
- *Is the chemical's function necessary for the product or process?* Certain functions might not be necessary to achieve product performance, such as antimicrobials in hand soaps or flame retardants in certain types of products. If that function is not required, it might be possible to eliminate the chemical of concern altogether.
- *Is the chemical of interest intentionally added, or is it an unintended by-product in the formulation?* If the chemical is an unintended by-product or contaminant, it serves no particular function, and the focus of the assessment might involve identifying ways to reduce or remove the contaminant from the formulation or identifying alternative chemicals that would not create specific by-products or contaminants. In that case, the assessment would focus on the function of the particular chemical resulting in by-product generation.

There are several ways to evaluate chemical function and application for the purposes of alternatives assessments, and there are numerous government and nongovernment options. Most government approaches consist of broad characterizations, such as surfactant or solvent. However, those characterizations might not provide enough detail for manufacturers to determine whether a particular alternative will work in their process or product. Manufacturers will want to

BOX 4-3

WHY FOCUS ON FUNCTION?

Alternatives assessments should consider the particular functions or "services" that a chemical provides in products and processes. This approach enables assessors to explore *how and why* a chemical is used rather than simply trying to find a chemical alternative to serve as a replacement. This approach can reduce the unintended consequences that might be associated with a "drop-in" substitute to replace a chemical of concern.

A focus on function provides an opportunity for government agencies and companies to screen chemical, material, and product or process redesign options in a comparative manner: by focusing on best-in-class options for a specific function and application. For example, a focus on the function of a solvent as a metal degreaser led the Toxics Use Reduction Institute to explore a range of options to meet that function, such as aqueous solvents, ultrasonic cleaning, and alternative metal-cutting methods, which removed the need for degreasing altogether. Likewise, alternatives to parabens as a preservative in a cosmetic product might include considering other chemical preservatives or entirely different ways of dispensing the soap (such as pumps) to avoid microbial contamination.

Focusing on function can provide opportunities for innovation in safer chemicals and materials. An understanding of a chemical's function can result in green chemistry attention on the molecular structures that give a chemical its particular physicochemical properties. In this way, chemicals that can serve the same function while minimizing potential toxicity can be considered. A broad focus can lead to materials and product or process design innovation. The connection between alternatives assessment and materials innovation is discussed in greater detail in Chapter 13.

define functions or applications narrowly, so that they can be analyzed more thoroughly; such analysis will lead to more actionable conclusions. However, the downside of such specificity, especially for government-facilitated alternatives assessments, is the need for multiple assessments for each particular application of a chemical rather than simply one assessment for the primary function.

TURI identified a number of functions and critical applications for five chemicals of concern for which alternatives assessments had been completed (TURI 2006b). TURI prioritized the functions and applications of the five chemicals (such as phthalates in flexible PVC sheeting) on the basis of key uses in Massachusetts and opportunities for substitution.

Thus, characterizing function and application, particularly for government alternatives assessments, provides an opportunity to focus the alternatives assessment on issues of greatest priority, such as exposure potential to sensitive populations, availability of potential alternatives, market or regulatory interest, or value to an entity, for a particular chemical of concern.

The outcome of this exercise is an evaluation of the function of the chemical of concern in a particular application or for placement on a list specifying functions of the highest priority for assessment. Not only does this step provide important input for identifying alternatives but can also provide important information for understanding potential hazards and exposures for the chemical of concern and potential alternatives.

Performance Requirements

Alternatives must meet the performance requirements of the original chemical formulation, material, product, or process, including compliance with applicable legal and customer requirements. Accordingly, once the functions and key applications of a chemical of concern have been identified, the performance requirements need to be identified as well. The purpose of defining performance requirements at the problem-formulation stage is to help identify viable alternatives and collect preliminary information for the performance evaluation and testing that might occur later in the alternatives assessment process (Step 9.2 in the committee's framework). Some legislation, such as the European Union's REACH, requires users to outline the full performance requirements at the problem formulation phase (ECHA 2011).

Although a more detailed performance assessment (including performance testing) generally occurs in later phases of an alternatives assessment, the assessor might wish to conduct an initial screen of alternatives against performance requirements to screen out those alternatives that clearly will not meet performance requirements. In fact, substantive performance testing for some established alternatives might already have been completed. Such screening can help focus the hazard and exposure assessments on the most technically viable alternatives. In some cases, it might be advantageous to complete even more detailed performance evaluations early in the alternatives assessment process; for example, when alternatives must meet certain specifications, or the list of potential alternatives is large. However, an alternative that does not meet certain requirements at this point should not necessarily be eliminated from consideration, although it might be assigned a lower initial priority. Correctly bounding the performance requirements increases the probability that the assessment process will find the most cost-effective, efficacious, and innovative solutions. The committee notes that defining and evaluating performance is an iterative process, and as noted, will need to be revisited later in the assessment process (Step 9.2). One approach to defining performance requirements is described in detail in the REACH framework (ECHA 2011) and also referenced by the IC2 framework (IC2 2013).

The committee's framework includes the following:

a. *Define specific function:* Although the function and application were characterized in the problem formulation step, the specific function should be defined in detail at this stage. For example, the general function of a substance might be as a solvent, but the specific function that it performs within a formulation could be to dissolve flux residue left behind from hand-soldering operations. Additional information, such as the type or chemical composition of flux residues, might be needed. The more completely the function can be defined, the easier it will be to set criteria to determine whether a potential alternative can be successful.*(*

b. *Identify relevant properties:* The relevant structures and physicochemical properties that determine the chemical's functions should be identified, if possible. In some cases, the properties that impart a specific function might not be fully understood.

c. *Define acceptability criteria:* It is important to specify the acceptability criteria for potential alternatives at the chemical level, the formulation or material level, the product level, or the process level, as appropriate. Acceptability criteria might include values or ranges of critical properties, such as boiling point, vapor pressure, or water solubility, that are determined on the basis of process or use conditions. It should be noted that a company might require a high level of specificity in its acceptability criteria, whereas a consortium, consultancy, or regulator might be satisfied with general criteria as long as they are sufficient to ensure basic functionality.

d. *Determine appropriate methods for testing alternatives against criteria:* In some cases, it might be possible to use established standards or test methods to evaluate criteria. For example, the efficacy of general purpose cleaners can be evaluated using the test method *ASTM G122–96(2002) Standard Test Method for Evaluating the Effectiveness of Cleaning Agents*, and a pass-fail criteria can include the stipulation that the product must remove at least 80% of the particulate or greasy soils (EPA 2012d). If standard methods are not available, qualitative methods might be required or specialized test methods might need to be developed to establish tolerance ranges.

e. *Identify regulatory, customer, specification, and certification requirements:* Certain types of products and materials might require specific performance levels to meet regulatory, specification (such as military specification), or other certification requirements. Those requirements and any accompanying test methods should be defined explicitly.

f. *Identify process or use conditions or constraints:* In addition to acceptability criteria, the process or use conditions required or expected during the performance of the function should be identified. They might include a specific temperature range; pH; purity, or presence of other chemicals; and other specific process constraints, such as drying time or process cycle time. The process and use information identified in this step might be useful in identifying potential exposure pathways.

The outcome of this exercise is a documented set of performance requirements for the particular function and application that the alternative will need to satisfy, as well as a plan for performance evaluation at the alternatives identification or performance assessment steps. The committee notes that it is important to not define the criteria too narrowly or too broadly. Defining criteria too broadly can lead to the selection of alternatives that fail to perform the central function. On the other hand, defining criteria too narrowly could lead to the rejection of alternatives that have markedly improved human health or environmental performance. These alternatives could be developed as suitable replacements, perhaps through other adjustments in the product, formulation, or process.

Human Health and Environmental Effects, Exposure Pathways, and Life Cycle Segments

Once the chemical function, application, and performance requirements have been identified, it is important to identify the human health and environmental effects associated with the chemical of concern. This information provides a baseline for comparison of the chemical with potential alternatives evaluated later in the committee's framework.

This step is also important for alternatives assessment planning in that it can help identify effects, exposure pathways, life cycle segments, and impacts of greatest concern for the chemical of concern. Once these features have been identified, they can be used as points of comparison between the chemical of concern and potential alternatives, which might exhibit similar hazard properties, exposures, or life cycle effects. These comparisons are appropriate because the use profile for the alternative and the chemical of concern are expected to be similar in the final product. Thus, this activity can help focus (or bound) the evaluations in Steps 5, 6, 8, 9.1, and 9.

The process for completing this step for the chemical of concern includes the following:

- *Characterization of physicochemical properties and hazards:* At the problem-formulation stage, it is important to develop a matrix of physicochemical properties and relevant human and ecological hazards for the chemical of concern, particularly those that have been identified as problematic. Additional details concerning relevant physicochemical properties and ecological and human health hazards to consider are discussed in detail in Chapters 5, 7, and 8 of this report.
- *Identification of use scenarios and exposure pathways:* It is important to know how the chemical of concern is used in a process or product to be able to identify its exposure pathways. Mapping the exposure pathways is designed to help in the interpretation of hazard data, not to curtail looking at hazards. That said, however, there may be some narrowing of focus in the hazard assessment. Expected patterns (acute vs. chronic) and routes (oral, dermal, inhalation) of exposure likely to be important can be identified given reasonably foreseeable exposure scenarios. A full exposure assessment is not needed at this stage; what is needed is enough understanding of exposure to determine exposure pathways of greatest interest for later assessment.

  Discussions with a variety of stakeholders, such as raw material suppliers, workers, communities, customers, and regulatory agencies, may assist in the identification of a variety of positive or negative exposure-related consequences, which may have been identified initially during the scoping exercise. For example, upstream consequences include those associated with the production, use, and storage of precursor chemicals and raw materials, and the production and use of energy and other materials. Other consequences include near-field exposures of workers along the production pathway, as well as the product's users; site-level or community-level exposures associated with upstream and product manufacturing facilities or at the point-of-use; and far-field exposures with potential impacts on distant human and ecological receptors from either upstream or downstream exposures.
- *Identification of life cycle segments that require additional consideration (life cycle segments of concern):* The purpose of this exercise is to identify and anticipate portions of the chemical of concern's life cycle that might need to be evaluated in Steps 8 and 9.1 and to make sure that the alternatives (identified in Step 7 of the committee's framework) also undergo this evaluation. The tasks that need to be completed are identifying concerns inherent to the chemical, such as toxicity of the building blocks and breakdown products, and those that are external, context-based concerns, such as energy and resource use and social impacts, over the chemical's life cycle. With that information in hand, it becomes possible to look at the alternatives in light of where important differences or trade-offs may be. A full life cycle evaluation is not needed at this point, because such assessments are costly. The goal is simply to identify areas of concern and to determine the focus of the assessment that will take place during Steps 8 and 9.

Some chemicals or chemical processes can result in the creation of by-products (or breakdown products) or involve other chemicals of concern during production of the final chemical. At this stage, such concerns associated with the "synthetic history" (intermediates, by-products, and breakdown products) of the chemical of concern should be identified because they will help in planning Steps 5, 6, and 8 in the committee's framework.

Specific chemicals or chemical processes also might have resource and energy impacts that are important to consider in the assessment, or there might be easily identifiable changes in potential life cycle impacts that need to be considered (for example, a change from a petroleum-based chemical to a biologically based chemical). Identifying life cycle segments of concern can help guide Life Cycle Thinking in Step 8.

The outcome of this step is a documented characterization of the chemical of interest that identifies its hazard profile, exposure pathways of concern, and anticipated life cycle segments of concern that should be evaluated in Step 8. This information might need to be augmented in later stages of the assessment as additional knowledge is gathered on potential alternatives. However, the goal at this stage is to have the information necessary to create a clear, focused plan for the assessment process.

Determining Assessment Methods

After human health and environmental hazards, exposure pathways, and life cycle segments of concern have been identified, decisions need to be made regarding the methods that will be used in the alternatives assessment. The methods should be clearly documented and include information on which assessment steps will be conducted, what hazard end points will be evaluated, what tools will be used to compare alternatives, and what approach will be used to address uncertainty. Some of the choices, particularly decision rules, are outlined in the Scoping exercise. Elements of the assessment, such as end points to examine and assessment steps to include, might need to be modified on the basis of knowledge gained throughout the assessment process. While the committee's framework is designed to be iterative and flexible, including flexibility in how each step is implemented depending on available resources, it does emphasize that documenting methodological choices must take place. This process is critical for minimizing concerns about whether the assessment has predetermined outcomes.

- *Assessment Steps:* Determining which framework steps—human health, ecological, exposure, performance, life cycle, economic or other evaluations—to include in the assessment should be identified at the outset. Making this decision can depend on the organization completing the alternatives assessment; the values driving the assessment (defined earlier); the use of the assessment (regulatory, non-regulatory, product development); issues identified in the assessment of hazards, exposure pathways, and life cycle segments of the chemical of concern; or knowledge about the nature of the particular product and chemical use. *At a minimum, Steps 1-8 should be included in each assessment.*

- *Tools to Evaluate and Compare Alternatives:* As noted in Chapters 5-8, there are a number of tools and approaches used in different frameworks to assess human and ecological hazards and intrinsic properties of alternatives. Although the tools are generally similar, there are some differences. In particular, specific end points to be evaluated and criteria for determining the degree of hazard might differ between frameworks, although many use decision criteria from the Globally Harmonized System of Classification and Labelling. The tools used in each framework might also differ in the data streams used to inform the assessment. Chapters 5, 7, and 8 provide guidance on end points and data streams to consider in the assessment process. Before conducting the assessment, the following decisions should be made:
    - which data streams and end points to evaluate,
    - how to compare alternatives (for example, qualitative vs. quantitative approach), and
    - how to present results (for example, numerical score vs. tabular or graphical format).

Making these decisions a priori will help to reduce bias in the assessment process. The committee notes that at a minimum, the physicochemical properties discussed in Chapter 5, comparative exposure discussed in Chapter 6, ecotoxicity discussed in Chapter 7, and the human health hazard end points discussed in Chapter 8 should be considered. Furthermore, how alternatives are compared will depend on the scope of the assessment. Comparing alternatives can be simple when the data are clear and there are not many options (IC2 2013). If many criteria are being considered or the data are not clear, the comparison becomes more complex.

- *Tradeoff strategies to determine which chemicals are "safer":* The definition of safer alternatives (Step 7) depends on the framework or tool used to evaluate and compare alternatives. Some, like the EPA's DfE framework, provide criteria for high, medium, and low scores for each end point, but there is no weighting of end points (EPA 2014c). In this instance, an assessor must determine what makes an alternative safer. The GreenScreen® tool, however, has benchmarks from one to four that are based on hazard and physicochemical properties, and they provide explicit weighting of which alternatives are safer (Atlee 2012). Other frameworks or tools implicitly weight certain hazards, such as human health, higher than other hazards, such as ecological.
- *Strategies for addressing uncertainty and data quality:* There will almost always be data gaps and uncertainty in an alternatives assessment. Chapters 5-8 provide some guidance on how uncertainties might be reduced through the alternatives assessment process. A variety of methods could be used to address data gaps. For example, certain data gaps can be addressed using models or alternative data streams. How such gaps are addressed can depend on the tools being used to evaluate and compare hazards and other attributes and decision rules established in the scoping process. In any case, it is important to document how data gaps will be addressed early in the assessment.

  Data quality is also addressed in several frameworks and tools. For example, the DfE framework has "data hierarchies" that indicate the types of data that are preferred in the hazard assessment process (EPA 2011a). The organization completing the assessment should outline early in the process what data will be used or preferred in the assessment (quantitative, qualitative, only lists, and government databases) and how data will be obtained.

The output of this exercise is a clearly documented, methodological plan for the alternatives assessment that will guide later steps. As noted, on the basis of the data being obtained, changes in methods might be warranted, but such changes should always be clearly documented.

IDENTIFYING ALTERNATIVES IN THE COMMITTEE'S FRAMEWORK

The committee's alternatives assessment framework includes a step (Step 3) that involves identifying alternatives. The purpose of this step is to identify a range of potential alternatives that meets a particular function in a product or process. In some cases, for example, if the number of alternatives is large and needs to be reduced, an initial screening based on goals, principles, decision rules, and performance criteria, as described in Step 2, can be completed. The goal here is to identify a range of viable alternatives and then to assess them through Steps 5 and 6 of the assessment.

Identifying a Range of Alternatives

For the purposes of the present report, the goal of the alternatives assessment is to evaluate safer alternatives for a particular chemical of concern for a particular function. In general, the initial alternatives identification should involve a broad range of stakeholders to ensure breadth and creativity of options. At this point, the alternatives identification should focus on available alternatives and those that might be on the horizon and highlight those that represent more than marginal improvements over the chemical of concern, given the costs associated with product or process reformulation. Options that seem unlikely should not necessarily be eliminated.

The breadth of alternatives to be considered in the assessment process should be made explicit in the scoping step. Often, an organization might want to evaluate only relatively simple chemical substitutes that do not result in substantial product or production process redesign requirements (known as drop-in substitutes). Such substitutions can be made more rapidly, often at a lower initial cost. In other contexts, an organization might want to consider greater chemical changes, including substantial product reformulation or redesign. A broad range of options can increase the complexity of the alternatives assessment because exposure pathways or hazard profiles of alternatives can be substantially different. Alternatives can be identified through a number of strategies, including review of scientific and trade literature and industry publications, interactions with suppliers, and engagement with experts in a company, government agencies, or technological institutes.

Initial Screening of Identified Alternatives

If the number of identified alternatives is large, it might be necessary to screen the list to a more manageable number for further assessment and potential adoption. Screening also can help prevent potentially regrettable substitutions for toxicity or performance reasons. That said, however, it might be useful to retain alternatives that appear to be improvements, particularly if few alternatives are available. What is important at this point is to identify those alternatives that clearly will not meet required functional, legal, or customer requirements. The screening process might also identify where there is a need for green chemistry and materials innovation (see Chapter 13). It is important to note that this initial screening is *not* a full assessment process, but rather a screen to limit the range of alternatives evaluated in depth in Steps 5 and 6 to a manageable size.

The first consideration in the screening process involves identifying those alternatives that might not be technically viable on the basis of performance. Although alternatives are often eliminated from consideration because of the potential to increase costs, at this point, cost should not be considered a determining factor. There are ways to reduce costs through process changes or purchasing agreements. Furthermore, although the unit cost of a chemical replacement might be higher, the comparison might not consider the range of cost reductions associated with an alternative chemical or material, including those related to durability, permitting, insurance, disposal, and liability. Those questions are more effectively considered in the economic and performance analyses (Steps 9.2 and 9.3) that occur after the comparative chemical hazard assessment.

The second consideration is based on toxicity or exposure concerns. This screening can be done rapidly by using authoritative lists or hazard classifications, as described in Chapter 8, but the listing criteria need to be transparent, understood by the assessor, and consistent with the criteria used to establish evidence of the health end point that the list is addressing. Several alternatives assessment tools include this approach as a screening step. Many countries and key stakeholders, including customers, have lists of chemicals that they choose to limit or ban on the basis of toxicity concerns, such as mutagenicity and reproductive toxicity, or PBT characteristics. Under the European Union's REACH legislation, the European Chemicals Agency (ECHA) has created a "very high concern" list of chemicals that will be highly regulated, making the import or use of them difficult (ECHA 2014a). Although such regulations have geographical limits, most companies today use global supply chains, and restricting the use of a chemical in one geographic region is also likely to affect other regions. As alternatives are being assessed, knowing what limitations (restrictions or exposure limits) already exist for the use of certain chemicals can help inform the assessment of alternatives.

The final consideration in the screening process involves reviewing goals, principles, and decision rules (identified in the scoping step), including the public commitments that a company has made that would affect products or chemicals they use or sell. For example, a company might have a decision rule to avoid all chemicals that are potential carcinogens or endocrine disruptors; any chemical meeting those criteria should be eliminated at this point.

It is important to recognize that this screening activity only eliminates clearly inferior or unacceptable alternatives on the basis of performance, toxicity, or exposure. It should result in a reasonable narrowing of alternatives to those that appear to be the most viable. Whatever the outcome, it is important to identify clearly the screening criteria. Alternatives eliminated from consideration at this step should be documented as a record of what was considered, in case they need to be reconsidered at later stages of the assessment (for example, in cases where new information regarding toxicity becomes available or where other alternatives are not viable).

5

Physicochemical Properties and Environmental Fate

Knowledge of the physicochemical properties of potential chemical alternatives is a requirement of the alternatives assessment process for two reasons. First, the inherent hazard of a chemical, such as its capacity to interfere with normal biological processes, and its physical hazards and environmental fate (degradation, persistence) are determined by its intrinsic physicochemical properties and the system with which it is interacting. For organic and inorganic chemicals, these intrinsic properties are determined by molecular structure, while for materials, they are determined by composition, size, structure, and morphology. Second, physicochemical properties can be used to eliminate from consideration chemicals that are likely to exhibit particular physical or toxicological hazards. As important as these data are, obtaining them is relatively fast and inexpensive, and can be readily done at the initial stages of the alternatives assessment.

This chapter provides a general background on physicochemical properties and briefly reviews experimental and computational methods that could be used to determine physicochemical properties. Current approaches for assessing physicochemical properties in several alternatives assessment frameworks are then discussed, followed by the details behind assessment of physicochemical properties and their relevance in predicting environmental fate and transport and human health hazards and ecotoxicity. Finally, the committee provides additional instructions on the implementation of Step 5 in its framework.

Box 5-1 provides a brief description of the elements of the committee's suggested approach.

PHYSICOCHEMICAL PROPERTIES OF INTEREST

For the purpose of this report, we broadly define physicochemical properties as physical properties, solvation properties related to interactions with different media, and properties or molecular attributes that define intrinsic chemical reactivity. The physicochemical properties of interest to chemical alternatives assessment can be used to identify physical hazards and to understand or predict a chemical's environmental fate, human toxicity, or ecotoxicity (see Figure 5-2). The committee cautions that given the active research in the field and the potential for special case concerns to arise for a given compound (e.g., atmospheric reactivity), the properties highlighted in this chapter

BOX 5-1
Elements of Step 5 (Assessing Physicochemical Properties)

The assessment of physicochemical properties is an early step (Figure 5-1) in alternatives assessment because physical hazards, environmental fate, and intrinsic human health hazards and ecotoxicity are directly related to a chemical's intrinsic physicochemical properties (Figure 5-1). Physicochemical properties such as those indicative of physical hazards could be used to eliminate particular chemicals from consideration and prioritize chemicals for further screening for human and ecotoxicological effects. A number of properties can be informative to alternatives assessment, as described in detail in this chapter, and a high-priority data set is also defined. Property data can be obtained from experimental or in silico (estimation) methods. In fact, state-of-the-art methodologies are making in silico methods increasingly reliable, low-cost approaches.

The suggested uses of physicochemical property data are:

1. To identify the potential for direct physical hazards posed by the chemical or material.
2. To determine the environmental compartment(s) into which the chemicals will partition.
3. To estimate the potential for bioconcentration and bioavailability.
4. To estimate the likely route(s) of mammalian exposure and bioavailability, and the likelihood for high aquatic toxicity.
5. To estimate the potential for inducing human toxicity.

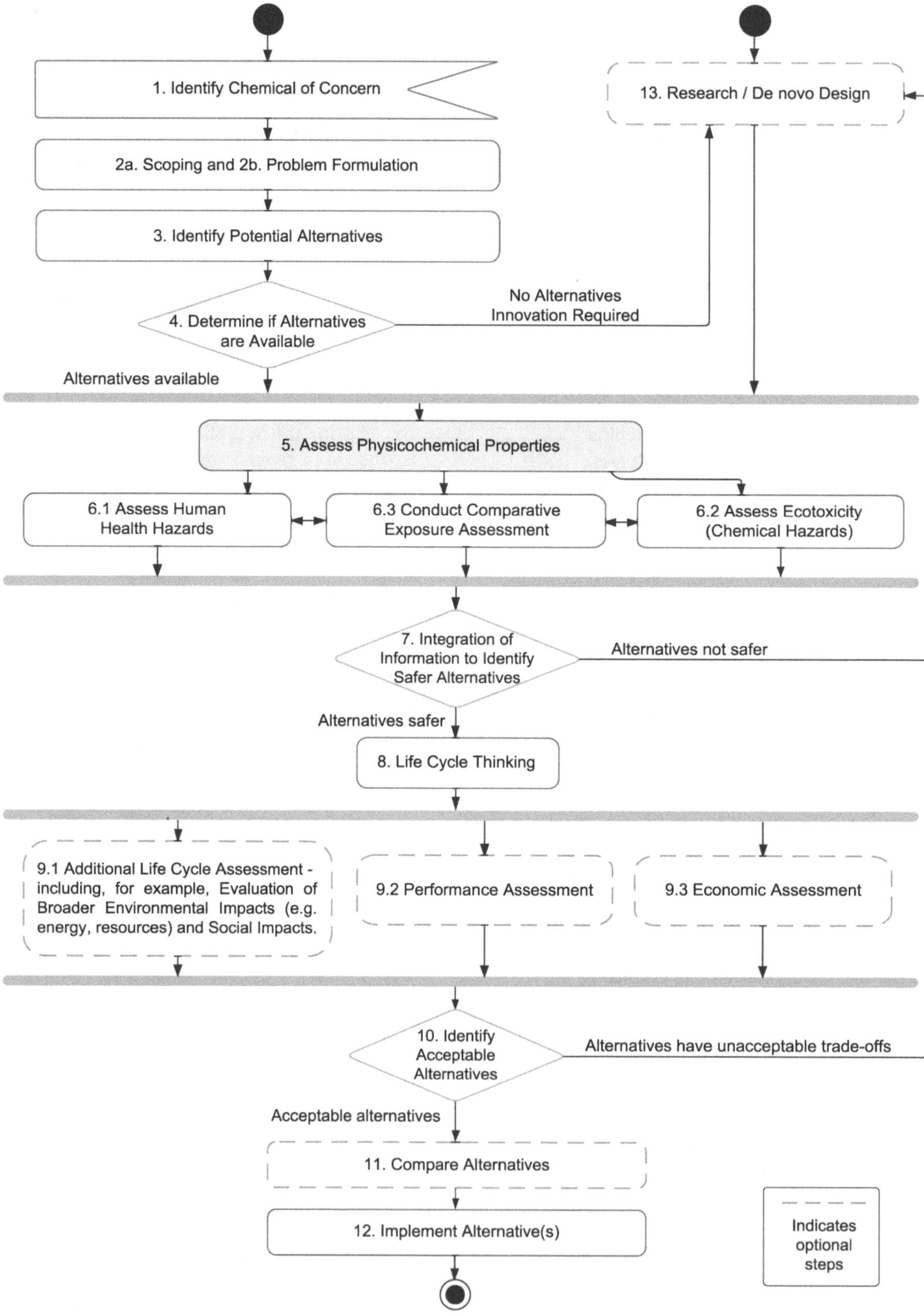

FIGURE 5-1 The committee's framework, with Step 5 highlighted.

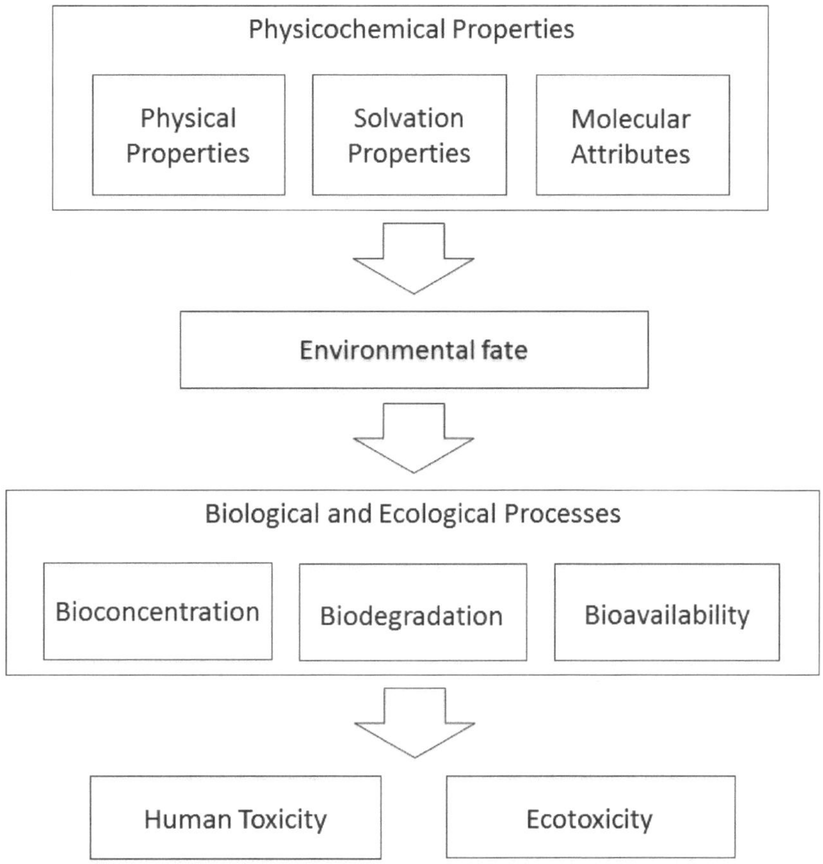

FIGURE 5-2 Scheme illustrating the relationships between the three primary types of physicochemical properties assessed in the committee's framework and their relationship to environmental fate, biological and ecological processes, and toxicity.

should be seen as general guidance, and care should be taken by the assessor to ensure that all appropriate physicochemical properties are identified for a given compound and system.

Physical Properties

Physical properties include freezing point, boiling point, melting point, infrared spectrum, electronic parameters, viscosity, and density. Some of these physical properties (e.g., electronic parameters, molecular weight, boiling/freezing point) are directly associated with environmental fate and health effects.

Solvation Properties

Solvation properties[14] describe a chemical's interactions with different phases and its partitioning between phases. Solvation properties of interest in alternatives assessment can be divided into three main types: (a) phase partitioning, (b) solubility, and (c) colligative properties:

- *Phase partitioning*. A partition-coefficient or distribution-coefficient is defined mathematically as the ratio of concentrations of a given compound across two mixed, immiscible phases at equilibrium. In the context of a chemical alternatives assessment, important partition coefficients are often measured in the liquid phase. Though partitioning can be measured across a range of solvents and phases, the phase partition coefficient most often encountered when assessing physicochemical properties is from a system where one solvent is water or an aqueous phase and the second is organic and hydrophobic, such as 1-octanol (i.e.,

[14] The terms *solvation properties* and *solution properties* are often used interchangeably. Solvation is the term used in this report.

octanol/water partition coefficient [K_{ow}] represented by *P*).[15]

$$\text{Partition coefficient } (P) = \frac{\text{concentration in organic medium}}{\text{concentration in aqueous medium}}$$

- *Solubility*. This chemical property refers to the ability of a given substance (the solute) to dissolve in a solvent. The primary measurement of interest in chemical alternatives assessment is solubility in water.
- *Colligative properties*. Colligative properties are properties of solutions that are not dependent on the chemical species but instead on the ratio of the number of solute particles to the number of solvent molecules in a solution. Examples of colligative properties include lowering of vapor pressure, elevation of boiling point, and depression of freezing point. Colligative properties generally do not play a significant role in alternatives assessments and are not discussed further in this report.

Molecular Attributes

The term *molecular attribute* is used to describe properties related to molecular shape and size. For the purposes of this report, the committee considers electronic parameters of molecules (e.g., frontier orbital energies and polarizability) that affect chemical reactivity as a type of molecular attribute.

Environmental Partitioning

In addition to the partition coefficient *P*, there are other media-specific partition coefficients that can provide valuable information about environmental fate, such as a chemical's phase partition coefficient in soil and water (K_d) and in water and air ($K_{w/g}$, Henry's law). As will be discussed in a later section, these coefficients provide insight into environmental partitioning of the molecule and the potential for bioaccumulation. As with other physicochemical properties, some of these values must be directly measured and some may be estimated.

MEASURED PHYSICOCHEMICAL PROPERTY VALUES

An extensive review of the experimental measurement of a chemical's physicochemical properties is beyond the scope of this report. Measured values of these properties often can be obtained from the scientific literature (Leo 1995). Some useful databases include: the National Institute of Standards and Technology Search for Species Data (NIST 2011) and the Syracuse Research Corporation's CHEMFATE Chemical Search database (SRC 2014). Since there is a wide range of environmental conditions of interest (especially temperature and pH), there are often no suitable literature values available. In those cases, direct measurement or estimation through computational approaches is required.

The OECD *Guidelines for the Testing of Chemicals* is a review of approximately 100 testing methods used by various governmental and non-governmental entities to identify and characterize potential hazards of new and existing chemicals (OECD 2014a). OECD test guidelines exist for the measurement of a variety of physicochemical properties, including K_{ow} and determination of pH, vapor pressure, density, water solubility, and melting and boiling points, among others. A number of comprehensive review texts have been authored on the measurement and estimation of physicochemical properties (Boethling and Mackay 2000). In cases where measurement is not feasible or is prohibited by cost, estimated parameters can be determined through a variety of methods, as discussed in the next section.

METHODS FOR ESTIMATING SELECT PHYSICOCHEMICAL PROPERTIES

This section briefly discusses the increasing number of computational, or in silico, tools available for estimation of the key physicochemical properties included in the committee's framework. These tools provide a rapid means for obtaining physicochemical data, often at a lower cost when compared with experimental measurement. A number of different software packages and algorithms are available for predicting physicochemical properties, and predictions are often in excellent agreement with experimentally-derived values. The user of such tools, however, must have a basic understanding of the inherent advantages and limitations of the various algorithms as they relate to the accuracy of physicochemical property prediction. Here we will briefly explore two broad categories of properties discussed in the chapter that are most amenable to accurate estimation—solvation properties and electronic parameters.

---

[15] In this chapter, *P* will be used interchangeably with K_{ow}, reflecting preferences in terminology across relevant fields.

Solvation Properties

Phase Partitioning

Molecular hydrophobicity (or lipophilicity) is expressed as *P* or *D* and is one of the most studied physicochemical properties in organic and medicinal chemistry. Log*D* is defined as the ratio of the concentration of compound in the lipid phase to the concentration of all species (ionized and un-ionized) in an aqueous phase at a given pH. This ratio is directly affected by the pH of the system; thus, this information is often included as a subscript, $\log D_{pH}$. Log*P* is defined as the logarithm of the ratio of un-ionized compound in each phase.

Log*D* for acids/bases can be readily calculated from log*P* when pK_a values are known. Thus, only methods for determining log*P* will be discussed here. Two types of in silico methods for estimating log*P* exist: those based on chemical structure, which are well established, and those based on spectroscopic data, which are fairly novel. There are five classifications of structure-based computational methods: whole molecule methods (which use only molecular parameters, such as size, polarizability and H-bond acceptor strength), atom-based, fragment-based, constructionist, and reductionist (Leo 2000). While some of these approaches use atomic- or fragment-based prediction algorithms, where a molecule is dissected into fragments (and its log*P* value is obtained by summing the hydrophobic contributions of each fragment), others use whole molecule attributes that take into account conformations (Meylan and Howard 1995; Muehlbacher et al. 2011). The most commonly used group contribution tools, such as ALOGP, CLOGP, ACD, and KOWWIN, have a coefficient of determination (r^2) in the range of 0.90-0.95 (An et al. 2014). Although these tools are very fast and accurate, these methods often show lower accuracy when externally validated (r^2 = 0.51-0.91). An and colleagues determined that this could be "due to limitations in the applicability domains to structures containing predefined fragments" (An et al. 2014). In particular, the authors identified concerns about the performance of compounds containing phosphorus, halogens, and other heteroatoms. They noted that there were clear disagreements between measured values of log*P* and those calculated by predictive programs. A detailed discussion of these nuances is available (Voutchkova et al. 2012). The algorithms based on spectroscopic data do not require knowledge of exact chemical structure (Voutchkova-Kostal et al. 2013). Although fairly new, they report performance on par with those of structure-based approaches, but their full applicability has not been determined (An et al. 2014).

Aqueous Solubility

Aqueous solubility is a direct measure of the hydrophobicity of a substance. The solubility equation developed by Yalkowsky can be used to estimate intrinsic water solubility at 25°C (logS) for structurally diverse organic substances (Ran and Yalkowsky 2001). This equation uses regression-derived correlation with log*P* and melting point (MP) for solids:

$$\log S = 0.8 - \log P - 0.01(\mathrm{MP} - 25)$$

Other factors that influence water solubility include temperature and pressure, neither of which is accounted for in this equation (Jorgensen and Duffy 2002). Another effect that should be considered arises from salinity ("salting-out"), which indicates that this equation is not appropriate for use with high-melting, non-ionic solids (Voutchkova et al. 2012).

pK_a

pK_a values provide insights into the lipophilicity and solubility of ionizable compounds. This, in turn, can be used to better anticipate and predict the compound's toxicokinetic behavior for processes such as gastrointestinal absorption, membrane permeability, protein binding, and metabolic transformations. Therefore, research has led to the development of computational tools for pK_a determination. As noted in the 2012 *Handbook of Green Chemistry* (Voutchkova et al. 2012):

> *In silico* pK_a methods are fast, cost-effective, and mostly reliable (some reporting correlation with experiment as high as 0.90) ... [T]hey can also provide structural assignment and identify which ionization center in the molecule corresponds to each pK_a value, and also predict the pK_a values of tautomers. Most of these methods use linear free energy relationships with Hammett σ and Taft σ^* constants for the calculation of microscopic and macroscopic ionization constants (Shields and Seybold, 2013). Some more fundamental approaches use semiempirical and higher level quantum calculations; however, these are problematic for larger systems, since

they require calculating very small differences in the energy of relatively large molecules (Shields and Seybold, 2013).... Importantly, as with all methods that require parameterization, the choice of *in silico* pK_a prediction tool should be guided by the type of compounds being analyzed, as every parameterization yields outliers (usually containing specific functional groups), and its range of applicability is limited by the training set used.

Molecular Attributes: Electronic Properties

Knowing the calculated electronic properties of molecules can be a useful part of a first-tier estimation of a chemical's reactivity with biological targets. For some end points, electronic properties have been shown to be helpful in identifying chemicals of high toxicity. These properties can be readily estimated with quantum mechanic calculations when the chemical structure is known. A multitude of electronic properties and molecular attributes have been used to describe biological activity of chemicals. Some of these relate to molecular size, shape, and volume, others relate to the distribution of electrons in the molecule, and yet another set is based on frontier orbital energies.

Properties that describe molecular size and shape include solvent accessible surface area, molecular volume, globularity, and ovality, and they can be related to bioavailability and reactivity. Accurate estimation of these attributes based on chemical structure necessitates prior optimization of the geometry via a conformational analysis and energy optimization.

Properties related to electron distribution are often related to chemical reactivity and biological activity (Voutchkova 2012). For example, molecular electronic dipole moments, μ, and dipole polarizabilities, α, are important in determining the energy, geometry, and intermolecular forces of molecules. Electric dipole moment μ_c is classically expressed as a sum of discrete charges, q_i, multiplied by the position vector, r_i, from the origin to the *i*th charge. Polarizability is the relative tendency of a charge distribution (ρ(r), an atom or molecule's electron cloud) to be distorted by an external electric field. Thus, the quantum method and the basis set[16] used impact the dipole moment and polarizability calculations.

Electronic properties based on frontier orbital energies are closely related to chemical and biological reactivity. Frontier molecular orbital (FMO) theory, pioneered by Fukui and coworkers (Fukui, et al. 1952), mathematically defined the role that frontier orbitals play on chemical reactivity. This theory is now well accepted in the field. In addition to the energies of the frontier orbitals (Highest Occupied Molecular Orbital [HOMO], Lowest Unoccupied Molecular Orbital [LUMO)], and the energy gap [ΔE] between the HOMO and LUMO orbitals), electronic properties can include chemical softness/hardness, chemical potential, and electrophilic index, to name a few.

Rather than attempting to provide a detailed description of these properties and their relation to biological activity, instead we illustrate the potential utility of one such property to alternatives assessment. This property is the HOMO–LUMO gap, which is a known measure of kinetic stability, such that a molecule with a small HOMO–LUMO gap (see Figure 5-3) is considered chemically reactive for covalent bonding (Kostal et al. in press). In the section of this chapter entitled "Use of Physicochemical Properties to Predict Aquatic Bioavailability and Toxicity," there is an example of the applicability of HOMO-LUMO gap for identifying chemicals most likely to exhibit high acute aquatic toxicity.

When calculating any electronic properties, the choice of quantum mechanical approach (i.e., semi-empirical, *ab initio,* and density functional methods) should be made judiciously. Recent advances have made density functional theory (DFT) methods comparable in accuracy to post-Hartree-Fock *ab intio* methods and often represent the optimal method of choice, especially for larger molecules. Semi-empirical methods can be accurate and are notably faster than *ab initio* or DFT methods; however, their performance is tied to the training set used in their development. Thus, these semi-empirical methods should always be benchmarked against experimental data or higher-level calculations for any given application.

[16]A basis set is a collection of vectors that defines a space in which a problem is solved. In quantum chemistry, the "basis set" usually refers to the set of (nonorthogonal) one-particle functions used to build molecular orbitals.

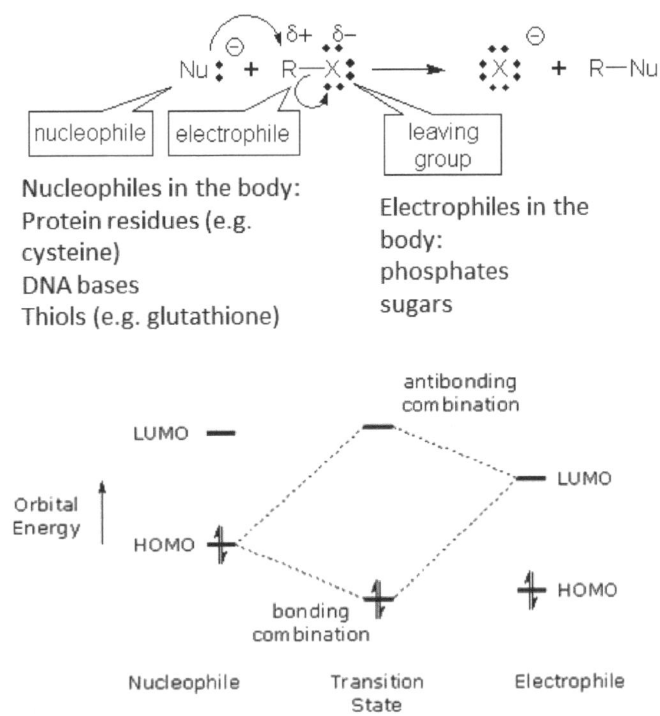

FIGURE 5-3 Relation of frontier molecular orbital energies to covalent interactions of nucleophiles and electrophiles, as illustrated with a generic nucleophilic substitution (Sn2) reaction. The reactivity of nucleophiles with HOMO energies close to the LUMO energies of the electrophiles will be higher than ones with larger differences, assuming steric effects are held constant.

PHYSICOCHEMICAL PROPERTIES IN OTHER FRAMEWORKS

Requirements for the collection and interpretation of physicochemical data are present in other frameworks, but specific guidance for the evaluation of these properties is not always included. For example, the disclosure of some physicochemical properties is required for registration under the European Union's REACH process (ECHA 2014c), but interpretation of those data is not.

Physical Hazards

The IC2 framework evaluates reactivity and flammability and uses the GreenScreen® methodology to categorize chemicals, which requires data on flammability and explosibility (Rossi and Heine 2007). The German Guide notes that physicochemical hazards may make certain chemicals difficult for workers to handle and pose safety hazards due to flammability and explosibility. It provides a chart with some guidance on categorization of different physical hazards based on common labeling standards.

EPA's DfE evaluates physical hazards using the United Nation's GHS, which is an internationally recognized structure for communication of a range of hazards (UNECE 2013a). In total, the GHS identifies 16 types of physical hazards. GHS also provides a structure for classifying these hazards, facilitating direct comparisons to be made across materials. Annex 8 of the 2013 edition provides an example of how to carry out a GHS classification (UNECE 2013b). The DfE framework lists several GHS categories, including those in the 2011 *Alternatives Assessment Criteria for Hazard Evaluation.* These criteria include explosibility, self-reactive substances, substances that on contact with water emit flammable gases, oxidizing gases, oxidizing liquids and solids, organic peroxides, self-heating substances, and corrosivity to metals as physical hazards of concern (EPA 2011a). These categories are explained in more detail in Table 5-1.

TABLE 5-1 GHS Criteria used by DfE for the Classification of Physical Hazards

Physical Hazards	Very High	High	Moderate	Low
Explosives	GHS Unstable Explosive	GHS Explosive Division 1.1 (Mass explosion hazard), 1.2 (Severe projection hazard), or 1.3 (Fire, Blast hazard or projection hazard)	GHS Explosive Division 1.4 (Fire or projection hazard), or 1.5 (may mass explode in fire)	GHS Explosive Division 1.6 (Extremely insensitive articles with no mass explosion hazard) or not classifiable as an explosive by GHS
Self-reactive Substances	GHS Type A (Detonates/ Deflagrates rapidly) or B (Liable to undergo thermal explosion)	GHS Type C (Possesses explosive properties) or D (Detonates partially when heated in confinement)	GHS Type E (Does not detonate when heated in confinement) or F (No effect when heated in confinement, not explosive)	GHS Type G (Thermally stable) or GHS not classified
Substances that on contact with water emit flammable gases	GHS Category 1 (In contact with water releases flammable gases which may ignite spontaneously)	GHS Category 2 (In contact with water releases flammable gases	GHS Category 3 (In contact with water releases flammable gases)	GHS not classified
Oxidizing Gases		GHS Category 1 (May cause or intensify fire; oxidizer)		GHS not classified
Oxidizing Liquids and Solids	GHS Category 1 (May cause fire or explosion; strong oxidizer)	GHS Category 2 (May intensify fire; oxidizer)	GHS Category 3 (May intensify fire; oxidizer)	GHS not classified
Organic Peroxides	GHS Type A (Heating may cause an explosion) or B (Heating may cause a fire or explosion)	GHS Type C (Heating may cause a fire) or D (Heating may cause a fire)	GHS Type E (Heating may cause a fire) or F (Heating may cause a fire)	GHS Type G (No hazard label) or not classified
Self-heating Substances		GHS Category 1 (Self-heating; may catch fire)	GHS Category 2 (Self-heating in large quantities; may catch fire)	GHS not classified
Substances corrosive to metal			GHS Category 1 (May be corrosive to metals)	GHS not classified

SOURCE: EPA 2011a and UNECE 2011.

TABLE 5-2 End Points, Thresholds, and Categories used to Evaluate Bioaccumulation Potential in Chemical Alternatives Assessment Frameworks Reviewed by the Committee. TURI's P2OASys worksheet returns numerical values based on a scale of 1 to 10 to represent relative hazard from low to high.

End Point	Framework	Threshold	Category
Log K_{ow}	DfE	< 2	Low
	IC2	≥ 5	Very high
		< 4.5	High
		4-4.5	Moderate
	TURI[a]	≥ 6	10
		< 6	8
		< 4	6
		< 2	4
BAF/BCF (Bioaccumulation Factor/Bioconcentration Factor) (mg/L)	DfE	> 5000	Very high
		1000-5000	High
		100 ≤ 1000	Moderate
		< 100	Low
	IC2	> 5000	Very high
		1000-5000	High
		500 ≤ 1000	Moderate
		100 ≤ 500	Low
		<100	Very low
	TURI[a]	≥ 1000	10
		< 1000	8
		< 200	6
		< 100	4
		< 10	2

[a]Category values calculated from the Pollution Prevention Options Assessment System (P2OASys) worksheet, September 2014. The P2OASys worksheet returns numerical values based on a scale of 1 to 10 to represent relative hazard from low to high.
SOURCE: EPA 2012; IC2 2011; TURI 2010

Solvation Properties and Molecular Attributes

Several reviewed frameworks provide an analytical system for assessing exposures on the basis of physicochemical properties or bioaccumulation.[17] For example, the IC2 framework lists a variety of physicochemical properties that should be considered when assessing exposure pathways, including: volatility/vapor pressure, molecular weight and size, solubility, log*P* (as K_{ow}), boiling point, melting point, density/specific gravity, pH, corrosivity, and dissociation constant. All but one of the alternatives assessment hazard classification schemes include a metric for bioaccumulation (see Appendix B for more information). Across a number of frameworks, the octanol–water partition coefficient log*P* or K_{ow} is used as an indicator of hydrophobicity. Table 5-2 shows the characterization and ranges that define classification scores for log*P* for three frameworks.

Among the frameworks, the potential for bioaccumulation generally is considered very high when log*P* exceeds 5 to 6 and generally considered low when the log*P* < 2. It should be noted, however, that a compound with a high log*P* value may be rapidly metabolized or degraded, and in these cases, would not bioaccumulate.

The DfE evaluates the ability of a chemical to bioaccumulate. When measured data are unavailable,

[17] "Bioaccumulation is defined as the accumulation of chemicals in the tissue of organisms through any route, including respiration, ingestion, or direct contact with contaminated water, sediment, and pore water in the sediment" (EPA 2000).

the DfE will consider the octanol-water (K_{ow}) and octanol-air (K_{oa}) partition coefficients. If the K_{ow} and K_{oa} have not been experimentally determined, then the DfE indicates that these values can be estimated using models, including KOWWIN and KOAWIN, available through EPI Suite[18] or SPARC.[19] Another appropriate method for determining these end points can also be used.

PHYSICOCHEMICAL PROPERTIES IN THE COMMITTEE'S FRAMEWORK

After reviewing the research literature and existing frameworks described in Chapter 2, the committee identified a high-priority data set of physicochemical property data. These properties are listed in Table 5-3, together with a brief description of the committee's rationale for their inclusion. In general, the committee selected those physicochemical properties that could support the following uses in an alternatives assessment:

- To identify the chemical or material's potential for posing a direct physical hazard.
- To determine the environmental compartment(s) into which the chemical or material will partition.
- To estimate the potential for the chemical to bioconcentrate[20] and/or be bioavailable.[21]
- To estimate likely route(s) of mammalian exposure and bioavailability, as well the likelihood for high aquatic toxicity.

This section describes in further detail how physicochemical properties can be used to inform an alternatives assessment with respect to evaluation of physical hazards, environmental fate, and human/ecotoxicological toxicity end points.

Physical Hazards

The first step in using physicochemical properties is to determine the likely physical hazards. For much of this report, the primary consideration of the hazards and impacts of a potential chemical substitution is directed toward the post-manufacturing, consumer, use, and end-of-life phases. When considering physical hazards in particular, this focus may shift, as many substances that are considered non-physical hazards at the consumer phase may pose greater risks at the manufacturing and transport stages.[22] The committee believes that for most cases, undesired physical hazard concerns that carry over to the consumer realm are most likely limited to flammability, gases under pressure, oxidizing liquids, and corrosivity to metals. There are, of course, rare cases where these properties might be desired in the final product (fireworks, lighter fluid, etc.).

Identification and classification of the physical hazards posed by potential alternatives allows for direct comparison and consideration of the risk and process safety management strategies that a company would need to develop when a given material or chemical alternative is chosen. As mentioned earlier, the GHS is a useful aid in evaluating and classifying physical hazards (see Table 5-1 for an example of its application to the DfE framework).

---

[18] Estimation Program Interface (EPI) Suite™. The EPI Suite™ is a Windows®-based screening-level tool developed by the EPA's Office of Pollution Prevention Toxics and Syracuse Research Corporation (SRC). This tool provides a suite of physical/chemical property and environmental fate estimation programs.

[19] SPARC (SPARC Performs Automated Reasoning in Chemistry). This EPA predictive modeling system is used to estimate chemical reactivity parameters and physical properties for a wide range of organic molecules. "SPARC is being designed to provide physical properties and chemical reactivity parameters describing factors for air, water and other environmental media needed to develop and apply models such as the Environmental Fate Simulator and Reaction Pathway Simulator" (EPA 2013b).

[20] Bioconcentration is a process leading to a higher concentration of a substance in an organism than in environmental media to which it is exposed (IUPAC 1993).

[21] In toxicology, bioavailability is that fraction of the total amount of material in contact with a portal of entry (lung, gastrointestinal tract, skin) that then enters the blood. In contrast, an ecotoxicologist may define bioavailability as that fraction of material solubilized in the water column under certain conditions of hardness and pH. An aquatic toxicologist might consider contaminants, which are soluble under specific stream conditions, to be bioavailable to fish or benthic organisms (EPA 1994).

[22] One example is household baking flour. On a shelf in a pantry, there is little physical hazard posed to the average home. However, the manufacturing process must be carefully managed to avoid the serious explosive hazard posed by the flammable particles suspended in air.

TABLE 5-3 High-priority Data Set of Physicochemical Processes and Rationale for their Inclusion

Type	Property	Rationale for Inclusion
Physical properties	Flammability	Associated with flammability hazard
	Corrosivity	Associated with ability to gradually destroy materials by chemical reactions
	Oxidizing ability	Associated with ability to give off oxidizing substances or oxidize combustible materials, increasing fire or explosion hazards
	Melting and boiling point	Impacts environmental fate and transport, as well as potential bioavailability
	Vapor pressure	Impacts environmental fate and transport, as well as potential bioavailability
Solvation properties	Acidity (pKa)	Determines ionization state in the environment as well as in biological compartments; ionization state in turn impacts other properties, such as water solubility and partition coefficients, which directly impact toxicokinetics
	Aqueous solubility	Reflects ability to partition into aquatic environment
	Octanol-water partition coefficient (log*P*)	Important determinant of human/mammalian oral and skin bioavailability; relevance to acute & chronic aquatic toxicity (narcosis) and directly related to bioconcentration
	Henry's law constant ($logP_{w/g}$)	Relevance to environmental partitioning and transport as well as human/mammalian alveolar absorption
Electronic parameters	Frontier orbital energies (HOMO, LUMO)	Reflects chemical reactivity with nucleophiles and electrophiles, which translates to reactivity with biomolecules in vivo
	Molecular electronic dipole moments, μ, and dipole polarizabilities, α	Important in determining the energy, geometry, and intermolecular forces of molecules, and are often related to biological activity
Inherent measures of environmental fate	Biodegradation	Indicator of persistence, and persistence is tied to ecotoxicity
	Bioconcentration factor (BCF)	Bioconcentration enhances the hazard potential of lipophilic chemicals; BCFs provide a comparative basis for assessing the potential for a chemical to have effects that resonate through the food chain.

Physicochemical Properties and Environmental Fate: Compartments of Concern

The second step in using physicochemical property data is to determine the environmental compartment(s) into which the chemicals will partition (environmental partitioning). The chemical's physical state, which can be predicted on the basis of melting point, boiling point, and vapor pressure, can indicate which environmental compartments—air, water, sediment, biota, soil— into which the chemical will partition. Highly volatile chemicals, for example, will escape from soil or water and primarily be present in the air. Conversely, chemicals with a high propensity to sorb onto organic carbon or move into lipid phases are likely to remain in soils or sediments or move into biota, respectively.

Aqueous solubility will provide information about whether a chemical will dissolve in water, a starting point for understanding its fate and transport into the water column or sediment. This understanding is further enhanced by knowledge of phase partition coefficient, log*P*. In general, chemicals with higher log*P* values will be more likely to cross into and be retained by biota, although there are significant exceptions to this rule (e.g., large molecules that cannot cross biological membranes). Chemicals with the propensity to environmentally partition into sediments will be more likely to sorb onto soils, so the soil-water phase partition coefficient (K_d) will be informative for both systems. For some chemicals, transformation processes need to be considered because transformed or metabolized products often have different physicochemical properties; thus, they may reside in different environmental compartments.

To obtain a sense of the escape potential for a given chemical, unit world models developed by Mackay and Paterson (1991) for organic chemicals and by Diamond et al. (1990) for metals provide a structural framework for determining potential chemical distribution based on intrinsic properties of fugacity (f), which is an inherent chemical property that governs the relative concentrations of chemicals in different environmental and biotic compartments. These models have been applied to ecological systems by Harvey et al. (2007) and Farley et al. (2011), who also provide examples of how to apply these model concepts to hazard assessments. Their application will be discussed in more detail in the next section.

Physicochemical Properties and Persistence, Bioaccumulation, and Biotransformation

Organisms take up and eliminate chemicals at different rates; when excretion or metabolic detoxification is slower than uptake, the chemical (or chemicals) accumulates in the organism, resulting in prolonged tissue delivery (Luoma and Rainbow 2005). Some chemicals increase in concentration at each level of the food chain; these chemicals are said to *biomagnify*. Because of the potential of some chemicals to biomagnify and persist in the food web, bioaccumulative substances require special consideration because they may pose a greater hazard than chemicals that are rapidly eliminated and do not accumulate.

Large chemical structural databases combined with recent developments in quantitative structure property relationships have greatly expanded the potential for rapid assessment of chemicals (Howard and Muir 2011). Thus far, efforts using these models have proven successful in identifying potentially persistent, bioaccumulative, and toxic (PBT) chemicals. (For more information about the application of physicochemical data to ecotoxicology, see Chapter 7).

Use of Physicochemical Properties to Predict the Persistence of Organic Chemicals

As defined by Pavan and Worth, "the *persistence* of a substance is the length of time it remains in a particular environment before it is physically transported to another compartment and/or is chemically or biologically transformed" (Pavan and Worth 2008). Most alternatives assessment frameworks consider persistence because molecules that persist will have increased concentrations, and possibly higher impacts, in environmental compartments and are more likely to bioaccumulate. For some classes of materials, it is possible to obtain useful, predictive information about potential persistence from physicochemical data, such as structural markers on the molecule and partition coefficients. For example, Howard and Muir (2010) screened more than 2,2000 commercial chemicals with in silico and expert judgment approaches and identified physicochemical properties that could be used to classify chemicals as persistent in the atmosphere (atmospheric oxidation half-life > 2 days), or potentially susceptible to long-range transport (log*P* > -5 and < -1).

The persistence of a chemical in the environment is often measured, or estimated, in terms of its *biodegradation*.[23] There are numerous modes of degradation that depend on the environmental conditions, types of microbes present, and the structure of the chemical. Degradation is typically quantified based on the extent of removal of dissolved organic compounds within an aqueous medium of a chemical and is expressed as a percentage of degradation in a given time.[24]

Degradation is usually a complex, multistep process that often produces chemical intermediates. These intermediates may present additional environmental hazards or persist if they are not readily degraded. Some transformations can increase the toxicity of the parent compound (e.g., methylation of mercury; photoinduction of polycyclic aromatic hydrocarbons [PAHs]), whereas other reactions may decrease the toxicity of a chemical. Biodegradation can also change the distribution of components within environmental compartments (e.g., due to ion formation from inorganic chemicals).

Chemical degradation, a subset of biodegradation, may involve a number of chemical reaction steps depending on the environmental conditions and the chemical structure (Khetan and Collins 2007). Chemical degradation processes include hydrolysis, photochemical transformations, and the action of microbial species (Khetan and Collins2007). The modes of degradation depend on the environmental compartment and conditions (pH, UV irradiation, microorganism population, etc). The diversity of conditions and chemical reactivities means that the results of degradation testing are sensitive to the conditions of the test. To complement those tests, researchers have developed some "rules of thumb" (see Box 5-2) to estimate or predict chemical degradation based on chemical functional groups and structure.

In addition, some models and databases have been developed to predict degradation rates (Arnot et al 2005). Examples of these models include:

BOX 5-2
Structural Attributes that Enhance Biodegradation

- Minimal number of halogens (especially F and Cl).
- Minimal chemical branching (especially quaternary C).
- Minimal number of tertiary amine, nitro, nitroso, azo, and arylamino groups.
- Minimal number of polycyclic residues (especially more than three fused rings).
- Presence of esters (including phosphonates).
- Presence of oxygen atoms.
- Presence of short linear alkyl chains (< 4 C) or phenyl rings that can act as sites for oxygenase enzyme activity.

SOURCE: Meylan et al. 2007; Howard and Muir 2013.

- Group contribution models that estimate and predict thermodynamic and other properties from molecular structures; for example, BIOWIN (Boethling et al 2004).
- Expert judgment criteria for biodegradability based upon "rule of thumb" models (Meylan et al. 2007).
- Degradation pathways model, including probabilistic models that calculate the probabilities of the individual transformations; for example, CATABOL (Dimitrov et al. 2007).

Several research challenges remain with respect to obtaining biodegradation data. These challenges include:

- Predicting degradation fragments.
- The need to develop more predictive structure/degradation relationships (SDRs) for parent chemicals and degradates.
- Predicting the rates of degradation for a new or previously unstudied chemical.

There are also a number of other factors that need to be considered when evaluating measured or predicted degradation data of organic chemicals. These include:

- Potential trade-offs between aquatic toxicity and degradation. Improving biodegradation often increases aquatic toxicity and may reduce durability.
- The initial stages of polymer degradation may make components more bioavailable until they are later degraded (Platt 2006). This can be a

[23] Biodegradation is the process by which microbial organisms transform or alter (through metabolic or enzymatic action) the structure of chemicals introduced into the environment (EPA 2012f).

[24] In the design of chemicals and components for formulating products, there is a tension between stability and degradation. During use, the chemical is expected to be stable. Upon release into the environment, the chemical should rapidly degrade. Ideally, it should form degradates that do not persist and are less toxic than the parent chemical. This type of strategy is one of the guiding principles of green chemistry (EPA 2014b).

major difference from the degradation of smaller molecules.

Use of Physicochemical Properties to Predict the Bioavailability of Inorganic Chemicals

Characterizing the lifetime of metals in the environment is difficult because the interactions are highly dependent on the characteristics of the environmental system where they are released. Leaching[25] and aging[26] are related to conditions within the soils and sediments, so predicting environmental hazard based solely on standard aquatic toxicity tests using dissolvable salts is not adequate. Therefore, the field of ecotoxicology is becoming increasingly reliant on sediment and soil toxicity test protocols that include leaching and aging steps or the application of bioavailability models to adjust data acquired under laboratory conditions to realistic conditions in soils, sediments, and water (Santore et al. 2002; Smolders et al. 2009).

Metals newly introduced into soils or sediments are more bioavailable than those that have aged for months to years. Metals are initially leached from soils or sediments, a process that happens relatively quickly (i.e., weeks to months), followed by a slow aging process (i.e., years), which results in decreased toxicity to sediment or soil organisms over time. Therefore, toxicity studies conducted with soil or sediment freshly amended with metal salts will result in effects at much lower concentrations than will be observed in real-world situations (Besser et al. 2011).

Aging occurs due to several different processes, including sorption to aluminum, manganese, or iron oxides and eventual incorporation of the metal ion into the crystalline structure of the mineral soil or sediment particles (Adriano 2001). The rate of chemical sorption to oxides, clays, other minerals, or organic matter is determined by the strength and number of negatively charged binding sites in the soil or sediment particle which, in turn, are influenced by the amount of aluminum, iron, or silicon present. Sorption reactions are reversible and highly dependent on pH, with higher rates of sorption occurring at higher pH, increasing bioavailability as pH decreases (i.e., becomes more acidic). Redox potential influences the bioavailability of cationic metals; highly insoluble sulfides of metals form under reducing conditions, such as those found in saturated soils or anoxic sediments. Therefore, the type of environment to which the metal is introduced also influences the degree of toxicity that might be expected, although this differs by metal. The strength of attraction between metal ions and charged sites is a function of the affinity of the metal to the charged site relative to its affinity for water molecules. Copper generally has the highest rate of sorption, followed in descending order by nickel, cobalt, lead, cadmium, and zinc. This order differs slightly for electrostatic binding to clays and other negatively charged particles, with nickel having the highest binding affinity and lead the lowest.

Binding affinity also influences the toxicity of cationic metals. For example, the gill of aquatic organisms is negatively charged and acts as another binding site for some metal ions (Playle 2004). Toxicity depends on the relative binding strength of the biotic ligand and other negatively charged particles in the water (e.g., organic matter, iron sulfides) and competition for the binding sites by other metals. The biotic ligand model can be used to predict toxicity for a given metal if the concentrations of other major cations are known (DiToro et al. 2001). This model adjusts values from standard toxicity tests to different types of aquatic environments and may affect the relative hazard of the different metals.

Anionic metals and metalloids such as molybdenum, arsenic, mercury, and selenium also bind to iron oxides, but binding decreases with increasing pH, which is opposite to what occurs with cationic metals. Therefore, toxicity of these metals differs substantially from that of the cationic metals in the same environment. Furthermore, methylation of metalloids plays a very important role in increasing their mobility and uptake as well as their ability to biomagnify in the food chain. Well-known examples of this phenomenon is the observation that methylated arsenic is less toxic than its inorganic form, while methylated mercury and organoselenium species are more toxic. Methylation is a biological process that occurs in bacteria, with the initial step occurring in sediments under reducing conditions (low oxygen) and the presence of high organic matter (Jonnalagadda and Rao 1993).

Because plants, invertebrates, and soil microorganisms interact with the soil or sediment pore water, the amount of free metal ions in solution is the most important determinant of toxicity. Plant roots may exude phytochelatins that

[25] Leaching is the process by which soluble materials in the soil, such as salts, are washed into a lower layer of soil or dissolved and carried away by water (USGS 2014).
[26] Aging refers to reduced bioavailability over time (Kelsey and Alexander 1997).

bind metals to either facilitate or exclude their uptake, while at the same time reducing pH of the soil or sediment to make nutrients more bioavailable (Pal and Rai 2010).

Cationic metals can occur as various ionic species, some of which are more soluble and therefore more bioavailable than others. For example, chromium is present in solution as Cr^{+6}, which is a highly bioavailable and toxic ion, while it is bound to soils and most sediments as Cr^{+3}, which is less toxic. Similarly, antimony trioxide (SbO_3) is highly insoluble, whereas antimony trichloride ($SbCl_3$) is not, which makes the latter less bioavailable. Toxicity studies with the soluble species of a metal, where the free ion is readily available, are useful for predicting effects to aquatic organisms, but generally are of little predictive value for soil or sediment organisms, largely due to the length of time needed for dissolution into the pore water and the confounding factors of pH and salinity from the added chloride (Smolders et al. 2009). In light of this complexity, no one physicochemical property or set of properties is currently adequate to define all toxicity concerns if metals are present in the structure of the compound and have the potential to become freely available during the degradation process. Toxicity testing and evaluation as described in the following chapters should be carried out to identify concerns related to the presence of metals in a compound.

Use of Physicochemical Properties to Predict Bioaccumulation

Bioaccumulation potential (B) is represented in most alternatives assessment schemes by the bioconcentration factor (BCF). The BCF is the ratio of the amount of chemical in an aquatic organism (usually fish) to the amount of chemical in the water under conditions of equilibrium. An alternative approach is to measure the bioaccumulation factor (BAF), which is the ratio of the amount of chemical in the fish to the amount in both food and water, expressed on a molar basis and frequently normalized to lipid content. Standard test protocols for these factors are available, but may be difficult to conduct and interpret due to several factors described in the literature (Fraunhofer Institute 2007).

Log*P* is a good surrogate for determining the extent to which a chemical would thermodynamically distribute between the lipids of biological organisms and water. In general, very lipophilic substances (ones with log*P* > 5) have the greatest potential to bioaccumulate. However, lipophilicity also affects whether a chemical will be taken up by the organism (i.e., its bioavailability). For example, chemicals with log*P* > 5 are primarily taken up from the diet, and the BAF is higher than the BCF. Chemicals with log*P* < 5 are primarily absorbed from the water, and the BCF and BAF are equal (Mackay et al. 2013). Aquatic organisms may need to be exposed to chemically treated water for 60 days or more before reaching chemical equilibrium. This is true for chemicals with slow excretion or metabolism rates, during which time dilution by growth generally occurs. While the BAF provides a more realistic measure of exposure to hydrophobic chemicals, additional uncertainty is introduced because the BAF includes partitioning of the chemical between water and food and simplifying assumptions about dietary preferences (Mackay et al. 2013). However, given that standard protocols have been developed to provide guidance for conducting BAF tests to enable comparability among chemicals, these data should be given preference over BCF values for estimating bioaccumulation potential in hazard classification and ranking. Note that for regulatory schemes where hazard classification is required (such as REACH or the GHS for transportation labeling), binning chemicals by whether they are non-accumulative (BCF < 2,000), somewhat accumulative (2,000 < BCF < 5,000) or very bioaccumulative (BCF > 5,000) is sufficient.

In the absence of measured BAF, it is theoretically possible to calculate the BAF from a measured BCF. Bioaccumulation for fish (BAF_F) is the bioconcentration factor based on freely dissolved chemical concentration (BCF_D) for its food items (phytoplankton such as algae) times the ratio of the uptake rate from the diet (K_D) and the uptake rate from the water via respiration (K_R). This is expressed as $BAF_F = (1 + K_D / K_R) \times BCF_D$. Furthermore, the tendency for a chemical to biomagnify can be quantified by the ratio of two trophic levels (BAF_2/BAF_1), with BAF_2 being a higher trophic level than BAF_1. However, because the diet for higher trophic-level species includes species that have a BAF_F, calculating the BCF_D can become quite complex.

A recent article addresses the question of whether BCF or BAF should be used to predict bioaccumulation potential. It concludes that for BCF and BAF values predicted by the EPA's EPI SUITE software, both BCF and BAF values provide comparable information (Costanza et al. 2012). The threshold values proposed by Costanza and coworkers (2012) are as follows:

- Not significantly bioaccumulative: BCF or BAF < 1,000.
- Bioaccumulative: BCF or BAF 1,000 and < 5,000.
- Highly bioaccumulative: BCF or BAF > 5,000.

Another approach for comparing bioaccumulation potential between chemicals is the use of chemical fugacity. Fugacity (f), expressed as Pascals (Pa), is an inherent chemical property that governs the relative concentrations of chemicals in different environmental and biotic compartments. Each type of media (air, water, lipid, biota) has an inherent fugacity capacity (Z) that defines the amount of a chemical fugacity that can be retained within that material, where Z is expressed as (mol/m^3 x Pa). Therefore, the ability of a chemical to bioaccumulate in any organism is a function of its chemical fugacity and the fugacity capacity of that organism for that chemical. Fugacity ratios between biota and their environment can be compared among chemicals to determine which chemical is most likely to bioaccumulate, or comparisons can be made between trophic levels to determine biomagnification potential (Burkhard et al. 2012; Mackay et al. 2013). Because fugacity capacity is a function of the Henry's Law constant and the log*P* for each chemical, these properties can be used to rank chemicals on their potential to bioaccumulate. The fugacity capacity for water, Z_{water}, is equal to 1/H, where H is the Henry's Law constant for the target chemical. Therefore, in a closed system, a chemical with a smaller Henry's Law constant will partition to a greater extent in water than one with a larger Henry's Law constant. For an organism, fugacity capacity is equivalent to Z_{water} times the *P* and percent lipid. Therefore, a chemical with a small Henry's Law constant and high log*P* will be most likely to bioaccumulate, and a comparative ranking scheme can be developed based on the ratio of these two parameters.

Current hazard classification and ranking schemes use BCF and BAF for aquatic organisms. Questions remain about whether aquatic BCF and BAF values are predictive for terrestrial organisms, where uptake into the food chain begins with movement of chemicals from soils into plants. Plant uptake of chemicals is highly related to soil sorptive properties of the chemical, solubility into soil pore water, and active uptake by plants. Terrestrial animals have different amounts of lipids than fish, making it questionable to directly extrapolate fish BAF values to birds and mammals. The relative fugacity approach described above, however, is equally applicable to terrestrial and aquatic systems, so it may form the basis of an approach for determining bioaccumulation (and biomagnification) potential in terrestrial systems.

Use of Physicochemical Properties to Predict Aquatic Bioavailability and Toxicity

Bioavailability is a measure of the amount of a chemical and the rate at which it crosses a barrier of the external environment and enters an organism's circulation. From there, the chemical can reach tissues in living systems and interact with cellular macromolecules. Adapted from the study of metals in the environment, chemical bioaccessibility, or environmental availability, can be defined as the amount of a chemical "in soil, sediment, water, or air that is available for physical, chemical, and biological modifying influences (e.g., fate, transport, and bioaccumulation)" (McGeer et al. 2004). For a chemical to exert a toxic effect, it typically must be bioavailable at a level that allows the chemical (or its metabolite) to reach a biochemical target, where it can exert its toxicologic effect. Blocking or reducing bioavailability is one potential means for reducing the intrinsic toxicity of a chemical (Voutchkova et al. 2010). While the lack of bioavailability is an indicator that the compound is likely to have low toxicity, high bioavailability does not suggest the compound is necessarily highly toxic.

Aquatic bioavailability: The scientific literature characterizes trends that allow comparative assessment of bioavailability in different species and through different routes of exposure. For example, in aquatic species, it is known that bioavailability is positively correlated with log*P* of the chemical, although the linearity of this relationship is not clearly defined (Pärt 1989). It is also known that aqueous solubility, molecular size, and ionization state also influence bioavailability. The log*P* at environmental or biological pH (i.e., logD) has been proposed as a measure that correlates with partitioning and ionization. The Biotic Ligand Model (Janssen et al. 2003) is useful when considering metal bioavailability to aquatic species as it relates competitive metal binding to ecotoxicological effects (Tessier and Turner 1996).

Aquatic toxicity. An example of how physicochemical properties can be directly used to estimate an ecotoxicological end point is acute aquatic toxicity. The physicochemical property limits listed in Table 5-4 are known to favor reduced acute and/or chronic aquatic toxicity. Meeting two or three property limits has been shown to substantially increase the probability that a chemical will have low

TABLE 5-4 Changes in Physicochemical Properties to Favor Reduced Aquatic Toxicity

Physicochemical Property	Changes
Molecular size and weight	Generally, as molecular weight increases, aquatic bioavailability and toxicity decrease. At MW > 1000 amu, bioavailability is negligible. Caution must be taken, however, to consider possible breakdown products that may have MW < 1000 amu and exert toxicity.
Octanol-water partition Coefficient (logP) and octanol-water distribution coefficient at biological pH ($logD_{7.4}$)	log*P* usually correlates exponentially with acute aquatic toxicity by narcosis for non-ionic organic chemicals up to a value of about 5-7. Chemicals with log*P* < 2 have a higher probability of having low acute and chronic aquatic toxicity (Voutchkova et al. 2011). For ionizable organic chemicals, $logD_{7.4}$ is a more appropriate measure: ionizable compounds with $logD_{7.4}$ < 1.7 have been shown to have increased probability of being safe to freshwater fish than those with $logD_{7.4}$ > 1.7 (Kostal et al. in press).
Water solubility	Generally, compounds with higher log*P* have lower water solubility. Very poorly water-soluble chemicals (<1 ppb) generally have low bioavailability and are less toxic.
ΔE energy [HOMO-LUMO]	The ΔE reflects broad chemical reactivity. It was recently reported that chemicals with ΔE > 9 eV (as calculated by semi-empirical methods) or > 6.5 eV (as calculated by DFT) are much less likely to be acutely or chronically toxic to aquatic species (Voutchkova-Kostal et al. 2012; Kostal et al. in press).

or no aquatic toxicity. This is one example of the use of global reactivity parameters to assess fundamental chemical reactivity that relates to biological activity, but other approaches may exist.

Use of Physicochemical Properties to Estimate Mammalian/Human Toxicokinetics

In addition to the use of physicochemical data to predict aquatic toxicity, these properties can also be used to estimate the toxicity of a given chemical in humans and other animals as they influence toxicokinetic and toxicodynamic parameters.[27] While the toxicodynamic interactions of chemicals are very challenging to relate to specific physicochemical properties, the influence of such properties on toxicokinetic behavior of chemicals can be more readily defined and used to prioritize the human health assessment of chemical alternatives.

The key toxicokinetic processes are absorption, distribution, metabolism, and excretion. The focus here is on the influence of physicochemical properties on the rate of absorption of a chemical into the bloodstream, its distribution to the organs and tissues, and its rate of elimination (clearance) of a compound.

The most prominent properties that have been shown to impact chemical toxicokinetics include:

1. molecular size and shape,
2. lipophilicity and hydrophobicity,
3. ionization potential or pKa, and
4. hydrogen bonding.

Physicochemical Properties That Influence Bioavailability in Humans

Chemicals that are highly bioavailable to mammals through particular exposure routes have also been defined by a set of property limits. These property limits were originally defined to assess the probability of drug candidates entering the human body, and are therefore highly dependent on the expected route of exposure. The property limits

[27] Note that "pharmacokinetic" and "pharmacodynamic" are often used interchangeably with the terms "toxicokinetic" and "toxicodynamic."

TABLE 5-5 Combinations of Property Limits Associated with Increased Bioavailability through the Four Main Routes of Exposure in Mammals

Exposure Route	Physicochemical Property	Property Limit
Ocular	Water solubility	Variable
	Molecular size	< 500 Da (corneal epithelium) < 10000 Da (conjunctival epithelium)
	Vapor pressure	< 0.0001 mm Hg
Oral	Molecular size	< 500 Da
	Log*P*	0-5
	Non-ionized at GI tract pH	-----
Respiratory (Lungs)	Particle size	< 5 µm
	Molecular size	< 400 Da
	Vapor pressure	< 0.0001 mm Hg
Dermal	Molecular size	< 400 Da
	Log*P*	0-6
	Presence of solvents	-----
	Ionization (polar, ionized)	-----

associated with increased bioavailability through the four main routes of exposure have been discussed in detail in the medicinal chemistry literature, and review articles are available (DeVito, and Garrett 1996; Voutchkova et al. 2010). These property limits are provided in Table 5-5 and are further discussed in Chapter 8. The inverse of these property limits is likely to increase the probability of minimal human bioavailability, but concrete studies to support this assertion are still lacking.

Ocular bioavailability: The topical delivery of pharmaceuticals for the treatment of the anterior segment of the eye (i.e., cornea, conjunctiva, sclera, anterior uvea), where the bulk of the research in this area has been done, has proven challenging, largely due to the complex structure and variety of clearance pathways and barriers that can reduce absorption and remove xenobiotics from the eye. For example, the flow of lacrimal fluid quickly removes most instilled compounds from the surface of the eye.

Mechanism of delivery and exposure most relevant for the consideration of chemical alternatives is that of direct absorption through the cornea or through systemic exposure; this is reflected in the values presented in Table 5-5. The vapor pressure of the material reflects the potential for gas-phase exposure to the compound. The importance of molecular size reflects the paracellular pore sizes in the corneal and conjunctival epitheliums, and lipophilicity appears to affect the route of entry into the body, whether through the cornea (reduced absorption of compounds with high lipophilicity) or the conjunctiva (where lipophilicity appears to play no role in absorption).

Oral bioavailability: As defined by Varma et al., "Oral bioavailability (F) is a product of fraction absorbed (F_a), fraction escaping gut-wall elimination (F_g), and the fraction escaping hepatic elimination (F_h)" (Varma et al. 2010). The property limits for oral bioavailability are well characterized. Lipinski identified four physicochemical properties that govern optimal oral absorption: molecular weight (MW) < 500 amu; octanol/water partition coefficient (log*P*) < 5; number of hydrogen bond donor atoms (HBD) < 5; and the number of hydrogen bond acceptor atoms (HBA) < 10 (Lipinski et al. 1997). Although there are numerous exceptions (Ganesan 2008), chemicals are generally less likely to have good oral absorption if they violate two or more of these physicochemical "rules."

Varma and coworkers (2010) also evaluated the physicochemical space for optimum human oral bioavailability. They showed that molecular weight, ionization state, lipophilicity, polar descriptors, and free rotatable bonds (RB) influenced oral bioavailability, stating that:

> These trends were due to a combination of effects of the properties on F_a and first-pass elimination (F_g and F_h). Higher [molecular weight] significantly impacted F_a, while F_g and F_h decreased with increasing lipophilicity. Parabolic trends were observed for bioavailability with polar descriptors. Interestingly, RB has a negative effect on all three parameters, leading to its pronounced effect on bioavailability (Varma et al., 2010).

Dermal bioavailability: Dermal or topical absorption predictive models have been in existence since the early 1990s, when Potts and Guy (1992) published a simple model that showed a relationship between the molecular volume or molecular weight and the lipophilicity of a chemical and its ability to permeate the skin. Although many other models have been proposed and published, most rely on related properties to determine the skin permeation rate. A framework incorporating the impact of exposure scenarios and application conditions on risk assessment of chemicals applied to skin is described in a number of key references (Ibrahim et al. 2012).

Respiratory bioavailability: Nasal uptake and regional deposition are influenced by the physical and chemical properties of the inhaled material, including water solubility, reactivity, and airborne concentration (Morgan and Monticello 1990). The pharmacokinetics of inhaled particles is also dependent upon physicochemical properties of the particles, including aerodynamic diameter (size) and solubility (Kreyling et al. 2013). The size of the particle will influence where it deposits within the respiratory tract; for example, particles under 1 μm penetrate to the alveoli and over 30 μm rarely progress farther than the upper respiratory tract.[28]

[28] Note that larger particles may not be inhaled, but upon deposition in the nose, mouth, and throat may still enter the body by mucociliary clearance and ingestion.

Knowledge of the particle size distribution of any powder, mist, aerosol, or other similar material is important for identifying hazards that should be eliminated or managed through the use of appropriate engineering, procedural, and personal protective equipment control at the sites of manufacture and use.

In addition to size, other physical and chemical properties can also influence transpulmonary transport (Holder 2012; Ibrahim and Garcia-Contreras 2013). These include molecular weight, melting point, boiling point, vapor point, molecular polarity, Henry's phase distribution, and the extrinsic properties of pressure (P) and moles (n). Localized tissue responses and respiratory tract absorption of deposited metals are also highly dependent upon chemical solubility, particle size, and surface area, which contribute to metal release from the inhaled particle (Kang et al. 2011; Oberdorster 1996).

Physicochemical Properties that Influence Distribution in Living Organisms

Volume of distribution (V_d): One important estimate of a compound's distribution that has been demonstrated to have a link to toxicity in animals is the volume of distribution, V_d. If a quantity of compound is introduced into the body, some amount will enter into the bloodstream and some will undergo different processes that remove it from the bloodstream, such as uptake by tissues and elimination from the body. V_d is defined as the theoretical volume of blood plasma required to achieve that concentration if no removal processes were occurring. If V_d is roughly equivalent to the total blood volume of the organism or individual, then no uptake is occurring. If V_d is higher than the total blood volume of the organism, then it indicates that some amount of compound has been lost from the bloodstream by those processes. The higher the V_d, the greater the distribution of the compound thoughout the body is likely to be. Those drugs that are lipophilic at pH 7.4 are likely to have higher values of V_d than those that are ionized or those that have a high affinity for plasma binding protein. The V_d directly influences the half-life of a compound, whereby large V_d leads to a longer half-life; that is, it prolongs the duration of exposure. The V_d has also been shown to influence the lowest observable adverse effect level (LOAEL). In rodent studies, a larger value for V_d generally results in a lower LOAEL (Sutherland et al. 2012).

Plasma protein binding (PPB): In general, xenobiotics within in vivo systems are either (i) bound to proteins and lipids in plasma (more commonly referred to as plasma protein binding [PPB]), (ii) bound to proteins and lipids in tissues, or (iii) unbound and free to diffuse among the aqueous environment of the blood and tissues (Smith et al. 2010). PPB strongly influences V_d and the half-life of chemicals in the body (Hollósy et al. 2006) because it is typically the unbound fraction of xenobiotics that interacts with protein receptors, forms DNA adducts, or interferes with a biological system in other ways to produce either a pharmacologic or toxicologic effect. Studies have shown that chemicals that interact with a protein receptor (e.g., the estrogen receptor) and are also highly bound to plasma proteins, will generally require higher doses to achieve the required free concentrations to elicit an equivalent response to a chemical that has a lower PPB level, provided the rate and fraction absorbed for both are equivalent. Physicochemical properties that influence PPB include lipophilicity, as measured by log*P*, and pKa. In general, chemicals with high lipophilicity and/or ones with acidic character will have a smaller unbound fraction, and thus a greater degree of PPB, than more hydrophilic or basic compounds (Vallianatou et al. 2013).

Physicochemical Properties that Influence Elimination/Clearance in Living Organisms

Clearance (CL) describes the rate of elimination of a given chemical to its concentration in plasma and is expressed as volume of distribution cleared per unit time. Total clearance describes the elimination of a chemical from the body without identifying the mechanisms involved (e.g., metabolism, urinary or biliary excretion, etc), but most chemicals are eliminated primarily via the liver and/or kidney.

Clearance is one of the most important pharmacokinetic parameters. It is affected significantly by the PPB of the chemical, because only the free fraction can be cleared. The clearance of the unbound chemical, CL_u, is independent of the PPB. Thus, CL_u only depends on chemical structure and physicochemical properties. For example, the rate of clearance is heavily dependent on the distribution coefficient of the chemical at biological pH (7.4), expressed as log $D_{7.4}$ (Zhivkova and Doytchinova 2013).

In sum, examining physicochemical properties can be used to help screen chemicals for their potential to induce human toxicity. For example, the lack of bioavailability and high clearance often indicate that the compound is likely to have low mammalian toxicity. However, high bioavailability and low clearance do not necessarily indicate that the compound is highly toxic. More retrospective and prospective analyses are needed to inform decisions about the use of materials that pose environmental risks. In some cases, development of specialized analytic methodology will be required. Continued assessment of known hazardous compounds will be important. For the present and for the immediate future, decisions will have to be made on the basis of limited data and information.

IMPLEMENTATION OF STEP 5 IN THE COMMITTEE'S FRAMEWORK

The implementation of Step 5 requires a comparative approach to the evaluation of the chemical of concern and its alternatives. In essence, information concerning the chemical of interest serves as a "baseline" for all subsequent comparisons. Completion of this step requires several broad activities, including:

a. Identification of the chemical of interest, chemical alternatives, and their most likely degradates or metabolites. Whenever possible, primary data about the identity and structure of the degradates and metabolites should be used. A variety of software tools can also be used to predict degradate and metabolite structures. Chemical identity includes the chemical name, chemical formula and structure, and whenever possible, the Chemical Abstracts Service (CAS) registry number.
b. Compilation of the minimal data set described in Table 5-3. Data should be compiled for the chemical of interest (serves as the baseline for subsequent comparisons), chemical alternatives, and their most likely degradates or metabolites. Physicochemical data to be collected and analyzed can be either measured or estimated values. Missing data should be clearly identified as such. All data sources, including software programs used to estimate physicochemical parameters, should be documented, and judicious awareness of the applicability domain of the estimation tool(s) should be used. The completed data set should be represented in a tabular or graphical display.
c. The compiled data should be categorized in such a way to determine the relative difference (such as high, medium, or low) between the physicochemical property of a chemical alternative and the chemical of interest. Widely accepted categorization tools like GHS available

for some physicochemical properties (e.g., flammability, corrosivity, oxidizing ability) should be used. The committee also provided categorization systems used by other frameworks for logP, vapor pressure, and several other physicochemical properties of interest. In some cases (e.g., aquatic solubility), the comparison of a physicochemical property is intended to identify potential differences in the environmental compartment(s) into which the chemical or material will partition. As a minimum, the identity of the environmental compartments of concern should be documented. In other cases, secondary end points (e.g., bioconcentration factor, or BCF) could be estimated from the physicochemical property data. Categorization schemes for BCF are also available in other frameworks and tools, such as GreenScreen®, and could be used with the committee's framework.

d. Some physicochemical data can be used to estimate likely route(s) of mammalian exposure and bioavailability, as well the likelihood for high aquatic toxicity. Information gleaned for physicochemical properties should be made available to members of the assessment team performing Step 6 (comparative exposure assessment, ecotoxicity hazard assessment, and human health hazard assessment).

e. Compilation of physicochemical property data may require an iterative approach. For example, the evaluation of degradates and metabolites may occur at later stages of the alternatives assessment process. Staging of effort may increase efficiency when a large number of alternatives are initially identified. In this case, some alternatives may be removed from consideration because of other factors (e.g., inherent toxicity).

f. It is not typically anticipated that a compound will be eliminated from consideration based on physicochemical properties alone. The exception to this would likely occur in the case where property data for a chemical reveal a high risk of physical hazards, such as flammability and explosibility, especially when these are not desirable properties of the alternative. Elimination of chemicals with undesirable physical hazards may be particularly critical if the consumer will be directly exposed to the chemical in question (as opposed to an intermediate in a production/synthesis process, which is only handled under controlled conditions).

6

Comparative Exposure Assessment

Exposure assessment is the process of considering and estimating the extent of exposure of human and ecological receptors.[29] *Comparative exposure assessment* plays an important role in the committee's alternatives assessment process in understanding the overall safety of alternatives (Figure 6-1). The committee's approach to exposure is to: a) consider the potential for reduced exposure due to inherent properties of the alternative chemicals; b) ensure that any substantive changes to the routes and any substantive increases in the levels of exposure are identified; and c) allow for consideration of the routes (dermal, oral, inhalation, etc.), patterns (acute, chronic), and levels of exposure (irrespective of any exposure controls) when integrating the evidence related to human and ecological toxicity among alternatives (Step 7 in Chapter 9).

In this chapter, the committee provides an overview of the approach to exposure assessment employed in other alternatives assessment frameworks. The committee then focuses on its framework and expands on the earlier discussion of exposure as it relates to scoping and bounding the assessment (Step 3; see Chapter 4). It is important to note that the consideration of exposure in the committee's framework is not to demonstrate "safe" levels of exposure. Instead, it is comparative and is focused on the intrinsic potential for exposure without physical or administrative controls. In this way, the committee's approach is different than most other approaches outlined in the frameworks reviewed.

The final section of the chapter goes through the sub-steps to be taken to complete Step 6.3, the comparative exposure assessment. Box 6-1 presents the elements of the committee's suggested approach.

EXPOSURE ASSESSMENT IN EXISTING FRAMEWORKS

The committee considered the role of exposure assessment in existing frameworks. The role of exposure and how it is determined varies significantly from framework to framework, and depends on several factors. These include the objective or focus of the alternatives assessment, regulatory requirements, framework policies and procedures, and how the alternatives assessment results are used. Many of the existing frameworks only consider exposure to a cursory degree, such as considering intrinsic properties that influence persistence or bioaccumulation. This is often because exposures of alternatives are assumed to be the same, or "substantially equivalent" to each other and/or an original chemical of concern. This assumption allows the user of these frameworks to primarily focus on reducing toxicological indicators of hazard. Therefore, when an exposure assessment is included, it may be used in a secondary role, to confirm that the alternatives that appear acceptable or preferable to the chemical of concern from a toxicological perspective are not clearly worse from an exposure perspective. Some frameworks (e.g., TURI 2006a) include exposure potential (environmental, occupational, and public health) in their preliminary prioritization of chemical alternatives. Information such as the mobility of the chemical for a particular use, and potential for user exposure when the chemical is in a product, is used to determine the exposure potential of a chemical. In addition, TURI uses occupational exposure limits as measures of acute toxicity in comparative chemical hazard assessments.

In some frameworks (e.g., BizNGO (Rossi et al. 2012), an exposure assessment is performed only after alternatives are first identified based on hazard assessments. In this instance, the exposure assessment may be initiated based on the results of applying Life Cycle Thinking (see Chapter 10), meaning that potential impacts to human health or the environment across the life cycle of the alternative are considered. If the exposure assessment identifies concerns, then a partial or full risk assessment would be conducted, (depending on resources) to assess health effects. Similarly, a full or partial life cycle assessment (depending on resources and needs) would be conducted to assess remaining environmental impacts. It is unclear, however, how

[29] Ecological receptors can include tissues, organisms, populations, communities, and ecosystems (EPA 2014d).

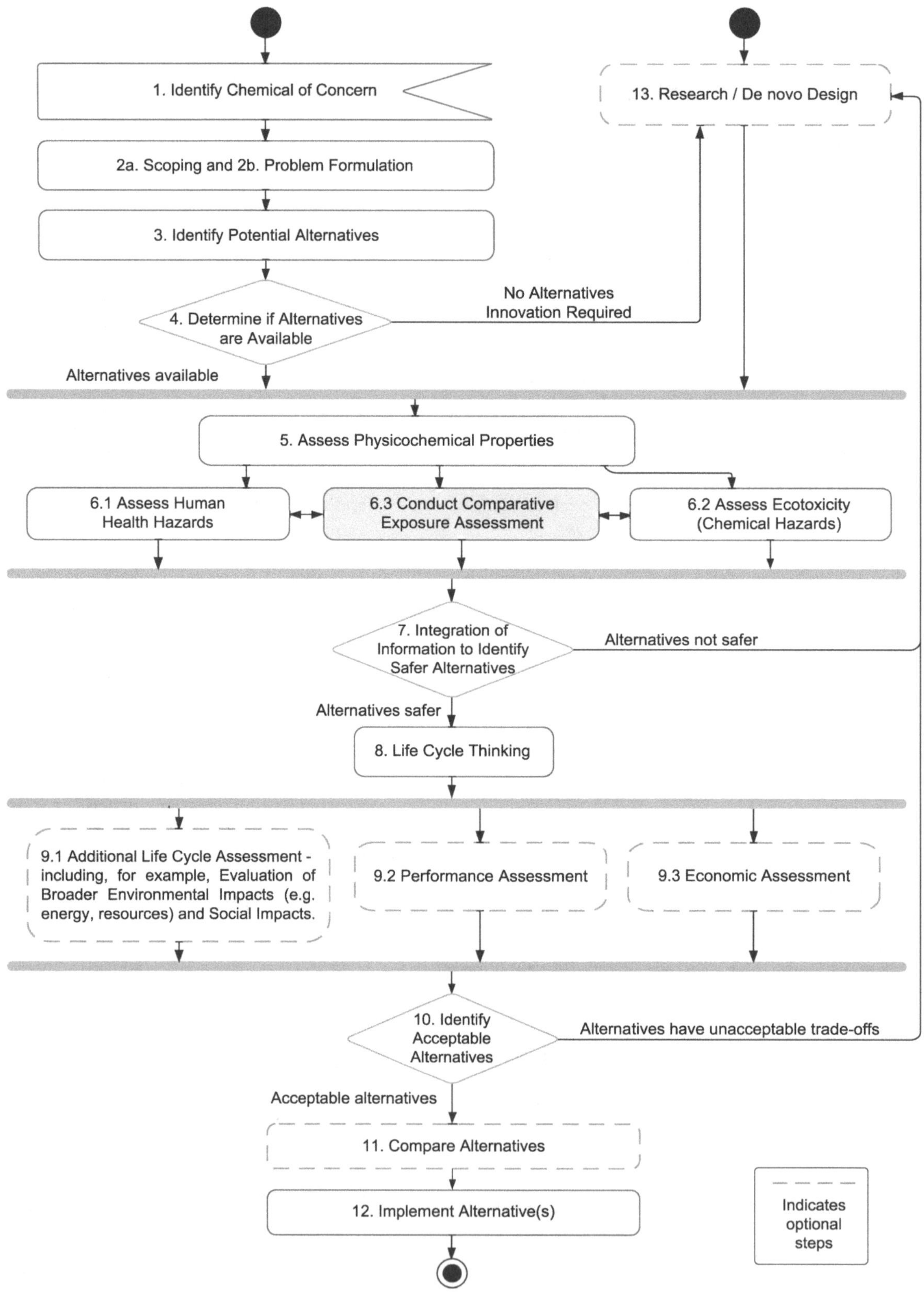

FIGURE 6-1 Committee's framework highlighting comparative exposure assessment (Step 6.3).

BOX 6-1

COMPARATIVE EXPOSURE ASSESSMENT AT A GLANCE (STEP 6.3)

- Determine if the alternatives would be expected to result in substantially equivalent exposures (Step 6.3). This can be accomplished by looking at outputs of simple exposure models (especially those considering estimates based on observed use patterns), comparing key physicochemical properties of alternatives (considered and compiled in Step 5), or, in some cases, applying knowledge about use scenarios and material properties.
    - If alternatives are substantially equivalent in their expected exposure, then the assessment can be mainly hazard based (i.e., based on inherent hazard).
    - If an alternative is deemed to have a substantially higher potential for exposure than the chemical of concern, then a more detailed exposure assessment may be appropriate. But a more detailed exposure assessment should only be performed if the toxicological and other advantages of the alternative are found, after analysis in later steps, to be attractive enough to warrant this additional effort.
- If the exposure potential of an alternative is preferable due to its inherent properties, this should be noted. It may add further weight to the choice of the alternative.
- To focus the consideration of alternatives on the inherent properties of substances, exposure estimates should be derived in the absence of assumptions about reliance on alternative-specific administrative, engineering, or personal protective equipment (PPE) controls.

For the required elements of Step 6.3, the exposure considerations are limited to the stage at which the chemical is used for human exposures and the use and disposal stages for ecological exposures. Broader upstream and downstream exposures that need to be considered may result from Life Cycle Thinking (Step 8) and life cycle analysis (Step 9.1).

these frameworks assess exposure and define and identify exposure concerns.

While exposure assessment is a module in the IC2 framework (IC2 2013), it is conducted after the hazard, performance evaluation, and cost and availability modules are completed. In the IC2 framework, chemical hazard reduction is viewed as a first step, and exposure is considered when examining potential trade-offs with identified alternatives.

Exposure also has a major role in the CA SPC (CA DTSC 2013b) and the REACH frameworks (ECHA 2011). For example, quantitative risk assessments[30] and, as a consequence, exposure assessments, are required components of alternatives assessment in the second tier of the REACH framework (after the first tier of replacing high-concern chemicals with those of lower hazard).

As this discussion shows, there is considerable variation in the way exposure is considered in existing frameworks. This variability may be partly explained by the principles that frameworks have adopted to guide the consideration of exposure in an alternatives assessment. Many frameworks have a *stated principle* to prevent harm by focusing first on inherent toxicity rather than relying on downstream controls of exposure to mitigate the risk. This approach is consistent with the industrial hygiene hierarchy of controls, which prefers to completely prevent exposure from a hazardous chemical, rather than control exposure, because exposure controls can fail (Schulte et al. 2013). In addition to the principles, other factors may limit the user from performing an exposure assessment. For example, some users may not have enough detailed knowledge about how downstream product developers or end users use the chemical. The added cost, time, or expertise requirements needed to perform an exposure assessment may also be a consideration for some users.

Despite the trend of many existing frameworks to only minimally address exposure, this approach may not be appropriate in some cases. Chemical alternatives can have different chemical structures that influence their toxicity *and* exposure. The presence of different functional groups and physicochemical properties may increase (or decrease) the likelihood of chemical exposure to humans or ecological receptors, thus negating benefits derived from selecting a chemical alternative on the basis of relative hazard alone. In many cases, the greater the difference in the chemical structure, the more likely that the exposures will not be equivalent.

The committee also observed that in some frameworks, the role of exposure differed between human health and ecological assessments—a difference that appears difficult to justify. Within the ecological component of alternatives assessment, the

[30]Within risk assessment, exposure assessment serves the function of providing an estimate of dose that, when combined with dose-response assessment, converts the potential for harm into a probability of harm.

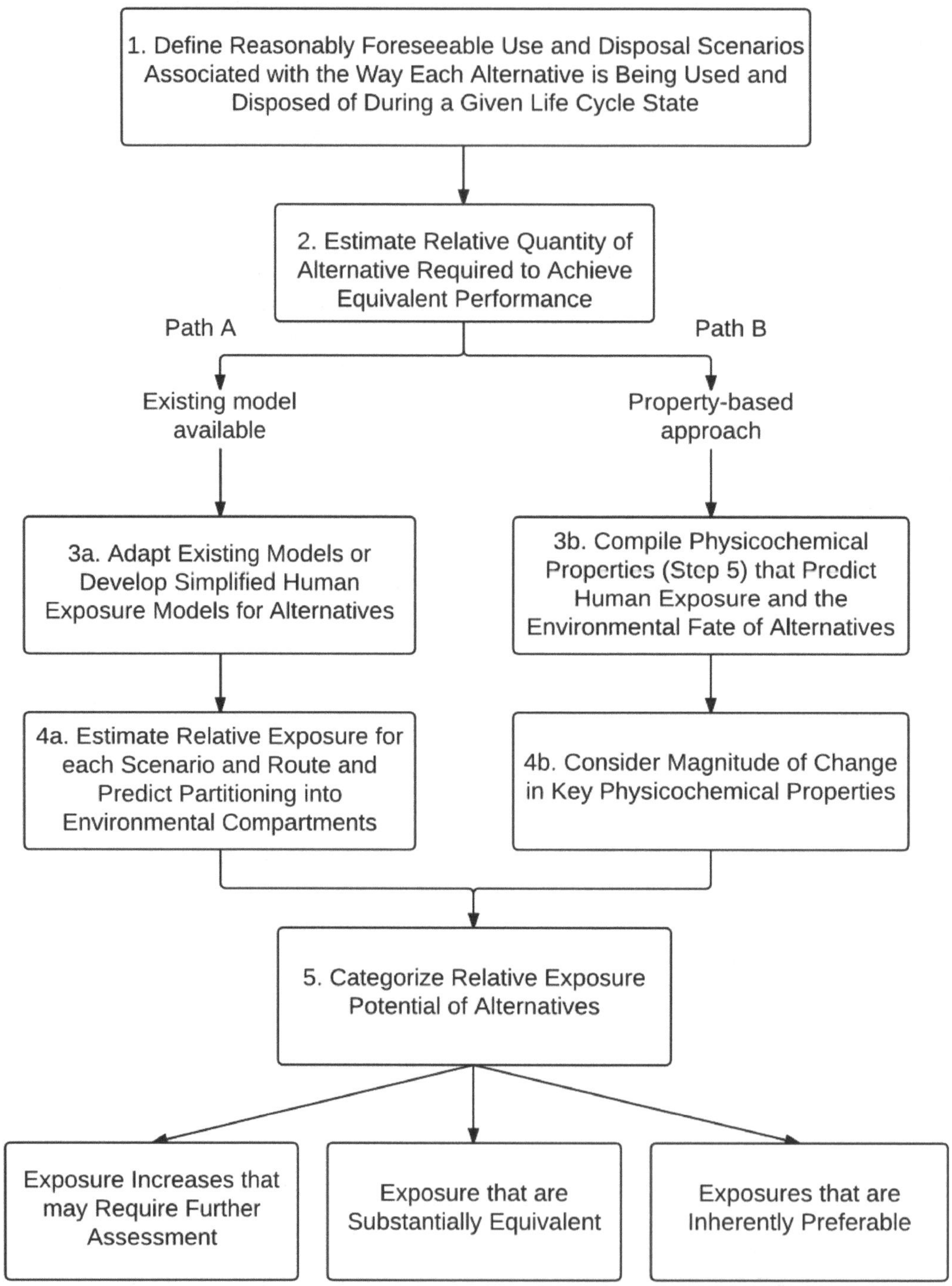

FIGURE 6-2 An approach to comparative exposure assessment within the committee's chemical alternatives assessment framework. Exposure potential could be derived from either the outputs of simple exposure models or the comparison of key physicochemical properties to arrive at one of three determinations with respect to an alternative, as compared to a chemical of concern.

extent that potential exposure of ecological receptors (given degradation, bioaccumulation, persistence, and other processes related to a chemical's fate in the environment) is considered stems from inherent properties of the chemicals, but consideration of the impact of these inherent exposure-related chemical properties on human health seems absent.

THE COMMITTEE'S APPROACH TO EXPOSURE ASSESSMENT

The role of exposure assessment needs to be carefully considered during an alternatives assessment. The committee notes that its role is very context-dependent, and may range from minimal to greater importance. In the following section, the committee describes its approach to exposure assessment in its framework, which is comparative and focused on the evaluation of intrinsic potential for exposure, in the absence of any physical or administrative controls.

To conserve assessment resources and still facilitate informed and efficient decision making, the committee's framework describes a staged approach to the assessment of comparative exposure, with exposure being considered to different degrees at different points in the framework:

- *In the problem formulation step (Step 2b, Chapter 4)*, the exposure pathways of the chemical of concern are considered early, during the problem formulation step, to focus the effort. Expected patterns (acute versus chronic) and routes (oral, dermal, inhalation) of exposure likely to be important were also identified during Step 2b, assuming there are intended and reasonably foreseeable exposure scenarios. Further, a qualitative consideration of a chemical's use (in a formulated or dispersive product, locked in a polymer matrix, etc.) and physicochemical properties provides important information in Step 2b.
- A *comparative exposure assessment (Step 6.3, described in this chapter)* to estimate relative exposure differences between potential alternatives and the original chemical of concern. This step can be done in parallel with the assessment of toxicological evidence for both ecological receptors and human populations. At this stage, the only human exposures that should be considered are those that occur while using the chemical of concern.[31] Ecological exposures to consider are those related to use and disposal, the areas of the most immediate interest. The procedure for completing the comparative exposure assessment is described in Sub-steps 1-7 of Step 6.3.
- *Additional exposure assessment consideration*, if concerns are identified when applying Life Cycle Thinking or examining the synthetic history of the alternatives in Steps 8 or 9.1.
- How to conduct a fuller, quantitative exposure assessment is explained in optional Sub-step 6.

Conduct Comparative Exposure Assessment (Step 6.3)

The committee's framework includes specific exposure considerations alongside the hazard assessment. The consideration of exposure assessment concepts at this point in the framework aims to determine whether exposure to alternative chemical(s) might be decreased or increased compared to the original chemical of concern. The committee's approach focuses on factors that are intrinsic either to the chemical alternatives or are inherent to the product into which the substance will be integrated. Therefore, extrinsic factors that may mitigate exposure (e.g., labeling, training, and a variety of engineering, administrative, or PPE controls) are not considered, which is consistent with the industrial hierarchy of controls (Schulte et al. 2013).

Figure 6-2 describes the committee's approach to comparative exposure assessment. This approach allows for the use of either available exposure models or comparison of critical physicochemical properties. If these are not readily available, other information on use and chemical and material properties can be used as a way to estimate the relative exposure potential of alternatives.[32] Each of

---

31 Or its disposal if it is being disposed of as it is used.

32 It is important to note that there are often significant uncertainties in exposure estimates that can lead to an underestimate of potential exposures. This can happen because of assumptions about the behavior of certain chemicals (how they might partition), misunderstanding of use scenarios (unexpected uses), or how the chemical might "escape" a particular matrix like a polymer. This does not, however, minimize the importance of considering potential exposure, but rather points to the importance of a broad exploration of exposure potential, as well as ensuring stakeholder involvement and a

the numbers in the diagram refers to a sub-step of Step 6.3.

All alternative chemicals need to go through Sub-steps 1 and 2 of the comparative exposure assessment (Figure 6-2). If existing models are available, the assessment follows Path A, Sub-steps 3a and 4a. If no models are available, then the assessment follows Path B, Sub-steps 3b and 4b. Both paths converge at Sub-step 5, from which it is possible to arrive at one of three possible outcomes shown at the bottom of the diagram.

Comparative Exposure Assessment

Sub-step 1. Define reasonably foreseeable use and disposal scenarios associated with the way each alternative is being used and disposed of during a given life cycle stage: During this step, a set of reasonably foreseeable use scenarios, such as how the chemical is used in cleaning products, fuels, cosmetics, or personal care products, as well as corresponding routes of human exposure, are identified for each alternative. These can be derived from knowledge about functional use, or the behavior of a chemical during a particular activity, such as manufacturing, and the application of a chemical in a process or product (for example, is the chemical bound in matrix or dispersive in its application?), as well as physicochemical properties. Stakeholders can be helpful in identifying these exposures in that they can provide important input and data that the assessor may not have access to.

Sub-step 2. Estimate relative quantity of alternative required to achieve equivalent performance: Given that the alternatives may have very different properties, it is reasonable to assume that the mass of each alternative required to achieve the same performance per unit of product as the chemical of concern may be highly variable. It follows that the relative amount of exposure to both humans and ecological receptors from the alternatives may partly depend on the amount of the chemical required to achieve the functional requirements identified in Step 2 (Chapter 4). Therefore, in some cases, the relative quantity required would need to be considered in more detail, as described in Sub-steps 3 and 4. In other cases, however, completion of these first two steps may be sufficient to determine if an alternative presents substantially equivalent exposure.

multidisciplinary approach to the exposure assessment process that enhances the information and input that goes into the assessment.

Path A

Sub-step 3a. Adapt existing models or develop simplified exposure models for alternatives: For some exposure scenarios, the chemical of concern identified in Sub-step 1 might have an established exposure model describing the range of expected human exposures expected during its use. Accordingly, it may be possible to modify the existing models to compare the relative exposure expected from the alternative chemicals based on their physicochemical or other properties. If an established model for the chemical of concern is not immediately available, a simplified human exposure model may be developed using a variety of modeling approaches. For example, publicly available exposure models that address common exposure scenarios and the associated routes of concern (e.g., dermal exposures from chemicals in contact with skin, inhalation of chemicals in indoor air) may be used (Delmaar et al. 2005). A wide variety of exposure models that address common exposure scenarios and exposure routes (OECD, 2012a) are also available. Exposure models from publicly available exposure estimates for similar uses and chemicals may be an additional source. For example, manufacturers have developed and submitted models and estimates under REACH.

Exposure modeling tools are often deliberately structured in tiers of complexity (Tier 1 being the simplest, Tier 2 more complicated, and so on), to accommodate variety in the amount and types of information available to the user. The simplest tier can be applied to determine if substantially equivalent, substantially more, or substantially less exposure levels could be expected from different alternative chemicals. Even simpler qualitative assessments of exposure may suffice in some cases where models are not available.

Sub-step 4a. Estimate relative exposure for each scenario and route and predict partitioning into environmental compartments: Depending on the results from the simplified exposure models, the relative human exposure (taking into account the relative quantity of the substance required to achieve the required function) can be estimated. Table 6-1 shows an example of the use of a simplified exposure model to compare two alternative chemicals with respect to exposure. The level of detail in modeling and characterization of relative exposure can be limited to the extent needed to classify the exposure as either substantially more, less, or equivalent (i.e., it may be sufficient to say that exposure will be at least 10 times more, due to the relative quantities needed, without the need to

TABLE 6-1 Comparison of Exposure Potential Using Simple Exposure Models. The ConsExpo model (Delmaar et al. 2005) used to compare hypothetical fragrance data.

Inputs:	**Fragrance A**	**Fragrance B**
Use frequency:	365 days/yr	365 days/yr
Product amount:	1000 mg	1000 mg
Weight fraction compound:	0.0001	0.001
Exposure duration:	10 hr	10 hr
Room volume:	50 m^3	50 m^3
Ventilation rate:	0.5 rooms/hr	0.5 rooms/hr
Inhalation rate:	5 m^3	5 m^3
Uptake fraction:	100%	100%
Body weight:	60 kg	60 kg
Outputs:		
Acute Internal Dose:	330 μg/kg	33μg/kg
Daily Chronic Dose:	330 μg/kg	33μg/kg

be more precise in the exact value of the relative exposure). The relative exposure assessment should also consider the potential for bioaccumulation or persistence of the chemical, as revealed by physicochemical properties. Steps 1-4, the comparison of potential exposure using simple exposure models, can be illustrated with an air freshener comparison. As shown on Table 6-1, Fragrance A is used in an air freshener. Fragrance B is less hazardous than Fragrance A, but more of Fragrance B is required to achieve the same effect. The ConsExpo model (Delmaar et al. 2005) was used to evaluate both exposures. Based on the ConsExpo model outputs (estimated acute internal dose and daily chronic dose), the assessor could determine whether the exposure potential of Fragrance B, because more must be used in the product, is substantially equivalent to Fragrance A, or whether the exposure differences need to be taken into account when considering hazard and other data.

Path B

Sub-step 3b. Compile physicochemical properties (see Step 5) that predict human exposure and the environmental fate of alternatives: For those exposure scenarios and routes for which there are no available models, the critical physicochemical properties can be considered to predict potential exposure. This sub-step relies on the information compiled during Step 5 in the committee's framework (see Chapter 5 for more details). The exposure scenario and route of exposure will most often indicate which of these properties will be more or less relevant to evaluating whether an alternative is likely to lead to substantially more, substantially less, or substantially equivalent exposures by each route and scenario.

Sub-step 4b. Consider magnitude of change in key physicochemical properties: Comparing physicochemical properties that relate to ecological exposures should result in a qualitative indication of each chemical's potential for partitioning to various media. The comparative ecological exposure assessment should begin by a direct comparison of those physicochemical parameters that are most likely to describe the persistence of a chemical in environmental media (e.g., K_{ow}); partitioning of a chemical into the environmental media (water, soil, sediment, air); and potential for bioaccumulation of the chemical into biological tissue through direct contact with environmental media or through food chain exposures (see Chapters 5 and 7 for more detail).

While alternatives are not expected to be identical, they may be considered substantively equivalent, by virtue of having broadly similar patterns and numerical values for various key properties. What level of change in a key property indicates a chemical as "better," "equivalent," or "worse" with respect to exposure should be determined in advance and may be established through expert judgment.

Box 6-2 provides an example of how to use relationships related to physicochemical properties

BOX 6-2
COMPARISON OF EXPOSURE POTENTIAL USING PHYSICOCHEMICAL PROPERTIES FOR DERMAL EXPOSURE

Chemical A is an antimicrobial incorporated into metal-working fluid. Chemical B is less hazardous than Chemical A, but differs in its physicochemical properties such that the amount of dermal exposure can be expected to be higher than Chemical A.

For a surface area of 1000 cm^2, an exposure time of 8 hours, and a body weight of 60 kg:

	Chemical A	Chemical B
Physicochemical inputs:		
Molecular weight:	150 Da	150 Da
Octanol-water partition coefficient:	$10^{0.5}$	$10^{2.5}$
Concentration:	0.01 mg/mL	0.01 mg/mL
Outputs:		
Permeability coefficient (k_p):	0.0005 cm/hr	0.0135 cm/hr
Predicted amount absorbed per kg bw:	0.0007 cm/hr	0.0180 cm/hr

The permeability coefficient and amount absorbed is derived from models for dermal absorption available in the literature (Potts and Guy 1992; Cleek and Bunge 1993; McDougal and Boeniger 2002). Based on these calculations, an assessor may determine that dermal exposure potentials of Chemicals A and B are not substantially equivalent and that this difference should be considered during the Integration step (Step 7) and may require a more complete exposure assessment.

to compare human exposure potential among alternatives.

Categorizing Exposure

At this point in the process, regardless of whether the alternative chemical has been assessed along Path A or Path B, it should now be categorized, as explained under Sub-step 5.

Sub-step 5. Categorize relative exposure potential of alternatives: The inference of exposure potential could be derived from either the outputs of simple exposure models or the comparison of key physicochemical properties to arrive at one of three determinations. The determinations are comparisons of the alternative to the chemical of concern or other baseline as follows.

- *Exposures that are substantially equivalent:* An alternative may be considered substantially equivalent in that the differences in exposure are considered to be minor, perhaps in comparison to what may be significant differences in hazard. The notion of substantial equivalence is not strictly defined and is considered to be context-dependent. The primary purpose of this determination is to simplify the subsequent assessment of alternatives so that the determination of relative safety of alternatives can be limited to a discussion of their relative hazard.
- *Exposures that are inherently preferable:* A second possibility is that the alternative is actually preferable to the baseline chemical due to its inherent properties or the specific way it is being used in a product. Inherently preferable exposures are those that substantially reduce the potential for human or environmental exposure. Alternatives with inherently preferable exposure profiles might be considered safer, especially if there are uncertainties in hazard or exposure potential. If any alternatives are preferable because of their inherent properties, that should be noted for further consideration in Steps 7 and 10 of the overall framework (Chapter 9).
- *Exposure increases that may require further assessment:* This refers to an alternative determined to have potentially higher exposures than the baseline chemical. If, after further steps and analysis are completed, the alternative is found to be preferable for other reasons (e.g., reduced hazard in Steps 6.1 or 6.2 or additional considerations from Step 8), it may be worthwhile to conduct further exposure assessment efforts to arrive at a more quantitative estimate. This optional analysis is described as Sub-step 6. It is not intended to be a requirement of the alternatives assessment process. Furthermore, because additional effort is required, it is to

be expected to be conducted after further analysis justifies the additional work.

Sub-step 6. Quantitative comparative exposure assessment (optional): This chapter has focused on qualitative comparative exposure assessment, but for a number of reasons, a more quantitative or expansive exposure assessment may be required. The reasons this may be needed are: 1) toxicity is similar enough that exposure is a tiebreaker (as explained in Chapter 9); 2) the alternative is considered favorable for other reasons than exposure; or 3) the implications of Life Cycle Thinking or analysis (Steps 8 and 9.1) expand the number of chemicals or chemical use patterns that need to be evaluated. As a result, the alternative chemicals may have to undergo assessment of use patterns and exposure pathways to further examine how exposure might change.

Quantitative comparative exposure assessment is not considered a simple task. A useful reference for exposure assessment is the report, "Descriptions of Existing Models and Tools Used for Exposure Assessment, Results of OECD Survey" (OECD 2012a), which includes a table of available exposure models and tools with descriptions and links for each tool. Table 6-2 is an excerpt from this survey. It highlights models that are useful when considering human exposure. They may be suitable in the committee's alternatives assessment process.

Another source of commonly used tools is the EPA's "EXPOsure toolBOX" (EPA-Expo-Box), which was publicly released in 2013 (EPA 2014d). EXPOsure toolBOX is a compendium of exposure assessment tools and contains links to guidance documents, databases, models, reference materials, and other resources, including an "Exposure Factors" module designed to facilitate use of the 2011 *Exposure Factors Handbook* data (EPA 2014e).

Integration of Exposure Assessment into Subsequent Steps

The result of the required sub-steps of Step 6.3 is to identify and categorize the potential exposure for each alternative, in a relative sense, as being a) substantially equivalent, b) inherently preferable, or c) potentially worse (higher). In most cases, Step 6.3 can help identify differences in exposure that should be considered when integrating information in Steps 7 (Chapter 9) and 10 (Chapter 11) of the committee's framework. If the extent, pattern, and degree of exposure are considered to be substantially equivalent between an alternative and the chemical of concern, then the determination of "safer" can be limited to the relative hazard of the chemicals. Where one or more of the exposure scenarios is inherently preferable due to intrinsic properties of the chemical or its integration into the actual product, then this can be noted as a further contribution to the relative safety of this alternative.

In the case where substantially increased human or ecological exposure is predicted for an alternative, then more detailed or rigorous exposure assessment may be called for. Rather than proceed immediately to a more complete exposure analysis, this assessment can be delayed until it has been determined that the alternatives in question have sufficient merit to justify the effort and the broader life cycle consequences have been explored in Step 8.

TABLE 6-2 Selected Human Health Comparative Exposure Assessment Tools

Model name	Owner	Description
ART (Advanced Reach Tool)	TNO (Netherlands Organisation for Applied Scientific Research)	Advanced worker inhalation exposure assessment.
CALENDEX	Exponent	Estimates human exposure to chemical residues in foods and home-based chemical treatments, such as pest control and turf treatments (subscription required).
CALTOX	Lawrence Berkeley National Laboratory	A risk assessment model that calculates the distribution of a chemical in the environment and the risk of an adverse health effect due to a chemical. It also evaluates the distribution among different environmental compartments.
CARES (Cumulative and Aggregate Risk Evaluation System)	US-EPA (U.S. Environmental Protection Agency)	Databases to evaluate potential risk from dietary, drinking water, and residential sources and from oral, dermal, and inhalation routes of exposure.
ChemSTEER (Chemical Screening Tool For Exposures & Environmental Releases)	US-EPA	Model for estimating (1) occupational inhalation and dermal exposures and (2) environmental releases to air, water, and land for chemicals during manufacturing, processing, and use.
CHESAR (Chemical Safety Assessment and Reporting)	ECHA (European Chemicals Agency)	REACH specific model to predict the concentration in environmental compartments, exposure of workers, and exposure of consumers via food and environment. Consumer exposure to be added soon.
ConsExpo		Exposure assessment of compounds in non-food consumer products.
E-FAST (Exposure and Fate Assessment Screening Tool)	US-EPA	Model for screening level estimates of chemical concentrations from releases to air, surface water, landfills, and from consumer products. Also estimates inhalation, dermal and ingestion potential dose rates and aquatic organism risks.
EMKG-EXPO-TOOL	BAUA (Federal Institute for Occupational Safety and Health)	Quantitative tier I assessment of occupational exposure (inhalation) to hazardous substances.
EUSES 2.1 (European Union System for the Evaluation of Substances)	EC-JRC (European Commission Joint Research Center)	A decision- support instrument to carry out assessments of the general risks of industrial chemicals and biocides posed by substances to people and the environment.
FHX (Farfield Human Exposure)	Trent University	Holistic fate and exposure model for chemical exposure assessment of humans of different age-classes.
G-CIEMS (Grid-Catchment Integrated Environmental Modeling System)	NIES (National Institute for Environmental Studies)	Assessment of compounds in environmental and human exposure.
Generic Exposure Scenarios	CEFIC (European Chemical Industry Council)	Tool (database) developed by sector associations to communicate generic exposure scenarios in that sector.
HERA (Human and Environmental Risk Assessment)	AISE/CEFIC (International Association for Soaps, Detergents and Maintenance Products)	Multiple human and environmental risk assessments on ingredients of household cleaning products according to HERA principles.

TABLE 6-2 (Continued)

Model name	Owner	Description
IGEMS (Internet Geographical Exposure Modeling System)	US-EPA	Includes models and data for ambient air, surface water, soil, and groundwater.
Industry Specific Generic Scenarios	US-EPA	Industry-specific methods and models for estimating occupational exposures and environmental releases for chemicals during industrial and commercial operations.
LIFELINE software suite	LifeLine Group Inc.	Addresses exposures that occur from the use of pesticides on agricultural crops and in residences, as well as pesticide residues that occur in water supplies (subscription required).
PROMISE (Probabilistic Methodology for Improving Solvent Exposure Assessment)	American Chemistry Council	Designed to evaluate exposures and doses from single or multiple uses of products that contain volatile organics; not a population- based model.
RAIDAR (Risk Assessment, Identification and Ranking)	Trent University	Holistic mass balance framework providing chemical exposure and risk assessments for humans and the environment. It is predominantly used as an evaluative model.
Risk Learning	AIST (Institute of Advanced Industrial Science and Technology)	Estimating human health risks of a specific chemical substance in environmental media (air, water, soil, etc.) or contact media (food, etc.) using carcinogenic risk and hazard quotient as risk indices.
RiskCaT-LLE (Risk Calculation Took for the LLE-based Risk Estimation)	AIST	Estimating human health risk as loss of life expectancy (LLE) from exposure to chemicals.
RiskOfDerm (Risk Assessment of Occupational Dermal Exposure to Chemicals)	TNO	Worker potential dermal exposure assessment.
SDA (Soap and Detergent Association)	Exponent	Exposure and risk screening methods for consumer product ingredients methodology for screening level exposure and risk assessments of chemicals used in consumer products, mainly laundry, cleaning, and personal care products.
SHEDS (Stochastic Human Exposure and Dose Simulation)	US-EPA	A probabilistic human exposure model. There are currently three versions of SHEDS. SHEDS-Multimedia version 3 / 4 is a probabilistic aggregate residential exposure model. The other SHEDS models address exposures to particulate matter (SHEDS-PM), air toxics (SHEDS-ATOX), and wood (SHEDSWood).
Stoffenmanager	TNO	Control banding for worker dermal and inhalation exposure and quantitative exposure assessment for worker inhalation exposure.
USES 4.0 (Uniform System for the Evaluation of Substances)	RIVM (National Institute of Public Health and the Environment)	Quantitative assessment of the risks posed by new and existing chemical substances, as well as agricultural and non-agricultural pesticides to people and the environment.

SOURCE: OECD 2012a.

7
Assessment of Ecotoxicity

This chapter begins with general background on ecotoxicology and then briefly reviews current approaches for comparative ecotoxicity assessments that are used in several alternatives assessment frameworks (see Appendix B for a more detailed description of approaches used in the alternatives assessment frameworks considered by the committee). The details behind the committee's framework concerning ecotoxicity assessment (Step 6.2) are then presented (Step 6.2, see Figure 7-1). Box 7-1 outlines the elements of the committee's suggested unified approach. Methods that could be used in ecotoxicity assessment are then discussed. Near-term and aspirational improvements, such as the use of adverse outcome pathways based on in vitro high throughput data and in silico read-across methods, are considered.

ECOTOXICOLOGY

Ecotoxicology is the study of how chemicals interact with organisms in the environment. Environments that are potentially at risk vary greatly and include marine and freshwater environments, terrestrial environments from the arctic to the tropics, and even the air where respiratory exposures and foliar uptake by plants can occur. Organisms at risk from chemical exposures include plants, fungi, and algae (primary producers); invertebrates (such as worms, bugs, beetles, and mollusks); fish; amphibians; reptiles; birds; and mammals.

There are an astonishing number of organisms in the world, representing close to 6.5 million species on land and another 2.2 million species in the oceans (Mora et al. 2011). Given this wide range of biodiversity, it is impossible to know everything about the potential ecotoxicological effects of chemicals. Instead, ecotoxicologists rely on a small set of indicator organisms and an understanding of how the physicochemical properties of compounds cause them to partition in the environment and organisms. Those model systems and approaches have provided toxicologists with a surprisingly robust ability to predict the relative hazard of different substances. Because the stated goals of most environmental assessments are primarily on the preservation of species and populations and less with individual organisms (with the exception of large charismatic species, such as bears, mountain lions, and most birds), the end points used most often in hazard assessments are survival and reproduction, with growth included as a surrogate for reproductive fitness in many species.

The ecotoxicology literature is heavily weighted toward aquatic systems, particularly freshwater organisms, because of the historical and ongoing use of water bodies for the discharge of various waste streams. However, land application of sludge, landfills, and terrestrial-based activities (such as mining, refining, and transportation), and air deposition can result in contaminated soils. Relative chemical hazards to terrestrial organisms do not necessarily follow the same patterns as that seen

BOX 7-1

ASSESSMENT OF ECOTOXICITY AT A GLANCE (STEP 6.2)

1. Review physicochemical data to determine into which environmental compartments the chemicals will partition.
2. Compile ecotoxicity data, paying particular attention to data for compartments identified in the first step. For missing data, estimate toxicity using read-across, QSAR, or other method.
3. On the basis of all available data for each alternative, categorize toxicity as high, medium, and low for each end point and include the uncertainty associated with each categorization. Include a narrative description of the data.
4. Create a visual display to show relative hazard in different environmental media (soil, water, sediment, air).

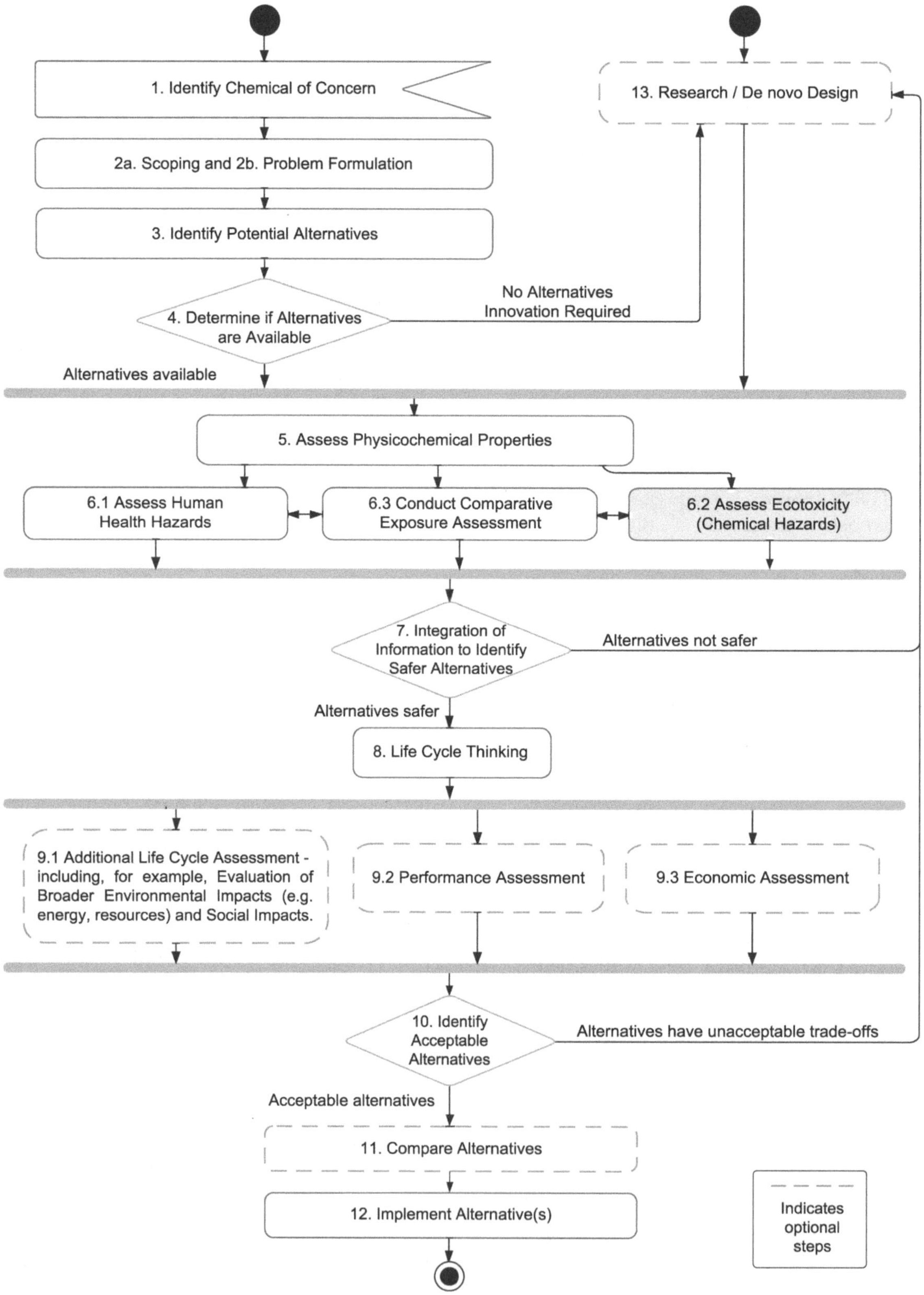

FIGURE 7-1 Committee's framework with the ecotoxicity assessment highlighted.

with aquatic organisms, necessitating separate testing and assessment schemes. Toxicity tests with plants and soil invertebrates are becoming more commonplace, thus reducing reliance on extrapolations from aquatic toxicity tests. In contrast, monetary and ethical considerations make it more difficult to conduct toxicity tests on terrestrial vertebrates.

Hazard classification schemes for environmental and ecotoxicological effects also include estimating the amount of bioaccumulation of a chemical within the food web and its persistence in the environment. These two attributes affect the amount and duration of environmental exposure and help predict which organisms are most likely to be affected (primary producers, invertebrates, or top predators). Those intrinsic chemical properties are discussed in more detail in Chapter 5.

ECOTOXICITY ASSESSMENT IN OTHER FRAMEWORKS

The committee considered how ecotoxicity was evaluated by the frameworks that it reviewed (see Chapter 2 and Appendix B). The goal of its analysis was to identify commonalities and distinct differences among the approaches for incorporating those hazards into the frameworks. In the context of this review, *ecotoxicity* characterizes potential adverse effects that a chemical causes to an aquatic or terrestrial receptor. That definition is used in various assessment methods, including many of the reviewed frameworks. Ecotoxicity is based on the toxicological properties of the chemical and the susceptibility of the organism. Ecotoxicity is distinguished from *environmental hazards,* which refer to potential adverse effects of the chemical that occur on larger (often geological or meteorological) spatial or temporal scales, such as global warming, ozone depletion, depletion of resources, or effects on indicators of sustainability. As noted above, the committee focused on ecotoxicity in this chapter.

The committee found that the frameworks display varying levels of specificity concerning the assessment of ecotoxicity. Some frameworks have protocols that reference analytical tools or methods that can achieve the ecological evaluations necessary for the relative ranking of chemical alternatives. Most of the protocols do not recommend any particular tool or even under what conditions one tool might be superior to another. Instead, the protocols provide comprehensive lists of methods, tools, and resources that the assessor might use within the context of the framework. Assessors are left to their own discretion in making a selection from among the often long lists of evaluative tools.

Other frameworks specify analytical tools that characterize the potential toxicity of a chemical's persistence, bioaccumulative properties, or environmental mobility through the use of compendia of such data or the application of extrapolations from molecular structure or measured properties. These frameworks usually develop relative rankings of alternatives on the basis of some color-coded system (such as red is more problematic than green) or a narrative classification (such as persistent, very persistent, and not persistent).

Table 7-1 summarizes the aquatic toxicity end points characterized in each framework that provides an analytical system for assessing aquatic and terrestrial hazards. As noted in the table, Design for Environment (DfE) characterizes acute aquatic toxicity on the basis of the concentration at which 50% of the organisms are affected (EC_{50}) or survive treatment (LC_{50}) Chronic toxicity is based on a no observed effect concentration (NOEC) or a lowest observed effect concentration (LOEC) over a series of treatments. The Interstate Chemicals Clearinghouse (IC2) framework also characterizes acute aquatic toxicity on the basis of EC_{50} or LC_{50} for tests of specific time frames. It does not provide a chronic aquatic toxicity characterization at early hazard assessment levels, but incorporates the DfE benchmarks through the application of GreenScreen® at later levels. The Toxics Use Reduction Institute (TURI) framework suggests using the Pollution Prevention Options Assessment System (P2OASys) tool, which includes four unequal ranges of LC_{50} test results for aquatic toxicity and aquatic plant toxicity separately.

This Pollution Prevention Options Assessment System only characterizes chronic aquatic toxicity for fish. The Guide on Sustainable Development only characterizes aquatic toxicity, with a cutoff based on a NOEC of less than 0.01 mg/L (considered nontoxic). The committee notes that the frameworks generally reference the United Nations Globally Harmonized System of Classification and Labelling of Chemicals (GHS) to characterize aquatic toxicity. Thus, there is little difference in characterizing and ranking aquatic toxicity among the frameworks; they all depend on the same underlying tool, the GHS, as the basis of characterization.

Only the DfE provides a characterization of terrestrial hazards, which is based partly on the EPA Office of Pesticide Programs' "Ecotoxicity

TABLE 7-1 Aquatic Toxicity End Points, Thresholds, and Categories Used in Alternatives Assessment Frameworks

	Acute Toxicity			Chronic Toxicity		
Assessment Framework	**End Point**	**Threshold (mg/L)**	**Category**	**End Point**	**Threshold (mg/L)**	**Category**
DfE	EC_{50} or LC_{50}	< 1	Very High	NOEC or LOEC	< 0.1	Very High
		1-10	High		0.1-1	High
		10-100	Moderate		> 1-10	Moderate
		> 100	Low		> 10	Low
IC2	96 hr LC_{50} (fish)	< 1	Very High	Recommends GreenScreen® at Higher Levels of Assessment		
	48 hr EC_{50} (crustacean)	1-10	High			
		10-100	Moderate			
	72 hr or 96 hr EC_{50} (algae or aquatic plants)	> 100	Low			
TURI[a]	LC_{50} (animals)	< 0.1	10	NOAEC (fish)	< 0.00002	10
		0.1-1	8		0.0002	8
		1-50	6		0.002	6
		50-1000	4		0.02	4
		> 1000	2		< 0.2	2
	LC_{50} (plant)	< 0.1	10			
		0.1-1	8			
		1-10	6			
		10-100	4			
		> 100	2			
Guide on Sustainable Chemicals	NA			NOEC	< 0.01	Not Toxic

[a]Category values calculated from the Pollution Prevention Options Assessment System (P2OASys) worksheet, September 2014. The P2OASys worksheet returns numerical values based on a scale of 1 to 10 to represent relative hazard from low to high.

SOURCES: Rossi et al. 2006; Reihlen et al. 2011; IC2 2013; EPA 2014c.

Categories for Terrestrial and Aquatic Organisms" (EPA 2014f). The system categorizes avian acute and chronic toxicity, acute toxicity for wild mammals, and toxicity for insect pollinators (see Table 7-2).

Overall, the frameworks provide relative ranks with an underlying assumption that execution of the framework will allow the user to select the safer chemical through a one-to-one comparison. The frameworks are not intended to identify a "safe" alternative per se, but rather evaluate whether the alternative is safer than the chemical of concern. In some cases, the safer alternative may have appreciable hazards that need to be considered.

COMMITTEE'S FRAMEWORK FOR ECOTOXICITY ASSESSMENT

The elements that the committee suggests for evaluating ecotoxicity are shown in Box 7-1. Once the appropriate environmental compartments have been identified using data on physicochemical properties (Chapter 5), ecotoxicity information for organisms associated with those compartments is assembled and compared. Unlike the existing frameworks, the committee's framework allows the analyst to focus on gathering ecotoxicity data for the ecosystem (aquatic, sedimentary [freshwater or marine], or terrestrial) of concern. Current

TABLE 7-2 DfE Ecotoxicity Categories for Terrestrial and Aquatic Organisms

Toxicity Category	Avian: Acute Oral Concentration (mg/kg)	Avian: Dietary Concentration (ppm)	Aquatic Organisms: Acute Concentration (ppm)	Wild Mammals: Acute Oral Concentration (mg/kg)	Non-Target Insects: Acute Concentration (µg/bee)
Very highly toxic	< 10	< 50	< 0.1	< 10	
Highly toxic	10-50	50-500	0.1-1	10-50	< 2
Moderately toxic	51-500	501-1000	>1-10	51-500	2-11
Slightly toxic	501-2000	1001-5000	> 10-100	501-2000	
Practically nontoxic	> 2000	> 5000	> 100	> 2000	> 11

SOURCE: EPA 2014f.

comparative hazard schemes are solely based on aquatic toxicity because of the large database of information. Aquatic toxicity tests are highly standardized, relatively straightforward to conduct, and have been in use for decades. Standardized sediment toxicity tests are available for a few organisms, but differences in bioavailability and organism survival in different sediment types complicates the testing methods and data interpretation (ECHA 2012, 2014d). Soil testing has become more prevalent, particularly with soil invertebrates and microbial function tests, and standard soils for comparative toxicity testing are well established. Higher order terrestrial organism tests with plants and vertebrate animals are more difficult to conduct and therefore data are less prevalent. However, high throughput in vitro studies coupled with adverse outcome pathways (AOP) appropriately predictive for species other than humans may be used in the future as a substitute for hazard comparisons or provide a basis for extrapolating aquatic toxicity data to other species.

The steepness of the slope of the concentration-response curves from the toxicity tests could also be considered in the assessment. Under certain exposure conditions, a steeper slope could indicate a greater hazard potential, as a small increment in chemical concentration will result in a large increased effect, whereas a shallow slope indicates that a greater amount of chemical in the environment may not substantially increase the effect level.

The analyst should gather all available data for the environmental compartment of concern, with no a priori prioritization of particular species (invertebrates vs .vertebrates vs. plants). Toxicity should then be categorized for each end point as low, medium, or high. For the purposes of chemical substitution, it is not necessary to be precise in such comparisons; the goal is to choose a chemical that has substantially less potential hazard, and the variability in the measurement end points across various species tests precludes precise comparisons. Cutoff values that could be used to help to categorize toxicity are shown in Table 7-2. Users of the committee's framework will need to exercise professional judgment since cutoffs in classification tools could result in the assignment of alternatives to different hazard categories (e.g., high vs. medium), when the actual difference in response can be toxicologically insignificant. In addition to categorizing toxicity for each end point, some indication (such as high, medium, or low) should be provided about the uncertainty associated with each categorization. These evaluations can be summarized in a table using a color-coded scheme. A narrative description of the data should be included (for example, if in vivo data are not available, how robust are the conclusions based on read-across, quantitative structure activity relationship [QSAR], in vitro, or other methods).

Visualization/Toxicological Priority Index (ToxPi)

One approach to visualizing the available data is to use the Visualization/Toxicological Priority Index (ToxPi) visualization software, which is illustrated in

Figure 7-2 (Reif et al. 2010; see Appendix C for additional information about ToxPi as well as discussions in Chapters 8, 9 and 12.). The "ToxPi visualization" is a visual representation of the relative magnitudes of the hazards (e.g., aquatic, sediment, and terrestrial). The width (in radians) of each slice represents the number of end points in each category, while the length of the slice indicates the overall degree of hazard. The distance of the slice from the origin (i.e., the radius) represents the potency (i.e., the distance from the origin is normalized to the maximum toxicity value; each equivalent is divided by the maximum, and the resulting values are summed). The aquatic, terrestrial, or sediment slices of the pie could be divided further if there are multiple endpoints represented (e.g., test end points such as mortality, growth, or species groups, such as invertebrate, vertebrate, or plant). Toxicity ranking based on high throughput suborganismal tests could be also included as a separate slice, with data normalized to the highest response value, as is done in the other slices. ToxPi describing ecotoxicity data can be displayed for each chemical under consideration, thus allowing for transparent comparisons across chemicals. Relative ranks for inherent hazards to aquatic organisms vs. sediment organisms vs. terrestrial organisms can be quickly visualized in this manner, or converted to ToxPi scores if desired, to aid in policy-dependent trade-offs of hazards to different ecosystems. The ranking of the compounds under consideration, accompanied by the confidence intervals (see Appendix C), can also be easily constructed to communicate the decisions made in the alternatives assessment. Ultimately, the ecotoxicity data may be combined with other available information (e.g., human health hazard, exposure, etc.) using ToxPi or other approaches.

METHODS FOR HAZARD DETERMINATION

Various methods can be used to obtain the data needed for the ecotoxicity assessment. The following sections briefly discuss the various methods.

Bioassays

Results of toxicity tests for aquatic, sediment, and soil or terrestrial organisms form the basis of most regulatory schemes for chemical registration or transportation, such as those for REACH and GHS. Primary measurement end points for acute and chronic exposures are survival, growth, and reproduction. As noted in Table 7-3, aquatic bioassays include water column (Daphnia; fish—freshwater and marine), sediment (Chironomus, Hyalella, oyster), and amphibians (FETAX; Frog thyroid assay).

Terrestrial assays include standardized studies for germination and growth of plants, various types of soft- and hard-bodied soil invertebrates, honeybees, and birds (see Table 7-4). Hazard determination for terrestrial mammals relies on data generated for human health assessments (primarily rodents, but some dog and nonhuman primate studies). Additional data might be available from livestock testing, although that testing is mostly limited to pesticides and pharmaceuticals. Soil microbial function tests are also available to determine chemical effects on respiration, decomposition, and nitrogen fixation. In general, information for toxicity to terrestrial organisms is sparse.

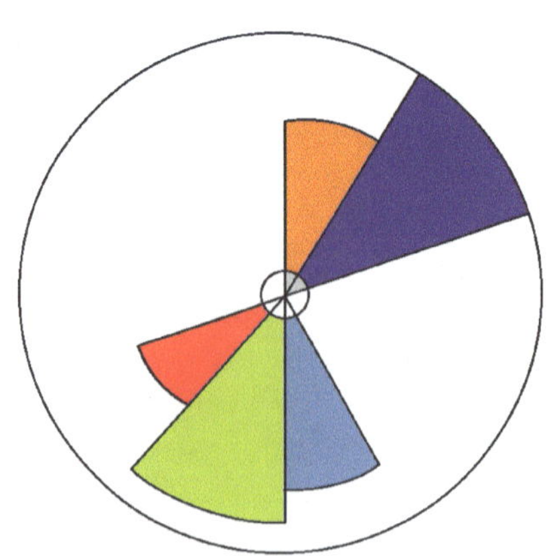

FIGURE 7-2 Illustrative ToxPi showing relative hazard to aquatic (dark blue), sediment (green), and terrestrial (red) organisms. Degree of persistence (orange) and bioaccumulation (light blue) are also shown here.

TABLE 7-3 Standardized Aquatic Tests for Ecotoxicity Properties

Media	Species	Guideline
Freshwater	Algae	OECD 201: Freshwater Alga and Cyanobacteria, Growth Inhibition Test OPPTS OPPTS 850.4500 - Algal Toxicity
		OPPTS 850.4550 - Cyanobacteria (Anabaena flos-aquae) Toxicity
	Fish	OECD 210: Fish, Early-life Stage Toxicity Test OPPTS 850.1400 Fish early-life stage toxicity test OECD 236: Fish Embryo Acute Toxicity (FET) Test OECD 212: Fish, Short-term Toxicity Test on Embryo and Sac-Fry Stages OECD 215: Fish, Juvenile Growth Test OPPTS 850.1075 Fish acute toxicity test, freshwater and marine OPPTS 850.1085 Fish acute toxicity mitigated by humic acid
		OECD 204: Fish, Prolonged Toxicity Test: 14-Day Study OECD 230: 21-day Fish Assay OECD 229: Fish Short Term Reproduction Assay OECD 234: Fish Sexual Development Test OPPTS 850.1500 Fish life cycle toxicity
	Invertebrate	OPPTS 850.1010 Aquatic invertebrate acute toxicity, test, freshwater daphnids
		OECD 211: Daphnia magna Reproduction Test OPPTS 850.1300 Daphnid chronic toxicity test
	Plants	OECD 221: Lemna sp. Growth Inhibition Test OPPTS 850.4400 - Aquatic Plant Toxicity Test Using Lemna spp OPPTS 850.4450 - Aquatic Plants Field Study
	Amphibians	OPPTS 850.1800 Tadpole/sediment subchronic toxicity test OECD 231: Amphibian Metamorphosis Assay
	Food web	OPPTS 850.1900 Generic freshwater microcosm test, laboratory
Freshwater sediments	Invertebrates	OECD 218: Sediment-Water Chironomid Toxicity Using Spiked Sediment OECD 219: Sediment-Water Chironomid Toxicity Using Spiked Water OECD 235: Chironomus sp., Acute Immobilisation Test OECD 225: Sediment-Water Lumbriculus Toxicity Test Using Spiked Sediment OECD 233: Sediment-Water Chironomid Life-Cycle Toxicity Test Using Spiked Water or Spiked Sediment OPPTS 850.1735 Whole sediment acute toxicity invertebrates OPPTS 850.1790 Chironomid sediment toxicity test
Marine water	Invertebrates	OPPTS 850.1020 Gammarid acute toxicity test
		OPPTS 850.1025 Oyster acute toxicity test (shell deposition)
		OPPTS 850.1035 Mysid acute toxicity test OPPTS 850.1350 Mysid chronic toxicity test
		OPPTS 850.1045 Penaeid acute toxicity test
		OPPTS 850.1055 Bivalve acute toxicity test (embryo larval
Marine sediments	Invertebrates	OPPTS 850.1740 Whole sediment acute toxicity invertebrates, marine

SOURCES: EPA Test Guidelines: EPA 2013c,d; OECD Test Guidelines: OECD 2014a.

TABLE 7-4 Standardized Terrestrial Tests for Ecotoxicity Properties

Media	Species	Guideline
Terrestrial systems	Birds	OECD 223: Avian Acute Oral Toxicity Test OPPTS 850.2100 - Avian Acute Oral Toxicity Test
		OPPTS 850.2200 - Avian Dietary Toxicity Test OECD 205: Avian Dietary Toxicity Test
		OECD 206: Avian Reproduction Test
	Plants	OECD 208: Terrestrial Plant Test: Seedling Emergence and Seedling Growth Test OPPTS 850.4100 - Seedling Emergence and Seedling Growth OPPTS 850.4230 - Early Seedling Growth Toxicity Test
		OECD 227: Terrestrial Plant Test: Vegetative Vigour Test OPPTS 850.4150 - Vegetative Vigor
		OPPTS 850.4600 - Rhizobium-Legume Toxicity
		OPPTS 850.4300 - Terrestrial Plants Field Study
	Honeybee	OECD 213: Honeybees, Acute Oral Toxicity Test OECD 214: Honeybees, Acute Contact Toxicity Test OPPTS 850.3020 - Honey Bee Acute Contact Toxicity Test OECD 237: Honey Bee (Apis Mellifera) Larval Toxicity Test, Single Exposure
		OPPTS 850.3030 - Honey Bee Toxicity of Residues on Foliage
		OPPTS 850.3040 - Field Testing for Pollinators
Soil	Invertebrates	OECD 207: Earthworm, Acute Toxicity Tests
		OPPTS 850.3100 - Earthworm Subchronic Toxicity Test
	Microbes	OPPTS 850.3200 - Soil Microbial Community Toxicity Test OPPTS 850.4900 - Terrestrial Soil-Core Microcosm Test

SOURCES: EPA Test Guidelines: EPA 2013d; OECD Test Guidelines: OECD 2014a.

In Silico Estimates of Ecotoxicological Hazard

Advances in in silico prediction methods through computational toxicology, computational chemistry and mechanistic toxicology often permit estimates to be made for untested chemicals, thus allowing data gaps to be filled. This approach is especially useful for an alternatives assessment, where a comparison between two or more chemicals is required. This section describes the in silico models used most commonly to fill such data gaps for ecotoxicology. Although there are a number of in silico approaches that can be used to fill ecotoxicity data gaps, challenges with accuracy and sensitivity of predictions remain.

Chemical Categories Approach, or "Read-Across"

One strategy for filling data gaps for a chemical of concern or alternative is evaluating hazard data pertaining to one or more structurally similar surrogates. According to the OECD guidelines (OECD 2007), this process is accomplished by grouping chemicals into "chemical categories," which

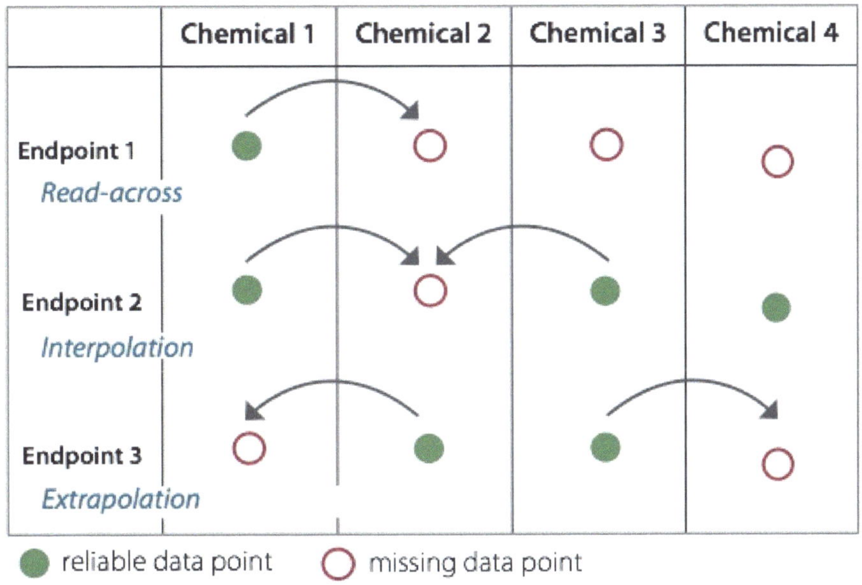

FIGURE 7-3 Schematic representation of the use of chemical categories to fill data gaps, enabling read-across from a data-rich chemical. SOURCE: Adapted from Worth (2008).

consist of chemicals that share a similar chemical structure or have physicochemical, toxicological, ecotoxicological, or environmental fate properties in common. As a result, it is assumed they are likely to have similar ecotoxicological hazards (see Figure 7-3).

The validity of this assertion, however, rests on how a "chemical category" is defined. The guidelines identified by EPA (2010a) and OECD (OECD 2007) for such groupings include the presence of common chemical functional groups, common breakdown products that might result in structurally similar chemicals, or common chemical classes or categories. The potential advantage of the approach is that it allows multiple chemicals to be assessed when only a few analogs have been tested, saving animals and costs. However, the major drawback is that the implied assumption that structural (and property) similarity is sufficient to impart comparative biological activity does not always hold, especially if the grouping rests only on structural similarity. Examples where the assumption does not hold can be found in the pharmaceutical industry, where a minor structural modification of an active pharmaceutical can result in order of magnitude differences in biological activity. If similarities in physicochemical properties are also a required criterion, the probability that chemicals in the same group will have similar biological activity will be increased; however, it is imperative that the properties used are mechanistically linked to the toxicity end point being predicted (see Chapter 5). Finally, the similarity within the category should be justified using a common mechanism or mode of action. The most widely used predictive tool for ecotoxicity using chemical categories is the OECD toolbox (OECD 2012b).

Quantitative Structure Activity Relationships (QSARs)

QSAR models provide estimates of a variety of ecotoxicity end points on the basis of chemical structure. Development of QSAR models for estimating ecotoxicity from chemical structures has advanced considerably (Cronin 2010; Hewitt et al. 2010). There are a number of QSAR tools that allow for a quick estimation of ecotoxicity and can be used by a non-expert. However, the resulting output can be misleading if the user is not trained in the appropriate application of such models. The major tools typically used are:

- Toxicity Estimation Software Tool (TEST) (EPA 2014g).
- Ecological Structure–Activity Relationships (ECOSAR) (aquatic toxicity) (EPA 2014h) based on structural fragments and logP. Validating

ECOSAR for three "valid" classes results in predictivity of at least 64%.

- OECD QSAR Toolbox (OECD 2012b).

There are several sets of criteria that can be used to assess the robustness of the QSAR models being used. The Setubal principles (Jaworska et al 2003) require a mechanistic basis, the availability of a training set, and validation. The OECD principles of validation require QSAR models to "have a defined end point, an unambiguous algorithm, a defined domain of applicability, appropriate measures of goodness-of-fit, robustness and predictability, and a mechanistic interpretation whenever possible" (Judson 2009). It is important to note that even the predictive ability of QSARs that meet the above criteria can be hindered by model training issues, such as domain applicability, overtraining, model bias, chance correlation, and overreliance on such testing methods as cross-validation.

In summary, QSAR models that are developed diligently and in keeping with established criteria for robustness can provide accurate predictions of ecotoxicity end points, if used astutely. However, they typically cannot be used to qualitatively assess whether a particular structural modification will result in a different toxicity profile.

Emerging Tools for Assessment of Ecotoxicity

There are several emerging tools that might eventually be valuable for assessing ecotoxicity and are discussed below. However, much research will most likely be needed before these methods can be incorporated confidently into alternatives assessment frameworks.

High Throughput Assays

The search for high throughput methods to predict toxicity to people has resulted in data generation that is directly relevant to the soil compartment. *Caenorhabditis elegans* is a small (about 1 mm long), free-living transparent nematode that lives in the soil in temperate regions. Its genome has been completely sequenced and the developmental fate of each cell is well known. *C. elegans* has been used for several decades as a model organism for many systems, including aging, neurobiology, and cellular differentiation, among others. Recently, it has emerged as a model for high throughput toxicological screens, including screening for genetic and molecular targets of new chemicals (Leung et al. 2008). Viability and behavior (such as locomotor activity) also are frequently reported. Such data could be added to information from standard test species of soil invertebrates (*Eisinia foetida, Folsomia candida, and Enchytraeus albidus*) to increase the range of data for assessing hazard to terrestrial systems.

The zebrafish (*Danio rerio*) is another species that is now being used in high throughput screening for chemicals. The embryo-larval bioassay was developed for use in preclinical screening of drugs because it is possible to visualize embryo development and there is a short time frame (4 days) from egg production to hatching (Fraysse et al. 2006). This test could provide a useful substitute for the longer fish reproduction studies traditionally conducted with rainbow trout (*Oncorhynchus mykiss*) and fathead minnow (*Pimephales promelas*). However, further comparisons of relative sensitivity of zebrafish with the standard test species need to be done before widespread acceptance of the data for predicting effects to aquatic organisms.

In vitro toxicity tests being developed as part of high throughput screening might also have application beyond human health (see Chapter 8) to inform users about potential adverse outcome pathways (AOPs) for other species. For example, EPA's ToxCast™ program has screened compounds using more than 700 biochemical- and cell-based assays (Kavlock et al. 2012). Although many cellular and subcellular systems are conserved across species, care must be taken when conducting cross-species extrapolations of AOPs to focus on commonalities in physiology and be aware of interspecies differences. Even some biological systems that are apparently well conserved across phyla can have differential sensitivities or outcomes depending on the chemical and species. For example, the endocrine system, including hormones and associated cellular receptors, is well conserved among vertebrates, but the same hormone might result in different outcomes, and receptor-binding affinities of a chemical will differ across species because of structural differences of the estrogen receptor. Rainbow trout estrogen receptors, for example, share only a 60% homology with the human estrogen receptor and have a 10-fold lower binding affinity for 17β-estradiol (Fairbrother 2000; Matthews et al. 2000). Furthermore, estrogen has different effects among the various classes of animals, suggesting that estrogenic chemicals would also result in different adverse outcomes. Oviparous (egg-laying) animals, for example, rely on estrogen for shell gland formation and oviduct development and the production of vitellogenin for deposition into the eggs; these are not seen in non-oviparous animals. Similarly, estrogen induces ovulation in mammals and fish, but not in birds, reptiles, amphibians, or invertebrates (Lange et al. 2002).

Prolactin is another hormone found in both mammals and birds, with different regulatory processes in each species. In mammals, it regulates lactation, while in birds, it induces broodiness and nesting. Some receptors and detoxification enzymes, such as the aryl hydrocarbon (Ah) receptors and cytochrome enzymes, seem to be more universal, while others, although nearly universal, have significantly different structures across species (for example, metallothionein). Oxidative stress and formation of free radicals is a common response to some toxicants, including many nanomaterials, and all cells (animal or plant) are responsive to subsequent changes in membrane permeability, gene activation, and enzyme activity. Huggett et al. (2003) summarized the receptor and enzyme expression assays that have been developed for fish and proposed a model for extrapolating toxicity end point values from human assays to fish.

ToxCast™ Phase I tested more than 300 chemicals, many of which are pesticides with ecotoxicology data available (Kavlock et al. 2012). The data can be used as a "training set" to develop predictive relationships between the ToxCast™ data and biologically relevant ecotoxicity outcomes for aquatic and terrestrial species, including plants. The 700+ chemicals tested in ToxCast™ Phase 2 can then be ranked more effectively for relative ecological hazard (Sipes et al. 2013; Wilson et al. 2014).

Predicting dose-response relationships, however, is difficult even for humans (Chapter 8), and currently is practically impossible to do when extrapolating from human fibroblasts, keratinocytes, or other cells to plant or animal species. Similarly, attempting to predict which organ system might be affected on the basis of cell culture responses is likely also impossible. Nevertheless, information currently available from ToxCast and other high throughput data should be able to at least group chemicals into yes-no categories regarding toxic potential for the different species groups (aquatic vs terrestrial), which would add significantly to hazard predictions, currently based solely on in vivo testing of three aquatic species (a fish, an invertebrate, and an algae). Additionally, in the absence of animal test data, information from the cellular or subcellular tests in ToxCast™ and similar programs can be used in a general hazard categorization and delineation of which system might be most affected. Because of the lack of information for cross-species extrapolation, however, this is not likely to differ from what would be done for human health hazard classification; therefore, the ranking would default to that discussed in Chapter 8 for human health effects. See the Glitazone case study discussed in Chapter 12 for an example on how high throughput data might be applied.

Design Guidelines

Another approach to fill data gaps and identify chemicals of concern is to use a rapid screening tool based on property-based design guidelines. The approach differs from QSAR in that rather than predicting a threshold of toxicity (such as an LC_{50} value), it predicts the probability that a compound with particular properties will exhibit ecotoxicity above or below a particular threshold. The approach can define both chemicals with a high probability to be highly toxic and those with high probability of being "safe" (that is, having low to no ecotoxicity on the basis of established thresholds). The distinction between such design guidelines and categorical QSAR models is that the latter use complex statistical approaches (such as random forest, neural network, and machine learning) to identify the classification algorithm, which typically renders the relationship between the descriptors and response undecipherable to the user. By contrast, the design guideline approach typically uses two to three mechanistically tied descriptors and a transparent statistical approach to derive the relationship between the descriptors and response variables.

An example of such an approach for acute aquatic toxicity is illustrated in Figure 7-4. By using two properties (one related to bioavailability and one to reactivity), this approach was shown to identify chemicals least likely to be of concern for acute aquatic toxicity (Kostal et al. in press). Compounds in the lowest toxicity category (colored green) are almost entirely confined to the quadrant of the plot defined by boundaries of $\log D_{o/w} < 1.7$ and $\Delta E > 6$ eV (Kostal et al. in press), where ΔE is the energy gap between the highest occuped molecular orbital and the lowest unoccupied molecular orbital (Hehre et al. 1986). In the Kostal et al. study (in press), only 1% of the compounds in the highest acute aquatic toxicity category (LC_{50} < 0.0067 mmol/L) are retained after filtering with these property limits.

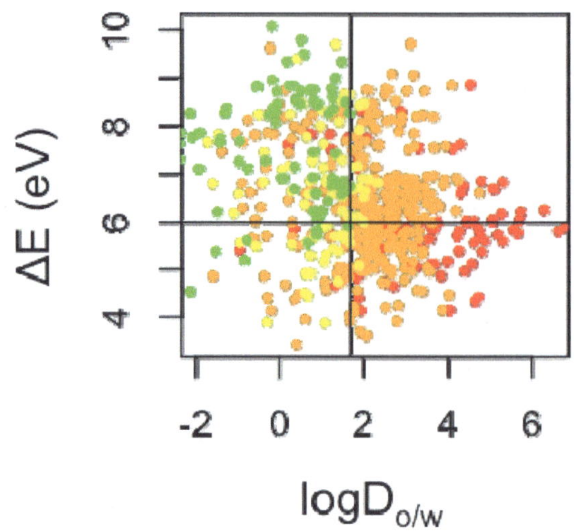

FIGURE 7-4 Scatter plots of the octanol-water distribution coefficient at pH 7.4 (log$D_{7.4}$) vs. ΔE (LUMO - HOMO energy gap, eV, determined by B3LYP/6-31+G(d)) for 555 compounds tested on 96-h toxicity assay of the fathead minnow. Compounds are colored by category of concern for acute aquatic toxicity as red - high concern (LC_{50}: < 0.0067mmol/L); orange - medium concern (LC_{50}: 0.0067-1.49 mmol/L); yellow - low concern (LC_{50}: 1.49-3.32 mmol/L); green - no concern (LC_{50} > 3.32 mmol/L). SOURCE: Kostal et al. in press. Reprinted with permission of PNAS.

A similar approach has been developed to identify chemicals of concern or those of no concern for chronic aquatic toxicity (Voutchkova-Kostal et al. 2012). A potential advantage of such methods is that they allow for intuitive comparisons between chemicals and inform the redesign of high-toxicity chemicals. However, a number of potential disadvantages also remain. For example, such approaches do not yield a discrete numerical threshold of toxicity, so if two alternatives are predicted to fall in the same quadrant, it is not possible to distinguish which has lower toxicity.

Quantitative Spectroscopic Data-Activity Relationships (QSDAR)

QSDAR models can provide estimates of ecotoxicity end points using an input of chemical spectra rather than structure (An et al. 2014). The spectroscopic data are used to generate descriptors, which are then fed into a quantitative model to generate a predicted threshold of toxicity. This emerging class of tools has a potential advantage over QSAR models in that it does not require knowledge of exact chemical structure. Therefore, in theory, these tools may be applicable to classes of chemicals, such as surfactants, that are found in mixtures with a variable structure. Thus far, only one example of such a tool exists for acute aquatic toxicity, and it uses nuclear magnetic resonance spectroscopic data (Voutchkova-Kostal et al. 2013). The accuracy of the model is closely comparable to the most robust QSAR models for that end point. However, QSDARs as a class of predictive tools still must undergo much further validation to establish wide applicability domains and the feasibility for estimating ecotoxicity of chemical mixtures.

8
Human Health

As noted earlier (Chapter 3), assessment of human health hazards should be included in each alternatives assessment. Human health hazard assessment of chemical alternatives is very similar to the hazard identification step of a traditional risk assessment. They are similar in that the types of adverse health end points and the sources of data for decisions are largely identical. Chemical alternatives assessments, however, typically use a comparative approach and are not meant to emulate formal dose-response or weight-of-evidence mode-of-action evaluations found in other chemical hazard assessments.

As shown in Figure 8-1, human health assessment is Step 6.1 of the committee's overall framework. Box 8-1 provides the elements of the committee's suggested approach.

TYPES OF DATA FOR HUMAN HEALTH ASSESSMENT

As illustrated in Figure 8-2, an implicit hierarchy exists with respect to the sources of data that are used in chemical assessments. Knowledge obtained from controlled clinical studies in humans is arguably the most desirable data for decisions on the potential for human health hazard. With the exception of pharmaceuticals, very few chemicals have this type of data available. Epidemiological studies of various designs are the next most useful data source because they examine whether there is an association between chemical exposure and human health.

The main advantage of these studies is that they involve humans; however, they are difficult to conduct and human evidence of chemical-induced effects, especially chronic effects, is rarely available. Data from experimental animal studies are often used to draw inferences about the potential hazard to humans when no adequate human data are available, yet the uncertainties associated with extrapolating the results from traditional toxicity studies in animals to humans are frequently poorly characterized. Other types of data, including results from studies in invertebrates, microorganisms, or in vitro experiments in animal or human cells, are also useful for traditional risk assessments. However, such data are most frequently used to determine the chemical mode of action. Traditionally, they have not been widely used to identify human health hazards beyond predicting specific hazard end points, such as genotoxicity, skin irritation, and eye irritation. Data from in vitro and in silico models, however, are, or will be, available for a far larger number of chemicals than experimental or epidemiological data will be (Collins et al. 2008). Thus, it is likely that at some point in the future, most decisions about environmental health protection will be made with in vitro and in silico[33] data and models, rather than traditional data (NRC 2007).

There are several different approaches for using the various levels of human health related data in an alternatives assessment:

a. Using traditional data, such as those that can be classified by the United Nations Globally Harmonized System of Classification and Labelling of Chemicals[34] (GHS) criteria. This is the typical approach used by alternatives assessments and is illustrated in the DecaDBE example in Chapter 12.

b. Using traditional data in combination with the use of new types of in vitro screening and in silico data as another type of primary data (for end points where this is deemed appropriate) or to fill data gaps.

---

33 The term in silico is used here to describe prediction and modeling (typically computational modeling) of effects based on information about a chemical's structure or physicochemical characteristics, including but not limited to structural alerts and structure activity relationship analysis.

34 GHS health hazards are agreed upon internationally for characterizing chemical hazards (76 Fed. Reg. 40850 2011). The Organisation for Economic Co-operation and Development High Production Volume Screening Information Data Set endpoints (OECD 2005), EU's Classification and Labeling and Packaging of Products regulation (EC 2011) and the U..S Occupational Safety and Health Hazard Communication Standard (77. Fed. Reg.17574 2012) are aligned with GHS. GHS criteria have been established for a number of human health end points.

c. A hybrid of a and b, which uses traditional data along with screening chemicals of concern with in vitro screening data and in silico modeling. This approach is illustrated in the glitazone example in Chapter 12.

There is interplay between health concerns, available data streams, and expertise that will contribute to the type of approach used. In any case, the type of approach used should be described in Step 2, as part of formulation and scoping. That said, the committee strongly supports a movement toward using in vitro screening and in silico data to fill data gaps when the necessary information is not available in the more traditional epidemiological and animal testing data. The committee points out later in this chapter that many high throughput in vitro assays may still have only limited applicability as primary data for predicting in vivo chemical hazards. The committee does, however, believe that the science will continue to advance in this area and that even today, there are opportunities to fill data gaps or screen for unexpected consequences using high throughput in vitro assays and in silico approaches.

To build on existing approaches, this chapter first describes how human health has been considered in existing alternatives assessment approaches and then describes the committee's framework for evaluating chemicals using traditional human health data in alternatives assessments. Second, the chapter provides more information on the state-of-the science of in vitro and in silico data, by health end point—showing where the science is in terms of predictivity and where the challenges remain. Third, the committee describes three scenarios of how novel in vitro and in silico data can be used in the context of chemical alternatives assessments and illustrates how visualization tools can inform the stakeholders about information available to help them make a decision about alternatives. Lastly, the committee offers advice on the path forward in using existing health data, as well as in vitro and in silico data, in chemical alternatives assessments.

BOX 8-1

HUMAN HEALTH ASSESSMENT AT A GLANCE

Phase 1. Evaluate the required health end points (identified in Steps 2) that must be addressed in the alternatives assessment

- Gather available data on health hazards associated with the chemical of interest and alternatives.
- Use authoritative lists to record previously identified health concerns.
- Use GHS criteria and hazard descriptors to the fullest extent possible to assess data, including potential effects on vulnerable populations, and classify the hazard data as indicating high (H), medium (M), or low (L) hazard. Indicate whether certainty of this classification is high, medium, or low.
- When conducting assessments of chemicals for reproductive toxicity and other health end points that require expert judgment to apply GHS criteria, use existing health hazard assessment guidance to ensure consistency and transparency.
- Where appropriate (e.g., for genotoxicity), use in vitro and in silico as primary data for an end point of concern (e.g., mutagenicity).
- Identify data gaps.

Phase 2. Develop strategies to address data gaps

- Use in vitro and in silico data and models to fill data gaps for an end point of concern (e.g., endocrine toxicity).
- Remaining data gaps should be classified as "No Data."

Phase 3. Develop a graphic or tabular display of health hazards associated with the chemical of interest and alternatives.

- Tabulation should include a placeholder for the full range of health end points typically considered in alternatives assessment, indication of which end points were considered, which end points warranted a H, M, L hazard level, which end points were based on novel in vitro or in silico approaches, and the certainty associated with each end point.
- ToxPi and similar approaches may be useful for visualizing novel high throughput data sets.

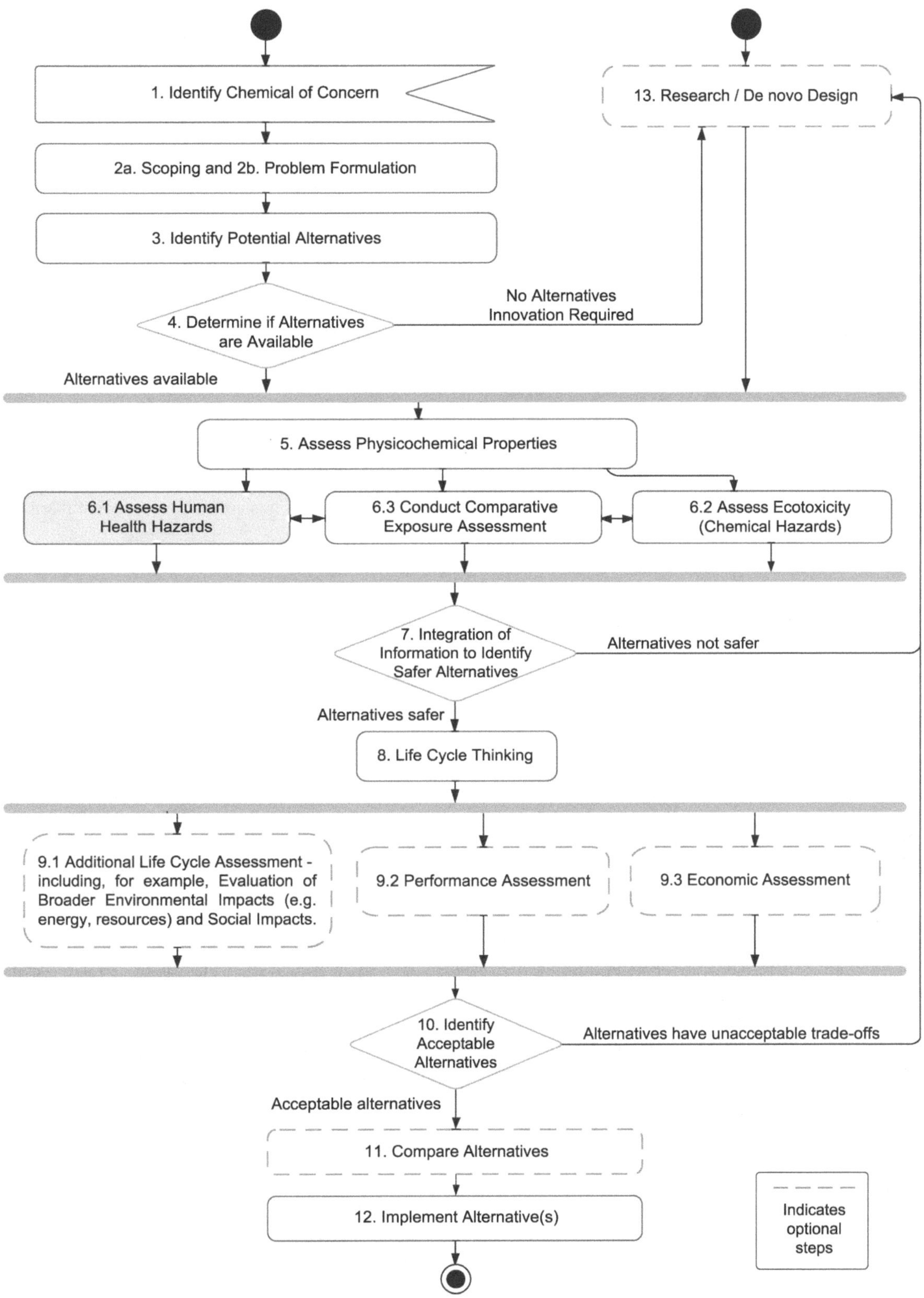

FIGURE 8-1 Committee's framework highlighting the human health hazard assessment.

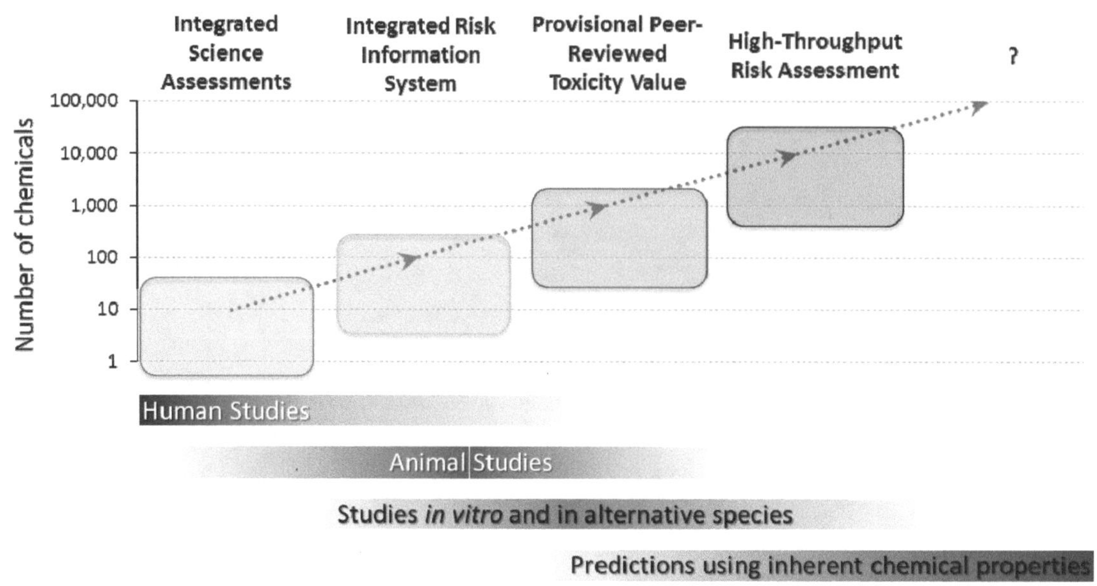

FIGURE 8-2 Decision contexts, data type, and data availability determine the type of human health assessment that can be performed on chemicals. The examples shown illustrate assessments performed by the EPA (EPA/NCEA).

HOW HUMAN HEALTH IS CONSIDERED IN EXISTING FRAMEWORKS

End Points Considered in Existing Frameworks

The committee considered the human health end points described in eight existing frameworks to compare current practices related to evaluating health hazards in chemical alternatives assessments and to inform the development of the committee's framework. Table 8-1 shows specific health end points; prioritized end points; the criteria and information sources the reviewed frameworks use to evaluate chemicals based on specific end points; and the types of data (e.g., human, animal, in vitro) upon which the criteria and source information are based. Appendix D provides more details on health end points and their evaluation in existing frameworks.

While the existing alternatives assessment frameworks are not identical, they contain common end points of concern, including carcinogenicity, mutagenicity, reproductive and developmental toxicity, endocrine disruption, acute and chronic or repeat dose toxicity, dermal and eye irritation, and dermal and respiratory sensitization. Several frameworks go further by identifying *priority* end points (e.g., carcinogenicity, mutagenicity/genotoxicity, reproductive toxicity, developmental toxicity, and endocrine toxicity). In determining which are "priority" end points, many of the frameworks use essentially the same rationale or basis—serious or irreversible health effects, or effects that may be transferred between generations and caused by low exposures to toxicants.

With two exceptions (endocrine activity/toxicity and epigenetic toxicity), the health end points in Table 8-1 align closely with the health hazards identified in the GHS. For example, the GHS defines acute mammalian toxicity as "adverse effects occurring following oral or dermal administration of a single dose of a substance, or multiple doses given within 24 hours, or an inhalation exposure of 4 hours" (UNECE 2013c). Regarding their acute toxicity, chemicals are classified into five hazard categories based on animal LD_{50} (oral, dermal) or LC_{50} (inhalation) values.

Endocrine toxicity is not included as a health hazard in the GHS. However, several frameworks identify endocrine-related health effects as an end point of concern. The criteria used in the frameworks vary because endocrine effects are not defined uniformly across frameworks. Data and authoritative lists are used to provide evidence of endocrine activity and/or disruption. For example, the DfE framework evaluates endocrine activity of chemicals, but does not characterize hazard in terms of endocrine disruption. On the other hand, the IC2 and BizNGO frameworks use criteria developed by GreenScreen® to evaluate chemicals for endocrine activity and assign hazard values based on adverse

TABLE 8-1 Health End Points Established by Other Frameworks

Human Health End Point (more detail in Appendix D)	Criteria and Information Sources Frameworks Use to Establish Evidence of the Human Health End Points		Types of Data Used to Establish the Health End Points			
			Human	Animal	Human & Animal (WOE)	In Vitro
Acute Mammalian Toxicity[1,2,4,5,6,7,8] Priority[7]	GHS criteria [1,2,4,5,6,8] Authoritative lists/databases[2,4] EU Risk phrases, Hazard statements[2,4,8] NIOSH[7]	MSDS[7] IDLH[7]; LD_{50}/ LC_{50}[7] OELs (NIOSH, OSHA, ACGIH)[7] HSDB[7]; RTECS[7]	OELs	OELs GHS 1-5; LD_{50}/ LC_{50}; IDLH	OELs	
Carcinogenicity[1,2,3,4,5,6,7,8] Priority[2,3,4,5,7]	GHS criteria[1,2,3,4,5,6,7,,8] Authoritative lists [2,3,4,6,7,8]	EU Risk phrases, Hazard statements[2,3,4,8]	GHS 1A	GHS 1B	GHS 1B, 2	
Mutagenicity/ Genotoxicity[1,2,3,4,5,6,7,8] Priority [2,3,4,5,]	GHS criteria [1,2,3,4,5,6, 7,,8] Authoritative lists [2,3,4,6,7,,8]	EU R-phrases, Hazard statements[2,3,4,5,6,7,,8]	GHS 1A	GHS 1B		GHS 1B; 2
Reproductive Toxicity[1,2,3,4,5,6,7,8] Priority [2,3,4,5,]	GHS criteria[2,3,4,5,6,7,8] EPA OPPT criteria (HPV)[1] REACH criteria (Annex IV)[1]	Authoritative lists[2,3,4,6,7,8] EU Risk phrases, Hazard statements[2,3,4,6,7,8] RTECS[7]	GHS 1A	GHS 1B	GHS 1B; 2	
Developmental Toxicity[1,2,3,4,5,6,7,8] Priority [2,3,4,5,]	GHS criteria[2,3,4,5,6,7,8] Authoritative lists[2,3,4,6,7,8] EPA OPPT criteria (HPV)[1]	EU Risk phrases, Hazard statements[2,3,4,6,7,8] REACH criteria (Annex IV)[1]	GHS 1A	GHS 1B	GHS 1B; 2	
Neurotoxicity[3] †Single Exposure (SE)[2,4] ††Repeated Exposure (RE)[1,2,4] Priority[3]	GHS criteria[1,2,4] Neurotoxicants (ATSDR; EPA IRIS)[3]	EU Risk phrases, Hazard statements[2,4] Authoritative lists/databases[2,4,]	GHS 1(a); 3; ATSDR; IRIS (SE) GHS 1(a) ATSDR; IRIS (RE)	GHS 1(b); 2 ATSDR; IRIS (SE) GHS 1(b); 2 ATSDR; IRIS (RE)	ATSDR; IRIS (SE) ATSDR; IRIS (RE)	
††Repeated Dose Toxicity[1,5,7]	GHS criteria[1,2,4] Authoritative lists [2,4]	EU Risk phrases, Hazard statements[2,4] EPA IRIS Reference Doses (RfDs)[7]	GHS 1(a) EPA RfDs	GHS 1(b) EPA RfDs	EPA IRIS RfDs	
Systemic Toxicity/Organ Effects[2,4,6,7,8] †Single Exposure (SE) ††Repeated Exposure (RE)	GHS criteria[2,4,6] Authoritative lists[2,4]	EU Risk phrases, Hazard statements[8] EPA IRIS Reference Doses (RfDs)[7]	GHS 1(a); 3 (SE) GHS 1(a) (RE)	GHS 1(b); 2 (SE) GHS 1(b); 2 (RE)	EPA RfDs	

TABLE 8-1 (Continued)

Human Health End Point (more detail in Appendix D)	Criteria and Information Sources Frameworks Use to Establish Evidence of the Human Health End Points		Types of Data Used to Establish the Health End Points			
			Human	Animal	Human & Animal (WOE)	In Vitro
Respiratory Sensitization[1,2,3,4,5,] Priority[3,5]	GHS criteria [1,2,4,5] Authoritative lists /databases[2,4]	EU Risk phrases, Hazard-statements[2,4] EU Annex VI Category 1[3]	GHS 1A; 1B			
Skin Sensitization[1,2,4,7,8]	GHS criteria[1,2,4,] Authoritative lists[2,4]	EU Risk phrases, Hazard-statements[2,4,8] HSDB[7]; Sax[7]; MSDSs[7]	GHS 1A;1B	GHS 1A; 1B		
Skin & Eye Irritation /Corrosivity [1,2,4,5,7,8] Respiratory Irritation[5,7]	GHS criteria[2,4,5,8] Authoritative lists [2,4,8] NIOSH[7] EPA Office of Pesticide Programs [1]	EU Risk phrases, Hazard-statements[2,4,8] HSDB[7] MSDSs[7]	HSDB; MSDSs; NIOSH	GHS 1, 2A, 2B		REACH skin irritation & corrosion tests
Endocrine Activity[1,2,4,7]/ Toxicity[3,5,6,7,8] Priority [2,3,4,5]	All available data[2,4,7] Authoritative lists[2,3,4,6,8]					
Epigenetic Toxicity [6]	No information provided		NA	NA	NA	NA

[1]DfE; [2]IC2; [3]CA SCP; [4]BizNGO; [5]REACH; [6]UCLA MCDA (Malloy et al. 2013); [7]TURI; [8]German Guide; †GHS Specific Target Organ Toxicity –Single Exposure (see Appendix D); ††GHS Specific Target Organ Toxicity –Repeated Exposure (see Appendix D).

Acronyms: OELs= Occupational Exposure Limits. AOEC=Association of Occupational and Environmental Clinics database of asthmagens. HSDB= Hazardous Substances Data Bank. CLP= ECHA's Classification and Labelling Inventory database. GHS 1A and 1B refer to GHS categories, which are explained in Appendix D. ACGIH = American Conference of Industrial Hygienist, ATSDR = Agency for Toxic Substances and Disease Registry, GHS = Globally Harmonized System of Classification and Labelling of Chemicals, HPV = High Production Volume, HSDB = Hazardous Substances Data Bank, IDLH = Immediately Dangerous to Life or Health, IRIS = Integrated Risk Information System, LC50 = Lethal Concentration that kills 50% of population, LD_{50} = Lethal Dose that kills 50% of population, MSDS = Material Safety Data Sheets, NIOSH = National Institute for Occupational Safety and Health, OELs = Occupational Exposure Limits, OPPT = Office of Pollution Prevention and Toxics, OSHA = Occupational Safety and Health Administration, RfDs = Reference Doses, RTECS = Registry of Toxic Effects of Chemical Substances.

Information in the table was obtained from a review of guidance documents, regulations, and other available information on the frameworks. Specifically, the guidance document for the hazard assessment tool, GreenScreen® (Clean Production Action 2013), is the source for information on the IC2 and BizNGO health end points. The California Safer Consumer Products (CA SCP) framework's health end points are the hazard traits that are used for listing a chemical as a "Candidate Chemical" or a potential "Chemical of Concern" in a priority consumer product (CA DTSC 2013c). The health end points for the UCLA Multi-Criteria Decision Analysis (MCDA) framework are the eight measures linked to the toxicity sub-criterion (associated with human health impacts) in a generic alternatives assessment model (Malloy et al. 2013). The REACH health end points are those specified in the REACH legislation (EC 2007). The health end points for the German Guide on Sustainable Chemicals framework (Reihlen et al. 2011) are based on the risk phrases and hazard statements used to identify high- priority chemicals for substitution ("red color code"). Two additional frameworks, the Lowell Center Alternatives Assessment Framework (Rossi et al. 2006) and the UNEP Persistent Organic Pollutants Review Committee General Guidance on Alternatives (UNEP 2009) framework, which are reviewed in other sections of the report, do not identify human health end points and are not included here.

endocrine-related health effects. Additional information about criteria the frameworks use to provide evidence of endocrine-related health effects is presented in Appendix D. Appendix D also discusses how other end points are characterized by the GHS classification scheme and their application in the GreenScreen® tool and DfE framework. Although GHS is widely used, different approaches have also been used to inform other alternatives assessment frameworks (e.g., TURI).

Information Sources Used by Existing Frameworks

Table 8-1 shows that frameworks use a variety of information sources, including authoritative lists and databases, to establish evidence of health end points when evaluating chemicals. Some of the frameworks specify review of all available, relevant information, including information obtained from searches of the scientific literature. For example, the EPA DfE framework uses primary data sources, public and confidential business information, expert predictive models, and other forms of expert judgment to characterize health hazards (Lavoie et al. 2010).

Authoritative lists are used extensively as the basis for alternatives assessments (i.e., as a reason for entry into Step 1 of the committee's framework). Authoritative lists, databases, and risk phrases[35] are also used to assess the health impacts of potential substitute chemicals. Several of the examined frameworks (e.g., IC2 and BizNGO) rely on the GreenScreen® chemical hazard assessment tool, which uses authoritative lists and databases to classify chemical health hazards. GreenScreen® defines "authoritative list" as "those lists developed by governmental bodies or government recognized expert bodies and include chemicals that are listed based on results from expert review of test data and scientific literature". The hazard lists that are required for a full GreenScreen® are called GreenScreen® Specified Lists (Clean Production Action 2012) and also include screening lists. According to GreenScreen®, "lists are identified as Screening Lists if they were developed using a less comprehensive review; or if they have been compiled by an organization that is not considered to be authoritative; or if they are developed using exclusively estimated data; or if the chemicals are listed because they have been selected for further review and/or testing, and result in a classification with a lower level of confidence."[36] Table 8-2 provides an example of how authoritative lists are used by the DfE framework and the GreenScreen® tool.

The IC2 and BizNGO frameworks also use authoritative lists (with GreenScreen®-assigned hazard levels) to establish evidence of the reproductive toxicity health end point.

Below are some of the lists included in their frameworks:

- *High Hazard* = NTP-OHAT (Clear Evidence of Adverse Effects-Reproductive);
- CA Prop 65 (known to the state to cause reproductive effects--male or female);
- H360F (may impair fertility);
- EU H360FD (may damage fertility); and
- EU 360Fd (may damage fertility).

Authoritative lists used by GreenScreen® are divided into A[37] and B lists. The A and B lists are

[35] Risk phrases were developed in the European Union (prior to adoption of the GHS Classification and Labelling System) to communicate risk. They are based on criteria that are essentially the same as GHS criteria and are being replaced by hazard statements based on GHS criteria. Both risk phrases and hazard statements are examples of authoritative lists.

[36] Although the types of lists the frameworks use are defined and explained in the GreenScreen guidance document (Clean Production Action 2013), some questions and issues remain. For example, it is not clear: (a) how lists are selected; (b) why some lists are used and others are not; (c) to what extent scientific rigor determines the level of confidence in lists; and (d) how the level of confidence in lists impacts the selection of safer alternative chemicals. The use of authoritative lists is discussed further in Appendix D.

[37] Authoritative A lists include: IARC Group 1 or 2A chemicals (carcinogenicity); EU CMR Category 1 or 2 chemicals [mutagenicity/genotoxicity]; and chemicals classified as H360F (may damage fertility), H360FD (may damage fertility; may damage the unborn child) and 360df (may damage the unborn child, suspected of damaging fertility) [reproductive and developmental toxicity]. Authoritative B lists include: IARC Group 3 chemicals (carcinogenicity); MAK Germ cell mutagens 1, 2, or 3a chemicals (mutagenicity/genotoxicity); chemicals classified as EU H334 (respiratory sensitization); and DOT Class 2,3 Group B chemicals (acute mammalian toxicity). Screening A lists include, predominantly, GHS lists of various countries, including Korea, Japan, Indonesia, and Australia for several end points (e.g., carcinogenicity, developmental toxicity, acute mammalian toxicity). Screening B lists include: WHMIS D1B chemicals (acute mammalian toxicity); OSPAR (endocrine disruptor); and MAK Pregnancy Risk Group D chemicals.

TABLE 8-2 Use of Authoritative Lists by the DfE Framework and GreenScreen® Tool

End Point/List Classification	DfE Classification	GreenScreen® Tool
Carcinogenicity		
NTP—Known to be human carcinogen NTP—Reasonably Anticipated to be a Human Carcinogen IARC Group 1—Carcinogenic to Humans IARC Group 2A—Probably Carcinogenic to Humans GHS H350—May Cause Cancer GHS H350i—May cause cancer by inhalation	Very High Hazard	High Hazard
IARC 2B—Possibly carcinogenic to humans EU CMR List Category 3—Cause for concern for humans owing to possible carcinogenic effects EU 351—Suspected of causing cancer	High Hazard	Moderate Hazard
Mutagenicity/genotoxicity		
EU CMR Category 1—Substances known to be mutagenic to man Category 2—Substances which should be regarded as if they are mutagenic to man EU H340—May cause genetic defects	Very High Hazard	

distinguished based on whether categories in the list translate directly into a single level of concern for a single GreenScreen® health end point or a single benchmark. In addition, the assigned health hazard level of an Authoritative A list cannot be modified using additional data; Authoritative B lists, however, can be modified. The confidence level is "high" for Authoritative A lists (Clean Production Action 2012, 2013). For Authoritative B lists, the confidence level is "low" in the current guidance document (Clean Production Action 2013), but is listed as "high" on the Specified List (Clean Production Action 2012).

The TURI framework also uses material safety data sheets (MSDSs), which in the future must be based on GHS criteria, as a source of information to compare the toxicity of chemicals [38] Another data source is the Hazardous Substances Data Bank (HSDB).[39] This database focuses on the toxicology of potentially hazardous chemicals and includes up-to-date abstracts of animal and human studies, including studies on the acute and chronic toxicity of chemicals. The abstracts undergo peer review before they are added to the database; however, the studies are not evaluated, and expert judgment is required to determine their relevance in providing evidence of chemical toxicity.

Use of Hazard Classification Levels in Existing Frameworks

To facilitate comparison of hazard levels across chemicals, some frameworks use hazard classification levels to describe information about the severity of the effect. Hazard classification levels are used most extensively in the DfE framework (Davies et al. 2013) and the GreenScreen® tool (Heine and Franjevic 2013).

For each human health end point considered in DfE and GreenScreen®, a descriptor is assigned based on criteria that constitute High, Moderate, or Low and, in some cases, Very High or Very Low (Davies et al. 2013). As discussed earlier, these

[38] Under the revised Hazard Communication Standard (77 Fed. Reg. 17574 2012), MSDSs will be renamed Safety Data Sheets, or SDS, and based on GHS criteria, which should make them a good information source. As of now, however, the committee's comparison of harmonized and un-harmonized chemical classifications for the acute toxicity end point in the European Chemical Agency (ECHA) Classification and Labeling Inventory Database, using the H330 and H311 hazard statements, showed a ten-fold difference in the number of classified chemicals (ECHA 2014e). This indicates that un-harmonized chemical classifications (by individual manufacturers and other safety data sheet preparers) may be inconsistent or inaccurate.

[39] Available at http://toxnet.nlm.nih.gov/cgi-bin/sis/htmlgen?HSDB

descriptors are usually based on GHS criteria. Table 8-3 shows acute mammalian toxicity levels by different exposure routes. It indicates that the hazard classification levels described by DfE can range from *Very High Hazard* = (Category 1 or 2) to *Low* (Category 5). Appendix D has a more detailed description of the hazard level classification systems in DfE and GreenScreen® for various end points. The hazard profile and assigned concern levels are ultimately reviewed by a group of experts before they are used in decision-making.

GreenScreen® uses a similar overall process of assigning High, Moderate, and Low classification levels. It groups human health hazard end points in the following way:

- Group I hazards can lead to chronic or life-threatening effects or adverse impacts that are potentially induced at low doses and transferred between generations. Group I end points include carcinogenicity, mutagenicity/genotoxicity, reproductive toxicity, developmental toxicity (including neurodevelopmental), and endocrine activity.
- Group II/II* hazards are additional end points that are necessary for understanding and classifying hazards (Heine and Franjevic 2013). Group II end points are acute mammalian toxicity, systemic toxicity/organ effects (single exposure), neurotoxicity (single exposure), and irritation/corrosivity for eyes and skin. Group II* end points include systemic toxicity/organ effects (repeated exposure), neurotoxicity (repeated exposure), and respiratory and skin sensitization.

Approaches to Handling Data Gaps in Existing Frameworks

Comparative chemical alternatives assessments, similar to more traditional human health risk assessments, are only as good as the data and information available. Identifying and addressing the potential impacts of health hazard data gaps in alternatives assessments is an important issue because it can help ensure that what are thought to be safer chemical substitutes do not subsequently pose health concerns. The extent to which data gaps and related issues are discussed in the frameworks varies widely.

The CA SCP framework does not require new data collection to address data gaps. Gaps in toxicity and health effects information are acknowledged in the German and TURI frameworks and noted in health hazard summaries, but are not discussed further. When primary data are not available or deemed inadequate, DfE has explicit procedures in place to assign a hazard concern level based on structure-activity relationship (SAR) considerations and professional judgment. These procedures ensure that all end points are covered so that the hazard profile can be completed. Similarly, GreenScreen® specifies review of at least one readily available suitable analog for each hazard end point for which data are missing on the parent compound, stating that expert judgment and estimated data from analog and SAR analyses may be used in lieu of measured data. If information is still deemed insufficient to provide any classification for a hazard end point, as is frequently the case, the end point is assigned a "data gap" or "no data" designation. For example, GreenScreen® states that a data gap exists when measured data and authoritative screening lists have been reviewed, and expert judgment and estimation such as modeling and analog data have been applied, and there is still insufficient information to assign a hazard level.

With regard to how any remaining data gaps are handled in the final analyses, a range of possibilities is described in Chapter 9. The UCLA MCDA framework evaluates the impact of data gaps in an alternatives assessment using multi-attribute utility theory and outranking. The GreenScreen® tool (and by extension, the IC2 and BizNGO frameworks) is the only example found to describe how data gaps are handled in the analysis. GreenScreen®'s procedure defines the minimum data requirements to achieve a given benchmark and describes the required data and permissible data gaps for each hazard end point category (Group I Human and Group II and II* are specified). The treatment of gaps, or failure to meet minimum data requirements, is negative as opposed to neutral and is benchmark-specific. For example, if a chemical meets Benchmark 2 based on hazard analysis but fails to meet the minimum requirements for this benchmark because of gaps in data, it is assigned an "unspecified" designation. If a chemical fails to meet the minimum data requirements for Benchmark 3 in the gap analysis, it is downgraded to 2_{DG}. No data gaps are allowed for Benchmark 4.

TABLE 8-3 Acute Mammalian Toxicity

	Very High	High	Moderate	Low
Oral LD_{50} (mg/kg)	≤ 50	< 50-300	> 300-2000	> 2000
Dermal LD_{50} (mg/kg)	≤ 200	> 200-1000	> 1000-2000	> 2000
Inhalation LC_{50} (vapor/gas) (mg/L)	≤ 2	> 2-10	> 10-20	> 20
Inhalation LC_{50} (dust/mist/fume) (mg/L/day)	≤ 0.5	> 0.5-1.0	> 1.0-5	> 5

SOURCE: EPA 2011a

HUMAN HEALTH IN THE COMMITTEE'S FRAMEWORK

For its recommended framework, the committee suggests that the assessment of human health hazards follow a similar process as that used by the frameworks described above, with these two modifications included:

- Consider the full range of scientific information to fill data gaps (see below).
- Continued focus on hazard as opposed to risk, except when directed otherwise by comparative exposure or decision rules.

The committee's framework for human health assessment would begin with the following end points, which are GHS health hazards, supplemented with endocrine activity that is not included in GHS at this point in time.[40]

- Acute toxicity
- Carcinogenicity
- Mutagenicity/genotoxicity
- Reproductive toxicity
- Development toxicity
- Respiratory sensitization
- Skin sensitization
- Specific target organ toxicity (single exposure):
    - Neurotoxicity
    - Respiratory irritation
- Specific target organ toxicity (repeated exposure):
    - Neurotoxicity
- Skin and eye corrosion/irritation
- Endocrine activity

The committee did not strictly define the above list as a minimum set of adverse health end points to be considered in alternatives assessment, but suggests that this list be used as the initial list for selecting end points in the problem formulation exercise in Step 2, with clear documentation of which end points were not considered in Step 6.1 of the assessment. Additional end points considered in the assessment also should be specified and clearly documented.

The committee advises using GHS criteria and hazard descriptors to the fullest extent possible in evaluating human health hazards, which is consistent with what is described in Chapter 7 for ecotoxicity. This approach is also consistent with several existing approaches that use GHS as their ultimate common denominator in human health assessment. The use of health end points that are aligned with health hazards identified in GHS ensures that assessments address internationally recognized chemical hazards. In addition, using GHS criteria enables alternatives assessments to use toxicity information on chemicals submitted as Screening Information Data Sets (SIDS) for the OECD SIDS program because GHS criteria include SIDS end points. EPA's Office of Pollution Prevention and Toxics is making non-confidential information submitted to this program publicly available (EPA 2007), enabling assessment of unpublished data, which is especially important for assessments of chemicals for which there are no published toxicity studies. This information should reduce data gaps. In addition, GHS alignment enables information from other resources, such as the ECHA Classification and Labeling Inventory Database and guidance documents, to be used in alternatives

[40] The descriptors for the Mutagenicity/Genotoxicity and Skin and Eye Corrosion/Irritation end points listed for the committee's framework do not conform to the GHS health hazard descriptors. The rationales for this are in Appendix D and would be included in the problem formulation section of the framework.

assessments. Another advantage of GHS alignment is that it links safer alternatives directly to workplace chemical hazards identified under the GHS-based OSHA Hazard Communication Standard.

While supporting the use of GHS criteria, the committee suggests the following refinements to the reliance on GHS criteria and hazard descriptors:

- Use available human data when GHS criteria indicate that they should be used.[41]
- Describe criteria used. When non-GHS criteria are used, explain the rationale.
- Align the description of GHS hazards to the GHS criteria. If there is a rationale for using a hazard description that is different from the ones used by GHS, explain this rationale and how to apply the criteria. An example of misalignment between GHS hazards and GHS criteria is use of the GHS criteria for "Specific Target Organ Toxicity (Repeated Exposure)" when referring to the health end point as "Systemic Organ Toxicity/Organ Effects (Repeated Exposure)."

The committee's framework advises using authoritative lists, as has been done by a number of existing approaches. The rationale for using such health end point-specific authoritative lists to compare chemical hazards is that it maximizes the use of existing evaluations of scientific information and helps ensure that alternatives assessments are efficient and based on consistent science. Assessments following the committee's framework would:

- Define "authoritative" lists.
- Describe criteria for which authoritative lists are used or not used in the framework.
- Include end point-specific authoritative lists of toxicants developed by government agencies that use human or animal data. For example, Agency for Toxic Substances and Disease Registry Minimal Risk Levels (ATSDR 2013), California Environmental Protection Agency Office of Environmental Health Hazard Assessment acute and chronic reference exposure levels (OEHHA 2014), and respiratory irritants identified by the National Institute for Occupational Safety and Health (NIOSH 2005), and others that include human data in the assessments.
- Ensure that the listing criteria are transparent, understood by the assessor, and consistent with the criteria used to establish evidence of the health end point that the list is addressing.
- Use authoritative lists only to identify hazards

Assessments following the committee's framework would consider existing health hazard assessment guidance to classify chemicals based on their end point effects, when GHS criteria require the use of expert judgment to establish a health end point. It is not clear whether some existing frameworks do this, but assessments following the committee's framework would do so in a transparent way when conducting de novo assessment to classify chemicals for reproductive toxicity and other end points. Box 8-2 describes examples of existing health hazard assessment guidance and includes EPA risk assessment guidelines for reproductive toxicity, neurotoxicity, and developmental toxicity.

Notably different from several existing frameworks, the output of the committee's framework would not include a "score" integrating human health data across health end points. Instead, the committee's framework would tabulate (a potential format is shown in Table 8-4) health end points, noting: which end points were considered; which of the typically assessed end points were not considered; indication of the hazard level suggested by the data (H,M,L); and an indication of the certainty of the data (known, limited certainty, highly uncertain). Gaps in data at this point would be clearly indicated.

The resulting tabulation makes no attempt to integrate information across health end point domains for three primary reasons: 1) there is no established consensus on which end points are of greater concern; 2) doing so unnecessarily carries forward the impact of benchmarking cutoffs; and 3) it is important to carry forward the certainty and the level of the hazard into the integration of other data in the decision-making step (Step 7). This approach is in contrast to the common approach of creating scores that integrate information and account for data gaps and uncertainty at this point in the process. While such approaches may be easy to use, they obscure information that should be considered across domains. Ideally, gaps would be addressed using novel high throughput in vitro data and in silico modeling, as described in the rest of the chapter.

---

41 In some existing frameworks, it is not clear if human data are used when prescribed by GHS criteria.

BOX 8-2
EXAMPLES OF EXISTING GUIDANCE FOR MINIMUM EVIDENCE REQUIRED TO ESTABLISH THAT SPECIFIC HEALTH HAZARDS DO OR DO NOT EXIST

Note: More specifics about minimal evidence requirements are described in the risk assessment guidelines.

- EPA's Risk Assessment Guidelines for Neurotoxicants (EPA 1998)

  Sufficient Experimental Animal Evidence/Limited Human Data
  The minimum evidence necessary to judge that a potential hazard exists would be data demonstrating an adverse neurotoxic effect in a single appropriate, well-executed study in a single experimental animal species.

  The minimum evidence needed to judge that a potential hazard does not exist would include data from an appropriate number of end points from more than one study, and two species showing no adverse neurotoxic effects at doses that were minimally toxic in terms of producing an adverse effect. Information on pharmacokinetics, mechanisms, or known properties of the chemical class may also strengthen the evidence.

- EPA's Risk Assessment Guidelines for Developmental Toxicants (EPA 1991)

  Sufficient Experimental Animal Evidence/Limited Human Data
  The minimum evidence necessary to judge that a potential hazard exists generally would be data demonstrating an adverse developmental effect in a single appropriate, well-conducted study in a single experimental animal species.

  The minimum evidence needed to judge that a potential hazard does not exist would include data from appropriate, well-conducted laboratory animal studies in several species (at least two) that evaluated a variety of the potential manifestations of developmental toxicity and showed no developmental effects at doses that were minimally toxic to the adult.

- EPA's Risk Assessment Guidelines for Reproductive Toxicants (EPA 1996)

  Sufficient Experimental Animal Evidence/Limited Human Data
  The minimum evidence necessary to determine if a potential hazard exists would be data demonstrating an adverse reproductive effect in a single appropriate, well-executed study in a single test species.

  The minimum evidence needed to determine that a potential hazard does not exist would include data on an adequate array of endpoints from more than one study, with two species that showed no adverse reproductive effects at doses that were minimally toxic in terms of inducing an adverse effect. Information on pharmacokinetics, mechanisms, or known properties of the chemical class may also strengthen the evidence.

IN VITRO DATA AND IN SILICO MODELS FOR CHEMICAL ALTERNATIVES ASSESSMENTS

Released in 2007 by the National Research Council (NRC), the report, "Toxicity Testing in the 21st Century (TT21C): A Vision and a Strategy" (NRC 2007), described the promise of high throughput in vitro approaches and in silico models in evaluating chemical safety. The idea that these novel approaches could replace animals in toxicity testing has been treated by some with skepticism and claims of unrealistic overreaching (Bus and Becker 2009; Meek and Doull 2009). But significant research investments have revealed numerous advantages to using high throughput methods in toxicology (Krewski et al. 2011; Kavlock et al. 2012) and led to the generation of a vast amount of data (Table 8-5) that are in the public domain and available for analysis and evaluation in hazard identification and dose-response assessments (Tice et al. 2013). Advances in molecular, cell, and systems biology, together with advanced analytical methods in biostatistics, bioinformatics, and computational biology, have led to toxicity testing now being routinely conducted in vitro by evaluating cellular responses in a suite of toxicity pathway-centric assays.

In silico approaches for predicting adverse effects have existed for more than 30 years, but research and development in this area has increased exponentially in recent years. Most in silico methods for toxicity prediction have focused on hazard identification; for example, determining whether a compound has properties associated with liver injury. The concept of chemical similarity has been used to develop a variety of methods to predict

TABLE 8-4 Hypothetical Tabulation Evaluations of Human Health Impact: One Potential Format.*

	Human Health Impacts																																		
Alternatives	Acute Toxicity			Carcinogenicity			Mutagenicity / Genotoxicity			Reproductive Toxicity			Developmental Toxicity			Specific Target Organ Toxicity (Single Exposure)—Neurotoxicity			Specific Target Organ Toxicity (Single Exposure)—Respirator Irritation			Specific Target Organ Toxicity (Repeated Exposure) - Neurotoxicity			Specific Target Organ Toxicity (Repeated Exposure)—Respiratory Irritation			Respiratory Sensitivity	Endocrine Activity	Skin Sensitivity			Skin and Eye Corrosion / Irritation		
	H	M	L	H	M	L	H	M	L	H	M	L	H	M	L	H	M	L	H	M	L	H	M	L	H	M	L			H	M	L	H	M	L
C																																			
A																												N C	N C						
B																																			

*High, medium, and low indicate health impact and relative uncertainty of each finding is depicted by colors (dark blue = known, light blue = limited certainty, pink = highly uncertain, gray = unknown (data gap), NC = not considered). C = chemical of concern.

chemical-induced responses based only on chemical structure. Both the simpler read-across analysis (Enoch et al. 2008; Hewitt et al. 2010) and more complex machine learning-based approaches (Voutchkova et al. 2010) can be easily adopted for the purpose of chemical alternatives assessment.

It is crucial that the next generation of alternatives assessment frameworks incorporate the use of in vitro and other high throughput assays—*toxicity pathway-centric assays*—into the assessment process. The question of how various types of human health assessments of chemicals, including chemical alternatives assessment, will be conducted once a proper suite of in vitro assays, in silico models, and other novel data streams become available has come to the forefront of the debate in the environmental health community, largely because the feasibility of obtaining complex data on hundreds, if not thousands, of chemicals became a reality in the past several years. A number of broadly applicable opinions have been voiced that, while unanimous in the overall conclusion that human and environmental health decisions will be made with new data, are somewhat different in how this information should be used and for what type of decisions (e.g., relative ranking/prioritization to select candidates for further traditional testing, or making choices about alternatives in the context of alternatives assessment).

Several approaches to using in vitro data in human health assessments impact the thinking about how to incorporate these data in alternatives assessment:

- Using in vitro data and in silico predictions in ways similar to current practices that rely on human and animal health end points, with the use of additional uncertainty or safety factors to account for in vitro to in vivo extrapolation (Crump et al. 2010). Crump et al (2010) reasoned that toxicity pathway-based models are unlikely to contribute quantitatively to decision making for several reasons, including that the statistical variability inherent in such complex models will hinder their ultimate utility for estimating small changes in response, and that such models will likely continue to involve empirical modeling of dose-response relationships.
- Using in vitro data and in silico models, coupled with estimates of population variability and uncertainty, to estimate the human dose at which a chemical may significantly alter a biological pathway in vivo. This dose is referred to as a *biological pathway altering dose* (BPAD) (Judson et al. 2011). This approach draws parallels between a chemical-associated perturbation of a pathway as observed in in vitro assays and a key event in the chemical's

TABLE 8-5 Toxicity and High Throughput Screening (HTS) Data amenable to the Evaluation of Human Health Hazard and the Relationship between Chemical Dose and Response.

ToxCast (Knudsen et al. 2013)	Quantitative (in most cases concentration-response based) information from a suite of diverse HTS cellular (viability, proliferation, or reporter gene), biochemical (enzymes and receptors), and zebrafish assays.	Phases I and II: 293 and 767 chemicals screened in 600+ assays Phase III: 1K+ chemicals screened in 100+ assays
Tox21 (Collins et al. 2008)	Ultra-qHTS (all data are collected for 8-15 concentrations of each agent) molecular, biochemical, and cell-based assays from a consortium of U.S. federal agencies.	8K+ environmental chemicals and drugs 50+ assays and 120+ endpoints
HTS Zebrafish (Truong et al. 2014)	HTS (concentration-response) of embryonic zebrafish for developmental, morphological, and behavioral end points	8K chemicals, including those screened by ToxCast and Tox21
PubChem (Wang et al. 2014)	A database of biological tests of small molecules generated through high throughput screening experiments, medicinal chemistry studies, chemical biology research, and drug discovery programs.	10K chemicals screened in up to 10K cellular, molecular or biochemical assays
Drug Matrix (Fostel 2008)	A large compendium of microarray data from in vivo (rat) exposures to various drugs and chemicals; profiling was performed on 9 organs.	658 drugs and chemicals 4.3K studies (dose, time, organ, etc.)/13K arrays
TG-Gates (Uehara et al. 2010)	Gene expression data from liver (rat) and cultured hepatocytes (rat and human) in dose- and time-dependent study design. Matching toxicity (pathology and clinical chemistry) data is also available.	170 drugs and chemicals 33K+ microarrays
CTD (Davis et al. 2013)	Manually curated chemical-gene/protein interactions, chemical-disease relationships, and exposure relationships (stressors, receptors, events, and outcomes) from published literature.	28K genes (change in expression information) 886K chemical-gene interactions
CEBS (Waters et al. 2008)	The Chemical Effects in Biological Systems (CEBS) database houses several types of study data from academic, industrial, and governmental laboratories.	6K chemicals and drugs 800+ molecular (mostly microarray) datasets
ToxRefDB (Martin et al. 2009)	Manually curated chronic toxicity data for a variety of organ systems in experimental animals from regulatory submissions to EPA.	474 environmental chemicals tested in guideline studies
NTP Toxicity data	Detailed toxicity data from bioassay (rat and mouse sub- and chronic regulation toxicity studies), CHO Cell Cytogenesis, Drosophila, Micronucleus, Mouse Lymphoma, Rodent Cytogenetics, and Salmonella assays on hundreds of environmental chemicals.	Close to 1K environmental chemicals Multiple doses, tissues, and end points

mode of action that may lead to an adverse health outcome. It offers an opportunity to not only compare alternatives with regard to the potential of human health hazard, but also take into account the quantitative and variability aspects of the underlying adverse effects.

- *A step-wise decision tree* that incorporates structure-activity relationship models, in vitro assays, toxicokinetic modeling, and short-term animal data into toxicity testing and risk assessment in an integrated fashion (Thomas et al. 2013). Tier 1 of this approach uses in vitro assays to rank chemicals based on their relative selectivity in interacting with biological targets that have been associated with known toxicity outcomes and to identify the concentration at which these effects occur. Reverse toxicokinetic modeling and in vitro to in vivo extrapolations (IVIVE) (Rotroff et al. 2010; Wetmore et al. 2012; Wetmore et al. 2013) are used to convert in vitro concentrations into external dose for derivation of the point-of-departure values. The latter can be compared to human exposure data or estimates (Wambaugh et al. 2013) to yield a margin of exposure (MOE).

BOX 8-3

IN VITRO TESTING BY END POINT

Genotoxicity - Direct

A battery of well-defined tests to assess a number of genotoxicity end points induced by direct-acting chemicals, such as point mutations, aneuploidy and chromosomal fragmentation, is necessary for regulatory consideration of drugs and other chemicals (Doak et al. 2012). Many OECD guideline protocols for genotoxicity assessment have been established and include the in vitro bacterial reverse gene mutation test (Ames; OECD 471), an in vitro mammalian cell gene mutation test (e.g., HPRT forward mutation assay, mouse lymphoma TK assay; OECD 476), and an in vitro mammalian cell chromosome aberration (OECD 473) or micronucleus (OECD 487) assays (Pfuhler et al. 2007). Despite concerns that such a battery of tests may result in a large number of false positives, it was shown recently that a combination of the Ames test and in vitro micronucleus assay can identify 78% of compounds known to be genotoxic in vivo (Kirkland et al. 2011). Standard OECD-approved assays for direct-acting genotoxicity are not meant for high throughput testing. Additional assays that can be used for screening of large chemical libraries are under evaluation. Additional assays in which large numbers of chemicals have been evaluated, without the advantage of established formal sensitivity or specificity of these assays, include a cell-based quantitative high throughput ATAD5-luciferase assay (Fox et al. 2012) and the induction of increased cytotoxicity in isogenic chicken DT40 cell lines deficient in DNA repair pathways (Yamamoto et al. 2011). While it is yet difficult to reach firm conclusion on the genotoxicity and potential tumorigenicity of a chemical using novel assays (Mahadevan et al. 2011; Benigni 2013), these experimental tools may be used in the context of a comparative assessment to provide a relative notion of safety among the alternative compounds being considered.

Genotoxicity-Indirect

A number of high throughput approaches are being considered in ToxCast and Tox21 programs. Despite concerns raised about the predictive nature of these in vitro (and rodent in vivo) approaches (Kleinstreuer et al. 2013; Corton et al. 2014; Rusyn et al. 2014), these assays likely will also prove useful for relative ranking of a chemical of concern and its alternatives. For example, a classification model that utilized in vitro screening data of 309 environmental chemicals in human constitutive androstane receptor (CAR/NR1I3), pregnane X receptor (PXR/NR1I2), aryl hydrocarbon receptor (AhR), peroxisome proliferator-activated receptors (PPAR/NR1C), liver X receptors (LXR/NR1H), retinoic X receptors (RXR/NR2B), and steroid receptors (SR/NR3) has been developed (Shah et al. 2011).

Endocrine Disruption

Endocrine disrupting chemicals have received heightened attention because of concerns that they may cause delayed reproductive and developmental effects in the general population (Birnbaum 2013). Concerns related to endocrine disrupting chemicals led to the EPA's development of an endocrine disruptor screening program (EDSP) and identification of chemicals that require screening. EDSP requires an initial screening battery (Tier 1) for the estrogen, androgen, and thyroid hormones, as well as steroidogenesis (EATS) pathways consisting of five in vitro (estrogen receptor binding—rat uterine cytosol, androgen receptor binding—rat prostate cytosol, estrogen receptor α transcriptional activation, recombinant aromatase, and steroidogenesis in H295R cell line), and six in vivo (rodent, fish and amphibian models) assays to evaluate a chemical's potential to interact with the endocrine system. High throughput screening assays that are not part of the Tier 1 panel in EDSP may have the potential for providing in vitro biological activity indicative of the potential to disrupt the endocrine pathways, as there are a number of assays that are highly relevant to EATS pathways (Martin et al. 2010; Huang et al. 2011) or nuclear receptor activation (see above). It has been suggested that such assays may assist in developing a prioritized list of chemicals for evaluation in the current Tier 1 battery or possibly replace in vivo assays in the long term. For example, a comparison of the results of in vitro screening of chemicals in a growth assay in the estrogen-responsive human mammary ductal carcinoma cell line T-47D with data from estrogen receptor binding and transactivation assays demonstrated that chemicals detected as active in both types of assays showed potencies that were highly correlated (Rotroff et al. 2013a). A follow-up study used high throughput screening assays for estrogen, androgen, steroidogenic, and thyroid-disrupting mechanisms to classify compounds and compare the results to in vitro and in vivo data from EDSP Tier 1 (Rotroff et al. 2013b). While it was reported that ToxCast estrogen receptor and androgen receptor assays showed significantly high concordance with the results of relevant in vitro and in vivo EDSP Tier 1 assays, no classification model could be developed for steroidogenic and thyroid hormone-related effects with the currently available ToxCast data.

Reproductive and Developmental Toxicity

Several recent studies have taken advantage of the available in vitro and in vivo information in the ToxCast Phase I chemical library and animal studies in the Toxicity Reference Database (Martin et al. 2009) to evaluate the utility of toxicity screening for predicting reproductive and developmental toxicity. These studies showed that ToxCast in vitro assay-derived information on steroidal and nonsteroidal nuclear receptors, cytochrome P450 enzyme inhibition, G protein-coupled receptors, and disturbances in cell signaling pathways could identify rodent reproductive toxicants with about 75% accuracy (Martin et al. 2011). Similarly, ToxCast in vitro assay-derived information on transforming growth factor beta, receptor signaling in the rat, and inflammatory signals in the rabbit can be used to classify compounds as developmental toxicants in the rat or rabbit with greater than 70% accuracy (Sipes et al. 2011).

Studies of chemical effects on zebrafish are also useful for predicting rodent developmental toxicity (60%-70% concordance) (Padilla et al. 2012; Truong et al. 2014). (While not technically an "in vitro" assay, zebrafish are high throughput animal models for development effects.)

Acute, Chronic, and Repeat Dose Toxicity

A variety of biochemical, molecular, and cellular assays are used in drug safety evaluation to identify potential unintended "off-target" effects that may result in adverse drug reactions (Kola and Landis 2004). A comprehensive profiling of compounds through a large-scale battery of experimental assays and in silico models is usually conducted, and many publications suggest that straightforward in vitro cytotoxicity assays are very informative of in vivo health hazard (Benbow et al. 2010; Greene et al. 2010a). The utility of a large-scale inference on the potential adverse drug reactions was recently demonstrated using prediction and testing of the reactivity of drug candidates toward a panel of 73 "receptors" that are known as side-effect targets (Lounkine et al. 2012). Importantly, human health hazard evaluation through these pipelines is not limited to a qualitative binary prediction of the potential to cause adverse drug reaction, but must also be accompanied by a quantitative prediction of the dose at which such effects may be seen. The latter is as, or even more, important for the estimation of the "safety margin" between the desired (i.e., therapeutic) and side effect (i.e., adverse) health effects.

Dermal Irritation/Sensitization

Predictive identification of skin sensitizers is now highly reliant on a range of in vitro approaches. A 2013 European prohibition on animal testing (Adler et al. 2011) of ingredients in cosmetics has led to a novel in vitro strategy that can reliably identify sensitizing chemicals and predict their relative sensitizing potential. Numerous in vitro approaches address key parameters of the sensitizing process (Gerberick et al. 2008; Vocanson et al. 2013). These include testing for the ability of chemicals to modify skin protein (e.g., by covalent binding), activate innate skin immunity, and promote skin emigration or surface/intracellular changes in dendritic cell phenotypes, or T-cell priming. Several recent tests have proven successful in correctly detecting large numbers of reference chemicals, sometimes with >85% of correlation to standard in vivo animal assays. Importantly, it has been found that none of these methods alone is able to detect all the sensitizers, and some of them are more likely to detect certain classes of chemicals (Vocanson et al. 2013). Nevertheless, in combination, they hold the promise that in vitro assays can detect chemicals with sensitizing properties. However, while appropriate in vitro solutions for the hazard identification step appear to be within reach, the field is now faced with the challenge of obtaining robust in vitro data on the potency of identified skin-sensitizing chemicals. The availability of such quantitative information may be crucial for an alternatives assessment, if the potential for human exposure varies widely among the chemicals being evaluated.

- Modification of methods under development by the EPA's Advancing the Next Generation of Risk Assessment program NexGen (EPA 2013e). The agency's draft approach to using in vitro data is based on the recognition that EPA deals with various decision contexts and that a "toolbox" of various NexGen methodologies could provide information and knowledge to support each of these decision contexts, from screening and prioritization, to limited and major scope assessments.
- Using the Adverse Outcome Pathway (AOP) concept to link molecular screening and mechanistic toxicology data to adverse effects of interest in assessments (OECD 2013b). The goal here is to reduce uncertainty by identifying key intermediate events and quantitatively linking them to adverse outcomes. An AOP should describe a sequential progression from the molecular initiating event to the cellular, organ, and organism response that underlies the in vivo outcome of interest (OECD 2013b). If an AOP accurately describes a sequence of events through the different levels of biological organization, it may be possible to determine which in vitro assays may be useful in identifying chemical effects or molecular initiating events. Conceptually, the AOP concept may therefore be very useful if a defined set of "adverse outcomes" to avoid in alternative selection are identified. Several AOPs for human health effects have emerged, including a well-developed one for skin sensitization (MacKay et al. 2013; Maxwell et al. 2014). Several additional AOPs are under development for mutagenicity, nuclear receptor-mediated non-genotoxic liver carcinogenesis, neurodevelopmental effects and

thyroid disruption, hematotoxicity, hepatotoxicity, and liver fibrosis (OECD 2013b).

Based on the above proposals for how in vitro data may be used to evaluate the potential for human health hazard, the committee suggests the following potential uses of high throughput in vitro data in alternatives assessment:

- Using in vitro data as primary evidence for an end point of concern (e.g., mutagenicity);
- Using in vitro data to fill data gaps for an end point of concern (e.g., endocrine toxicity);
- Using in vitro data to screen out possible unintended consequences of data-poor chemicals

These uses are consistent with an emerging structure for how in vitro and other high throughput assays, as well as in silico model-based predictions, may be used in the broader context of risk assessment (EPA 2013e,f; Thomas et al. 2013). In vitro and in silico approaches that address human health hazard are described in the next section.

In Vitro Approaches for Evaluation of Human Health Hazards

Using in vitro data in an alternatives assessment is conceptually similar to using in vivo animal data. For example, animal studies are used to make predictions of the potential for health hazards in humans, whereas in vitro assays are used to assess whether chemicals may perturb certain biological pathways. The committee did not undertake a comprehensive review of which health end points can now be assessed using novel in vitro data as primary data in the same way more traditional types of data are used. Similarly, a complete discussion of the strengths, limitations, and predictive value of in vitro tests is beyond the scope of this committee. The committee acknowledges that scientific input will be necessary to determine the breadth of assays that may be required to adequately assess a chemical of concern and alternatives. This could involve the identification of one or more assays to assess different endpoints of interest. In Box 8-3, the committee describes the state of the science of high throughput in vitro toxicity assays for several end points. Box 8-4 describes the committee's thinking on how to consider the relationship between the in vitro concentration used in such assays and the in vivo dose that elicits adverse health effects.

In Silico Approaches for Evaluation of Human Health Hazards

In silico models exist for a variety of human health end points, but the accuracy of these predictions can vary dramatically. The accuracy of in silico toxicity predictions is typically measured through internal and external validation of the model using data sets of known experimental activity. Internal validation is used during development to show that statistically derived models are robust, but this type of validation provides little information about the ability of the model to predict the activity of compounds outside the training set (Tropsha et al. 2003; Gramatica 2007). External or prospective validation is the gold standard method for evaluating model performance, but results have proved to be context dependent and difficult to generalize beyond the data training set. Furthermore, in silico prediction of a variety of toxicity end points has been limited by the quantity and quality of data available in the public domain for model development. In addition, most in silico approaches do not identify the dose at which effects are likely to happen.

In Box 8-5, the models and approaches available for predicting genotoxicity, carcinogenicity, skin sensitization, reproductive and developmental toxicity, and hepatotoxicity are discussed briefly. There are many commercial systems available, and a discussion of their strengths and weaknesses is beyond the scope of this chapter. Rather, the committee provides a survey of approaches and, where appropriate, provides illustrations of their use in toxicology. While Box 8-5 looks at approaches by end point, Box 8-6 discusses specific chemical structure and physicochemical properties that influence toxicity.

BOX 8-4

RELATIONSHIP BETWEEN IN VITRO CONCENTRATION AND IN VIVO DOSE THAT ELICITS ADVERSE HEALTH EFFECT

Most, if not all, novel high throughput toxicity screening assays evaluate the relationship between dose of the chemical and response of the assay. Dose-response relationship data are becoming a source of increasingly accessible information for evaluating the potential human health hazard. This information may be useful even in the context of chemical alternatives assessment. Potential dose metrics or points of departure that can be applied to high throughput toxicity data include:

- Chemical concentration that elicits a 50% effect (EC_{50}) in the assay (Neubig et al. 2003; Huang et al. 2008; Xia et al. 2008). The limitations of the EC_{50} approach in the analysis of in vitro screening data have been addressed by (Sand et al. 2012).
- Binary "active/inactive" classification of responses in each assay (Shukla et al. 2010). This approach facilitates concordance analysis with in vivo toxicity outcomes that are also frequently binary.
- The use of logistic curve modeling to fit the concentration-response relationships that may not reach the maximum effect (Sirenko et al. 2013).
- One standard deviation-based benchmark concentrations (BMCs) (Sirenko et al. 2013).
- Benchmark dose-transition (BMDT), which represents the dose where the slope of the dose-effect curve changes the most (per unit log-dose) in the low dose region (Sand et al. 2012).
- Lowest dose at which the signal can be reliably detected (Sand et al. 2011).

Several of these methods rely on statistically based approaches. Human health risk assessments, including alternatives assessments, will likely be improved if these approaches also consider inter- and intraspecies adjustments and biological considerations relating to the assessed in vitro end points (Chiu et al. 2012).

Even though concentration-response data are routinely collected in most in vitro assays, it has been repeatedly noted that in vitro toxicity screening-derived points-of-departure are not directly useable in assessment decisions (e.g., comparative analysis) unless they are converted to in vivo dose equivalents (Blaauboer 2010; Basketter et al. 2012; Blaauboer et al. 2012; Thomas et al. 2013; Yoon et al. 2014; Groothuis et al. in press). The relationship between in vitro concentrations and the concentration of the chemical in the blood/target tissue in vivo, however, can be complex and dependent on variables that are not captured in screening assays. The high throughput screening data do not account for pharmacokinetic factors, such as bioavailability, clearance, and protein binding, which can significantly influence in vivo toxicity and, depending on the assay, may not account for metabolism.

Computational in vitro-to-in vivo extrapolations (IVIVE) use data generated within in vitro assays to estimate in vivo drug or chemical fate. IVIVE is increasingly being used to predict the in vivo pharmacokinetic behavior of environmental and industrial chemicals (Basketter et al. 2012). A combination of IVIVE and reverse dosimetry can be used to estimate the daily human oral dose, called the *oral equivalent dose*, necessary to achieve steady state in vivo blood concentrations equivalent to the point-of-departure values derived from the in vitro assays (Rotroff et al. 2010; Wetmore et al. 2012,2013). Incorporation of pharmacokinetic and exposure information enhances the use of high throughput in vitro screening data by providing a risk context (Judson et al. 2011; Thomas et al. 2013), but more research is needed to produce such information. Consideration of alternative dose metrics instead of nominal concentrations is needed to reduce effect concentration variability between in vitro assays and between in vitro and in vivo assays in toxicology (Groothuis et al. in press). The quantitative IVIVE efforts will add information critical to interpreting the biological relevance of exposure scenarios (Wetmore et al. 2012).

BOX 8-5

IN SILICO PREDICTION BY END POINT

Genotoxicity

In silico prediction of genotoxicity has been a major research focus since the initial publication of structural alerts for DNA reactivity (Ashby and Tennant 1991). Access to large public domain data sets has helped stimulate progress and has resulted in a fair degree of success in the prediction of genotoxicity, particularly in the prediction of the Ames salmonella assay for mutagenicity by in silico models (Naven et al. 2010; Lynch et al. 2011). The overall concordance between the predictive tools and the assays they are designed to predict ranges between 70% and 85%. It is worth noting that these values are close to the inter- and intra-laboratory reproducibility of the Ames assay, reported as 87% (Kamber et al. 2009). However, the sensitivity of the in silico model—its ability to accurately predict an Ames positive compound—can vary much more dramatically, from up to 85% for public domain data sets to just 17% for some proprietary (e.g., pharmaceutical) data sets (Hillebrecht et al. 2011). This variability in sensitivity may result from the fact that few active pharmaceutical ingredients contain the classical DNA-reactive functionsal groups that are a common cause of genotoxicity.

Commercially available software packages for conducting in silico predictions—such as Derek for Windows[55] (DfW; Marchant et al. 2008), MC4PC (Saiahhov and Klopman 2010), and Leadscope Model Applier (LSMA; Valerio and Cross (2012))—are commonly used in the pharmaceutical industry for the prediction of genotoxicity and other toxicological end points. Other readily available systems like Toxtree (Benigni et al. 2010) are also being evaluated for their usefulness. Their comparative performances have been extensively reviewed and published (Hillebrecht et al. 2011; Sutter et al. 2013), but it is clear that no single system performs significantly better than any of the others. Although other models exist for the prediction of chromosomal aberrations, such as clastogenicity and anugenicity, these systems are generally less accurate than the other modeling tools and are not commonly used in industry settings.

Carcinogencity

Various methods for structure-based prediction of carcinogenicity have been developed over the last several decades, including some commercial applications, such as Derek, Case Ultra, Leadscope Model Applier, ToxTree, and OncoLogic. The value of these methods for predicting carcinogens has been limited by lack of public data availability and the complexity of the end point itself. Carcinogenicity can occur through genotoxic and non-genotoxic mechanisms. Most structure-based approaches are able to predict DNA-reactive genotoxic compounds (as discussed above). Some systems, such as Derek, contain structural alerts specifically targeting certain classes of non-genotoxic carcinogens. Other predictive packages, such as Case Ultra, do not always differentiate between these two classes in their predictions.

Two prospective exercises conducted by the National Toxicology Program (NTP) in the 1990s evaluated the performance of computational models for carcinogenicity. In these exercises, the NTP invited interested parties to publish model predictions on a set of chemicals that were scheduled for testing in the NTP's two-year rodent bioassay. Once the tests were completed, the in vivo results were compared to the predictions. The carcinogenicity of the first set of 65 chemicals was reasonably well predicted with computational models, achieving between 50%-65% accuracy. However, the second set consisting of only 30 chemicals was not predicted as well by the in silico systems and tended to over-predict non-carcinogens as carcinogenic (Benigni and Giuliani 2003). Ongoing effort to predict carcinogenicity through structure-based approaches continues with some recent examples from Fjodorova et al. (2010) and Kar et al. (2012).

Reproductive and Developmental Toxicity

"Developmental and Reproductive Toxicity (DART) occurs through many different mechanisms and involves a number of different target sites, making it very difficult to predict this end point" (Wu et al. 2013). Most of the published QSAR development has been done through collaborative projects with the computational toxicology group within the U.S. Food and Drug Administration (FDA), using data collected from preclinical and clinical data submitted by pharmaceutical companies. Matthews et al. (2007) reported the use of computational QSAR approaches to predict male and female reproductive toxicity, fetal dysmorphogenesis, functional toxicity, mortality, growth, and newborn behavioral toxicity. Matthews reported high specificity (i.e., the number of correctly predicted negatives) and positive predictive value (i.e., the number of correct positive predictions when compared to the total number of positive predictions) of greater than 80%. However, the sensitivity (i.e., the number of correctly identified positive compounds) was often less than 50%. Unlike the NTP carcinogenicity exercises, to date there have been no published prospective tests of performance of these DART models, so their accuracy compared to a set of novel compounds cannot be ascertained. Published models are available in commercial packages such as Case Ultra and Leadscope Model Applier. In addition, Derek Nexus also contains some structural alerts for DART effects that have been developed as part of acollaboration with Pfizer Inc., although these alerts and their respective performance have not been formally published.

Wu et al. (2013) recently published "an empirically based decision tree for determining whether or not a chemical has receptor-binding properties and structural features that are consistent with chemical structures known to have toxicity for DART end points." As with the above models and structural alerts, the performance of this decision tree has not been independently assessed, so its performance for truly novel chemical series that have not been previously tested may well be limited.

Skin Sensitization

Skin sensitization is primarily driven through hapten reactivity, which supports a central role for chemical reactivity in allergic sensitization (Vocanson et al. 2013), as well as skin permeability and metabolic activation. This requirement for chemical reactivity makes the prediction of skin sensitizers more feasible, and there has been substantial progress in this area. Structural alerts for skin sensitization have been implemented in Derek, ToxTree, and other systems; the relative performance of these approaches has not been extensively reviewed, but external validation studies do point out limitations in applicability and low external predictivity (Teubner et al. 2013).

Predictive tests for allergic contact dermatitis (ACD) have also been developed. ACD depends on the intrinsic capability of the chemical to cause skin sensitization and the ability of a chemical to penetrate viable epidermis. Numerous QSAR methods that predict ACD for specific chemical classes or non-congeneric data sets have been published (Deardon 2002; Guha and Jurs 2004; Sutherland et al. 2004). Factors that affect the ability of chemical to be absorbed into the epidermis are discussed in more detail in Chapter 5.

Respiratory Sensitization

Respiratory sensitization is an important disease (Mekenyan et al. 2014), but there are "no validated or widely accepted models for characterizing the potential of a chemical to induce respiratory sensitization." While efforts to model respiratory sensitization in silico have been hampered by an incomplete understanding of immunological mechanisms, structural alerts for this end point have been developed (Agius et al. 1991, 1994; Enoch et al. 2012). Typical structural alerts "have been encoded into the Derek Nexus knowledge based expert system developed by LHASA Ltd. Other efforts have focused on establishing statistical QSAR models; examples include those first derived by the developers of MCASE, Jarvis et al. (2005) and more recently by Warne et al. (2009), who investigated the use of pattern recognition methods to discriminate between skin and respiratory sensitizers" (Mekenyan et al. 2014).

Despite the lack of a universally accepted test method, REACH regulations and others still require the assessment of respiratory sensitization as part of a risk assessment. The REACH guidance describes an integrated evaluation strategy that includes a consideration of well-established structural alerts and existing read-across, QSAR, and in vivo data. As with many other toxicological end points, there has been no published comparison of these methods for prediction, so it is difficult to draw conclusions on the relative merits and accuracy of the models.

Hepatoxicity

Drug-inducted liver injury (DILI) is a major concern in the pharmaceutical industry and has led to the withdrawal of a significant number of marketed drugs (Holt and Ju 2006; Kaplowitz 2005). Adverse effects range from hepatic enzyme elevations to liver failure (Zimmerman 1999; Williams 2006) and are often difficult to predict in the preclinical stages. As a result of this interest, numerous in silico approaches for predicting hepatotoxicity have been developed. These approaches range from structural alerts associated with DILI (Hewitt et al. 2013) to QSAR methods (Chen et al. 2013). The mechanisms involved in DILI are often complex, making accurate prediction of this end point using QSAR and other computational approaches challenging. Most of these methods claim to have a sensitivity and specificity between 65%-70% depending on the method and data test set; however, no independent evaluation has been published, so true head-to-head performances are difficult to ascertain.

BOX 8-6

OTHER CHEMICAL STRUCTURE AND PHYSICOCHEMICAL PROPERTIES THAT INFLUENCE TOXICITY

While structures can certainly not predict all chemical activity, physicochemical and structural characteristics have been used for predicting toxicity end points. Changes in key physicochemical properties, such as pKa, lipophilicity, and polar surface area, can lead to dramatic effects on the toxicity of a chemical, either through influencing pharmacokinetic properties, such as clearance of the compound, or its ability to interact with a biological system in the form of pharmacological interactions and/or non-specific protein binding.

The effect of physicochemical properties on bioavailability is discussed in more detail in Chapter 5. Other physicochemical and structural characteristics that have been developed for predicting toxicity endpoints include:

- pKa and LogP (or calculated LogP, cLogP), which correlate with mitochondrial uncoupling for certain classes of chemicals (Naven et al. 2013).
- The energy gap between the highest occupied molecular orbital (HOMO) and lowest unoccupied molecular orbital (LUMO), which is associated with a chemical's ability to absorb light and phototoxicity (Peukert et al. 2011).
- A basic center and one or more lipophilic chains in a compound is associated with a higher likelihood of inhibiting an ion channel important to cardiac cells action potentials (Schmid et al. 2003).
- Structures indicating the potential for generating reactive metabolites (Stepan et al. 2011).
- Amphiphilicity, pKa, and lipophilicity (LogP), which have been used to predict the likelihood of phospholipidosis (Goracci et al. 2013).
- Acidity or basicity of the molecule as an indication of propensity to interact with different classes of receptors such as cyclooxygenases and the nuclear hormone receptors (acidic molecules) or the aminergic G-protein coupled receptors (basic molecules) (Parker at al. 2014).
- High lipophilicity, low polar surface area and low passive permeability are associated with induction of endoplasmic reticulum stress, which has been linked to drug-induced toxicity (Koslov-Davino et al. 2013).
- Structures predicted to have endocrine effects, using various QSAR methods to predict chemical docking with cellular targets (Vuorinen et al. 2013).
- LogP and topological polar surface area may be related to a chemical's ability to cause *in vivo* toxicity at low plasma exposures, cytotoxicity, and off-target pharmacological effects (Benbow et al. 2010, Greene et al. 2010b; Wang and Greene 2012).
- Higher volume of distribution, when combined with in vitro indications of greater cytotoxicity in hepatocytes, is associated with a reduction in the lowest observable adverse effect level (LOAEL) (Sutherland et al. 2012). LogD and pKa predictions based on structure are useful in predicting volume of distribution and plasma protein binding (Lombardo et al. 2002). Other structural predictors of the steady state volume of distribution and clearance (Gombar and Hall 2013) are also helpful.
- Lipophilicity (or LogP) and polar surface area descriptors are strongly correlated with chemical clearance (Hsiao et al. 2013).
- Higher LogP, low polar surface area, and pKa have been associated with higher target promiscuity (a wider area of nonselective interaction with biological targets) (Seidler et al. 2003).

Use of Novel In Vitro Data and In Silico Models as Primary Evidence

The NTP *Report on Carcinogens* identifies that among the most crucial information considered in the evaluation of human cancer hazard of chemicals is that obtained from "studies on genotoxicity [ability to damage genes]" (NTP 2011). Genotoxicity may be either the result of interaction between a chemical and DNA (direct genotoxicity) or action between a chemical and DNA regulatory elements (indirect genotoxicity). A number of in vitro assays are available to test for the potential of a chemical to be genotoxic. Due to the complexity and diversity of mechanisms that may lead to DNA damage (and the mutagenic events that follow), a battery of in vitro tests is needed to establish a chemical's genotoxic potential. In vitro mutagenicity data is included in

GHS as a primary data type (UNECE 2013c). More details are in Appendix D, but primary data are sufficient for considering a chemical as a GHS Category 2 Suspected/Possible Germ Cell Mutagenicity. Other (i.e., not mutagenicity specific) in vitro testing for genotoxicity is described in Box 8-3.

Use of Novel In Vitro Data and In Silico Models to Fill Data Gaps

High throughput in vitro data can also be used to fill certain primary data gaps for particular health end points. For example, ToxCast in vitro assay-derived information on steroidal and non-steroidal nuclear receptors, cytochrome P450 enzyme inhibition, G protein-coupled receptors, and disturbances in cell signaling pathways may identify rodent reproductive and developmental toxicants (Martin et al. 2011; Sipes et al. 2011), potentially showing a path toward replacement of more traditional (Clode 2006) uterotrophic assays and the Hershberger assays found in tiered endocrine assays.

How novel in vitro data can be used to address gaps in the human health data in an alternatives assessment was illustrated by Russell Thomas of EPA's National Center for Computational Toxicology (Thomas 2014). Thomas used the information from two case studies by Martin et al. that analyzed the potential use of ToxCast data (Martin et al 2012) for tiered testing. While these case studies were designed to examine the use of in vitro data to determine which chemicals warranted further reproductive toxicity testing in animals, Thomas showed how the concept applies to alternatives assessment as well, in that those chemicals with higher or lower probability of exerting reproductive effects can be identified. This concept is explained in further detail in Box 8-7.

Use of Novel Data to Screen out Possible Unintended Consequences in Data-Poor Chemicals

The third use of novel high throughput in vitro data suggested by the committee is to screen out unintended consequences. Some companies in the pharmaceutical industry are using this information in this way. In vitro high throughput screening and toxicogenomics appear promising in the screening of data-poor chemical alternatives for biological activity that may contribute to hazard identification. The information about potential modes of action can support the transition toward an integrated testing and assessment strategy. The pharmaceutical industry routinely uses mechanistic in vitro tests and high throughput screening to look for unexpected safety issues with potential drug candidates. As a result, the industry brings valuable experience tempered with some caution about setting overly high expectations for in vitro toxicity testing technologies (MacDonald and Robertson 2009).

In addition to screening for unintended and unexpected consequences by looking at mode- of-action information, it is also possible to use high throughput in vitro screening to look for evidence of nonselective chemical activity at low concentrations. That is, use the data for screening out chemicals that have general indicators of potential toxicity, even if the specific toxicities and their modes of action are not identified. This concept is based on the observation that while the batteries of in vitro toxicity assays utilized in the Tox21 and ToxCast programs provide a broad biological profile of the potential proximal biochemical and cellular targets for a chemical, the majority of environmental chemicals being tested likely act via nonselective interactions with cellular macromolecules (Thomas et al. 2013). Because most high throughput assays incorporate an extensive concentration range usually spanning several orders of magnitude, it is frequently observed that many biological targets are "engaged" at, or near, concentrations that result in cytotoxicity (Martin et al. 2010). Thus, one use of high throughput assay-derived information may be for separating chemicals into either those that cause toxicity primarily through nonselective interactions with cells and cellular macromolecules or those that act through more selective interactions (e.g., receptor-mediated chemicals). Specifically, it has been shown that promiscuity across multiple pharmacological targets at a concentration of 10 μM can lead to an increased likelihood of observing toxicity in vivo at low exposures (Hughes et al. 2008; Wang and Greene 2012). In the context of alternatives assessment, if alternatives under consideration exhibit varying levels of "selectivity," a compound with higher "selectivity" may be considered as lower risk for additional "off target" effects and thus be assigned a higher relative rank. Clearly, if a chemical is considered to be selectively active against a pharmacological target, whether intended or otherwise, the implications of this specific pharmacological activity in relation to potential safety concerns should be considered. For example, agonism of the 5HT2b receptor has been implicated in causing cardiac valvulopathy.

It is also possible that high throughput in vitro data could be used to screen out unintended consequences associated with particularly susceptible subpopulations of people. In general, existing alternatives assessment frameworks do not consider human variability in their analysis, but because this variability underlies differences in how people respond, addressing these differences is a key consideration in human health assessments for chemicals (NRC 2009). By using and expanding upon current analytical methods, these assessments may take advantage of novel in vitro data to better characterize and quantify variability in susceptibility (Zeise et al. 2013). Approaches that are now possible include using large-scale in vitro screening (Rusyn et al. 2010) in human cell lines obtained from genetically diverse subjects (Durbin et al. 2010; Welsh et al. 2009; Wheeler and Dolan 2012). The utility of such in vitro models to toxicology, especially for exploring the extent and nature of genetic components of inter-individual variability in pharmacodynamics, was recently demonstrated (O'Shea et al. 2011; Lock et al. 2012). The extent of inter-individual variability in response that was observed for different chemicals in in vitro assays could also be compared with previously collected sets of in vivo human pharmacodynamics variability data.

Limitations of Using In Vitro Data in Alternatives Assessments

High throughput in vitro screening to identify chemical hazards and prioritize chemicals for additional in vivo testing is an area of heightened scientific inquiry and regulatory scrutiny. While the promise of the novel in vitro assays and statistical methods is difficult to underestimate, their predictive power or classification accuracy is still not clear. It has been observed that the findings from in vitro assays may not provide more information than that of the chemical structural descriptors (see below), and aggregating the assays based on genes or pathways may even lead to reduced predictive performance (Thomas et al. 2012). Significant potential biases in the estimates of the performance of the classification models have been noted, though this point is still a subject of active debate (Dix et al. 2012; Knudsen et al. 2013). **Because the current high throughput in vitro assays may still have only limited applicability for predicting in vivo chemical hazards, the committee believes that high throughput in vitro assays with limited or uncertain predictivity should generally only be used in alternatives assessments to fill data gaps or screen for unexpected consequences, except as described earlier for certain mutagenicity and endocrine/reproductive toxicity assays.**

Limitations of Using In Silico Approaches in Alternatives Assessments

When using in silico methods or read-across approaches to infer toxicological activity, there are two main limitations that need careful consideration in the assessment:

- Measures of chemical similarity and their appropriate application to the effect being predicted; and
- The reported applicability domain of a prediction and hence the reliability of the prediction being made.

These aspects of computational models are briefly summarized in the following sections. Recent reviews by Patlewicz et al. (2013) and Modi et al. (2012) are available for more information.

Limitations of Chemical Similarity and Read-Across Approaches

Similarity in chemical structure is often used in read-across and QSAR models to identify chemical structures with known activities that could be used to infer the activity of a molecule with unknown activity. In other words, the idea is to infer that the less understood chemical will produce an equivalent test result as the more well-understood one (OECD 2014b). This approach presents the dilemma of how to define what is similar and what is not. Defining chemical similarity has been debated for several decades, and no one method for applying or presenting read- across concepts has been agreed upon, despite the frequent use[42] of grouping and read across to satisfy information requirements under REACH, legislation that has been enacted in Europe. The method used is often case dependent; for example, in genetic toxicology, when a chemical bears the same structural alert as an Ames negative comparison compound (and no other alert) in the same position and environment and has a similar molecular weight, then the chemical is considered

[42] For example, more than 20% of high production volume chemicals submitted for the first REACH deadline relied on read-across for hazard information on a number of toxicity end points necessary for registration (ECHA 2011).

BOX 8-7

USE OF IN VITRO HIGH THROUGHPUT SCREENING DATA TO IDENTIFY CHEMICALS WITH HIGHER OR LOWER PROBABILITY OF EXERTING REPRODUCTIVE EFFECTS—INFORMATION USEFUL FOR FILLING DATA GAPS

In this example, high throughput in vitro screening data (36 assays) from the ToxCast database were compared to the in vivo data from the ToxRef database to identify the in vitro assays that best predict in vivo results. The information about how activity of 56 chemicals (ToxCast Phase I and II chemicals, including plasticizers or other industrial chemicals) in these 36 in vitro assays was mapped onto the 8 pathways most predictive of in vivo assays identified (pathways illustrated here as pie slices for PPARα, AR, etc.). Strategies for predicting in vivo toxicity outcomes from in vitro data are detailed in Figure A and in Martin et al. (2012).

Based on the available data, there are numerous ways to visualize and compare the profiles of the chemicals. No one way is considered the preferred method. The Toxicological Priority Index (ToxPi) software (described in detail in Appendix C) is a prioritization support tool that incorporates diverse chemicals' bioactivity profiles and other relevant data (Reif et al. 2013). It is used here to visualize the outcome of the analysis and then rank the chemicals according to their impact on the eight pathways. ToxPi is a dimensionless index score that enables integration of multiple sources of evidence on exposure and/or safety, which is then transformed into transparent visual rankings to facilitate decision-making. The rankings and associated graphic profiles can be used to prioritize resources in various decision contexts, such as testing chemical toxicity or assessing the similarity of predicted compound activity profiles.

Figure B shows these results as ToxPi for 12 of 52 chemicals analyzed, where each slice represents multiple assays associated with a given gene pathway, and the chemicals are ordered based on the ToxPi score. For each slice, distance from

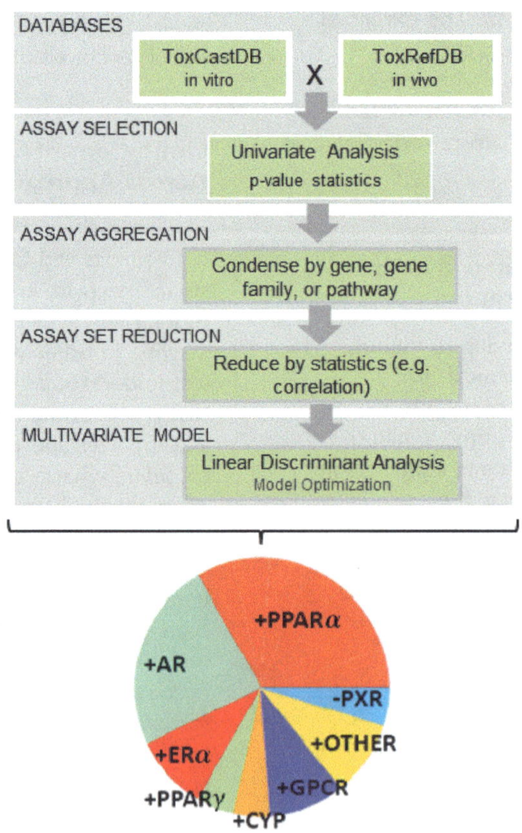

FIGURE A General workflow for developing statistical classification models for in vivo toxicity outcomes based on in vitro toxicity data. The example of the classification model for reproductive toxicity is for the rat (Martin et al., 2011). SOURCE: Thomas 2014. Reprinted with permission of the author.

the origin (center) is proportional to the normalized value (e.g., assay potency) of the component data points. The width (in radians) shows the relative weight of that slice in the overall ToxPi score calculation. Values closer to the unit score (equal to 1) translate to higher potency, or greater pathway perturbation relative to other chemicals in the analysis. Conversely, values closer to the origin (equal to 0) translate to lower potency and lesser pathway perturbation across the corresponding domains. Values at zero (i.e., slices not extending at all from the origin) translate to "inactive/no activity."

A judgment is made about which gene pathways should contribute most to the ToxPi score. These weighting factors, as well as the assays used, are described in Martin et al. (2011). The model shown here, however, used data from human cytochrome P450 assays instead of rat cytochrome P450 assays, and the Bioseek assays were removed because the data were not publicly available at the time of press. For each chemical, weighted combinations of data were combined from multiple data streams, with relative scores shown in ToxPi profiles as slices based on one or more components. Martin et al (2011) explains this process in more detail.

In this analysis, the chemicals with high ToxPi scores may be considered as representing a higher degree of reproductive health hazard. Indeed, a number of phthalates and other chemicals that have been associated with adverse reproductive health effects are at the top of the list. Among "alternatives," some compounds demonstrate low relative scores, at least for the type of information used to develop the classification model. In an alternatives analysis, the chemicals with higher ToxPi scores would be assumed to represent a greater hazard.

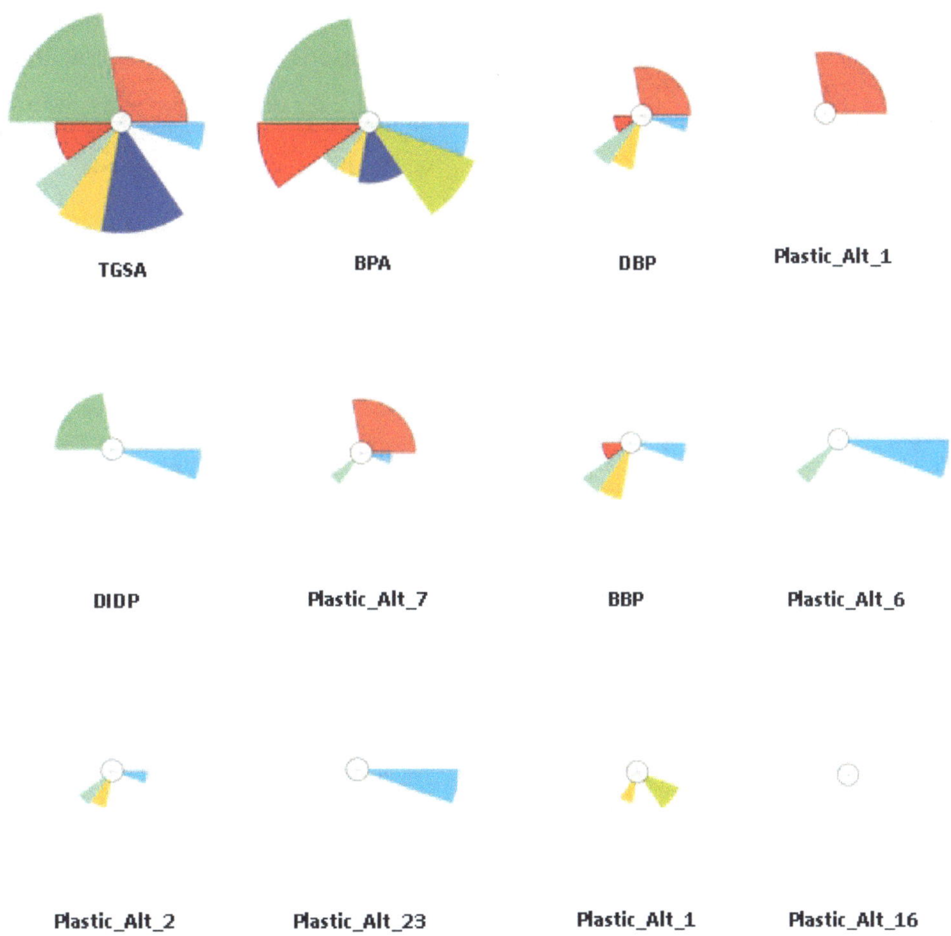

FIGURE B ToxPi profiles (ToxPi GUI v. 1.3) were used to prioritize chemical alternatives using 36 assays and pathways associated with reproductive toxicity (Reif et al. 2013). A subset of these prioritized chemicals is shown here. Some of the compounds were evaluated as potential alternatives and labeled "Plastic_Alt_*n*."

likely to be negative in the Ames test as well. In this case, no further testing is generally needed (EMA 2014).

In a read-across assessment, a chemically defined category of known adverse activity is represented by compounds that have common structural features and exhibit similar trends in their physicochemical properties. Generally, the presence of a common biological or chemical behavior is associated with a common underlying mechanism of action (e.g. alkylating compounds). This categorical approach provides the basis for identifying trends in properties across the category of compounds, resulting in the possibility of extending the use of measured data to similar untested chemicals. These estimates of biological activity may be considered adequate, without further testing, for regulatory purposes (e.g., classification and labeling and/or impurity hazard assessment for classification with respect to toxicity potential). Enoch et al (2011) have provided a description of chemical category function (Sutter et al. 2013). However, the standardization of this approach for defining structural similarity on the basis of a chemically-defined class of known biological actives is much more difficult when the mechanisms of action are both diverse and complex.

In QSAR approaches, the definition of structural similarity is crucial to the final result of an in silico prediction (Naven et al. 2012). The assessment of chemical similarity usually begins with a quantitative description of the molecular structure or fingerprint" (Sutter et al. 2013). Comparisons between structures are then performed using one of a variety of indices that have been developed; for example, Euclidean distance measures or maximum common substructures. However, similarity is a multidimensional concept, and the similarity between two compounds can be difficult to determine or set guidelines for. For instance, compounds (1) and (2) in Figure 8-3 are similar in that they both have the same molecular formulas ($C_6H_5NO_2$), yet their chemical structures bear little resemblance. They have different electron delocalization properties or aromatic behavior, physicochemical properties, and most importantly, probably dissimilar biological properties. Likewise, glucose (3) and galactose (4) share structural similarities but have very different pharmacologic properties. Many methods to measure the structural similarity between two compounds have been developed, but the more relevant question to consider is whether structural similarity an important factor for the toxicological end point being studied.

This is because minor modifications to the structural alert can significantly influence toxicological activity, yet major modifications to the periphery of the chemical structure may have little impact on activity so long as the structural alert remains intact. When assessing the relevance of a prediction of activity, it is not enough to ask how similar the query compound is to other inactive compounds. It is also important to identify the features of structurally alerting, active compounds that would attenuate the activity and to assess if these features can be adequately extrapolated to the compound being studied.

Limitations of Defining Applicability Domains

OECD guidelines currently recommend that QSAR models should define the domain within which the predictions of a model can be deemed reliable—the applicability domain (AD). Many methods exist for defining the AD of a QSAR model, and they have been extensively reviewed (Dragos et al. 2009; Hewitt and Ellison. 2010; Ellison et al. 2011). The AD of a model can be broadly described using two non-exclusive terms: (a) the region of chemical or response space relating to the model training set and (b) the region of a chemical or response space where a model makes an acceptable prediction error. For toxicological end points like mutagenicity or the uncoupling of oxidative phosphorylation, which are dependent on the presence or absence of structural alerts, the less applicable the concept of similarity becomes.

In the first definition (a), the underlying assumption is that those predictions based on interpolation from data in the training set are generally more reliable than those based on extrapolation. The second definition (b) is based on the assumption that by assessing where compounds are predicted most accurately, we can gain valuable information. The thinking is that inevitably, a subset of the training set will be incorrectly classified, casting doubt on the reliability of the predictions based on similarity to these compounds. In addition, definition (b) does not automatically assume that predictions for compounds that are considered dissimilar to the training set are unreliable (Dragos et al. 2009).

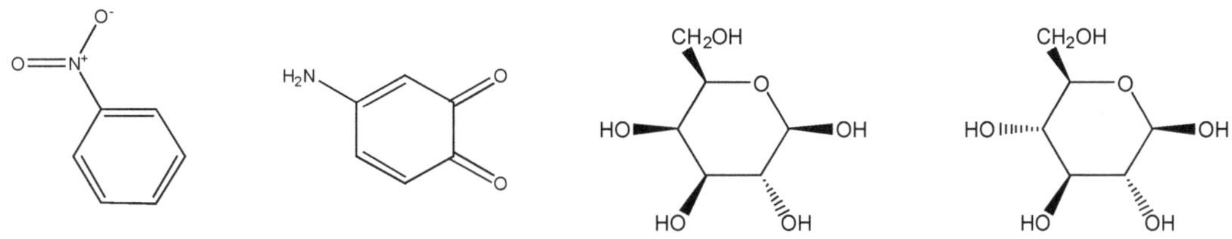

1 nitrobenzene

2 4-amino-2-benzo-quinone

3 glucose

4 galactose

FIGURE 8-3 Selected chemical compounds to illustrate structural similarities and differences.

Defining the AD of any model is difficult and presents challenges to the end user about whether a prediction is reliable or not. In addition, although the scope of structural alerts can be used to define their AD, this provides little information to a user when alerts are not matched to the compound in question. Expert systems that rely on structural alerts do not have a model training set per se, as the alerts are often based on disparate data sources, such as toxicity data, information pertaining to the biological mechanism, and knowledge of chemistry and reactivity, which are synthesized into the development of an alert. Furthermore, not all data are publicly available; thus, current approaches cannot reflect this expert knowledge and often require a complete model training set.

Most of the methods for defining ADs have been trained to reduce the error in continuous output QSARs where the assay data provides homogeneous responses; for example, for LogP values or an experimentally derived IC_{50} for protein inhibition. It should be noted that there is a distinct gap between the *applicability* of ADs of chemicals producing homogenous responses to categorical models based on assays that generate a more diverse range of outputs, such as carcinogenicity or reproductive effects. Because a large number of available QSARs are categorical, and their use in alternatives assessment is likely even if the prediction outcome is of active/inactive type, the assessors should have confidence that the alternatives they are evaluating can be classified using a particular QSAR, or fall outside of the AD of that model. There are a few exceptions, but generally it has been shown that there is only value in using an AD to qualify confidence in a positive response, rather than as a prediction for absence of activity.

IN VITRO AND IN SILICO DATA INTEGRATION: AN OPPORTUNITY FOR NEW INSIGHTS

Most of the current computational tools used in toxicity assessments, including chemical alternatives assessments, rely either on chemical or biological data. Specifically, structure activity-based computational approaches attempt to predict the toxicity for a particular health end point from chemical structure alone, whereas novel biological data-based bioinformatics approaches do not usually take full advantage of the inherent structural features of chemical molecules. Incorporating these features may improve the accuracy of the prediction or increase the confidence of the assessor. Integrative chemical-biological modeling may both improve the prediction accuracy and uncover insights previously invisible to either informatics discipline alone.

Using only chemical or biological modeling is unlikely to take full advantage of the richness of the modern data streams, which effectively capture chemical-biological interactions. Few integrative studies, however, have been reported; their paucity is stemming from the lack of both suitable data and integrative methods. Several recent reviews (Rusyn et al. 2012; Valerio and Choudhuri 2012) have proposed general schemes to integrate cheminformatics and bioinformatics for improved understanding of chemical effects on biological systems. A simple means of integrating the disparate data streams that may be available for an alternatives assessment is to apply existing statistical methods to both chemical and biological types of molecular features. Another way is to merge chemical models with biological models. Other approaches may be less straightforward. These include strategically combining chemical structures and biological assays such that the two data sources compensate for each other's shortcomings and the complementary information between them is maximally used.

While the approaches to integrative analyses of chemical and novel biological data streams are still maturing, it is clear that a multidisciplinary systems approach is the best available solution for translating molecular and preclinical insights into practice and guarding against unwanted outcomes of chemical use. In addition to addressing the issue of data quality, further gains in methodological innovations and cohesive integration of the various disciplines will be necessary. Starting such multidisciplinary efforts is unlikely to occur organically and will require deliberate efforts to foster a collaborative environment. As more data come online and advances in assay technologies reduce experimental variability, we expect integrative approaches to play a greater role in toxicology.

IMPLEMENTATION OF STEP 6.1: HUMAN HEALTH ASSESSMENT IN THE COMMITTEE'S FRAMEWORK

Box 8-1 at the beginning of this chapter provides a summary of how Step 6.1 in the committee's framework should be implemented. Additional illustrations of the committee's approach are also presented as two case studies in Chapter 12.

Specific advice for the completion of Step 6.1 includes:

- The assessment should focus on health hazards as opposed to health risks. Risk includes exposure, which is considered elsewhere in the framework.
- The health end points to be addressed in an alternatives assessment should be specified in Step 2. The set of human health hazard end points discussed in this chapter is the suggested initial list. there should be clear documentation of which end points were not considered, and which points, if any, were added.
- Use health end point-specific authoritative lists to identify previously identified health concerns.
- Collect available data, including by conducting scientific literature searches on health hazards associated with the chemical of interest and alternatives. Use the GHS criteria and hazard descriptions, when available, to assess available information for the required end points.
- Existing hazard identification guidance (e.g., as described in EPA risk assessment guidelines) should be used to ensure consistency and transparency when conducting de novo assessments of chemicals for reproductive toxicity and other health end points that require use of expert judgment.
- Assign hazard designations to criteria and authoritative lists to facilitate their use in comparing the health hazards of chemicals and selecting safer alternatives.
- Gaps in data for required end points should be addressed with qualitative and quantitative predictions based on high throughput in vitro data and in silico modeling, when available. The large-scale governmental efforts to collect novel data streams through Tox21 and ToxCast programs will provide critically important information that should inform gap analysis.
- Equally important is to evaluate which computational modeling approaches may serve as acceptable substitutes for alternatives assessment in lieu of additional data collection. In this regard, a closer examination of the read-across assessment framework in development by the European Chemicals Agency may provide additional guidance to inform user implementation of these approaches in alternatives assessments.

FUTURE RESEARCH NEEDS

Implementation of the committee's framework will require the ongoing development of tools and methods and regulatory guidance by the scientific community. This need is especially acute with the use of emerging in vitro and in silico data from high throughput toxicology programs (e.g., Tox21 and ToxCast programs). The committee anticipates that these data streams will provide critically important information that can be used to fill data gaps when traditional data from human and experimental animal studies are lacking. In addition, there is a need for scientists and regulatory agencies to determine which high throughput toxicology assays, end points, and model systems are most informative in assessing the human health hazard types used in chemical alternatives assessment. Once these decisions are made, then development of well-accepted classification schemes for these high throughput and in silico data, analogous to the GHS system, would enhance the use of this information.

In the case of the Tox21 and ToxCast programs, there is also a need for the development of data mining tools (e.g., user-friendly dashboards and software) that will enable stakeholders to access the novel data types that have been already collected and support comparative analyses in a transparent and statistically rigorous way. The EPA is uniquely positioned to demonstrate leadership in incorporating novel data streams and modeling outcomes on human health into alternatives assessment. Specifically, EPA can help develop best practices by determining which information available through the Tox21 and ToxCast programs are most informative in assessing human health hazards and how and when such data may be incorporated in alternatives assessment. Such model efforts could demonstrate how decisions based in part on high throughput data could be formulated and communicated to the stakeholders. In addition, EPA may want to consider assisting parties interested in alternatives assessment in their interpretation and use of results from innovative toxicity testing methods.

CONCLUSIONS

Alternatives assessment of human health hazards is critical not only within the committee's framework, but also in most other frameworks because it is central to determining whether an alternative meets the criteria for being considered safer than a chemical of concern. Identification of human health hazards will require evaluation of multiple data streams, including human epidemiologic or experimental studies in animals, and will be increasingly dependent on the use of novel in vitro and in silico approaches. It is important to keep in mind that in keeping with the committee's approach to the task (see Chapter 1), required tools were not specified as part of this step. The committee found that most frameworks rely heavily on GHS criteria for evaluation of human health hazard, which are generally thought to be acceptable for this purpose. **However, the committee strongly encourages users with adequate scientific resources to move beyond relying solely on traditional types of data associated with GHS or other benchmarking approaches toward data from novel in vitro and in silico approaches. This is especially true as the development and application of tools to assess and integrate novel data streams into the alternatives assessment process evolves.**

In Chapter 12, the committee provides one case example in which these novel types of data were used to evaluate and rank several alternatives. As can quickly be discerned from this case study, the application of novel data streams in particular will require expertise in computational modeling, molecular toxicology, and other scientific disciplines that may go beyond the capabilities of some existing assessment teams. This dilemma is not unique to the application of high throughput data. Indeed, the committee found that interpretation of traditional toxicology data even when using the GHS and GreenScreen® tools can remain a challenge even for experienced toxicologists. For example, the committee found that some differences in end points and descriptor language exist between the GHS hazard categories and the reviewed frameworks. In addition, cutoffs in classification tools used in some frameworks could result in the assignment of alternatives to different hazard categories (e.g., High vs. Moderate) when the actual difference in response can be toxicologically insignificant. Users of the committee's framework will need to exercise professional judgment so that they do not discount possible beneficial alternatives or adopt others that may have unintended health consequences.

1

9

Integration of Information to Identify Safer Alternatives

Any process of choosing among alternatives either explicitly or implicitly integrates the findings from a variety of sources, including human health and ecological assessments. This process therefore requires judgment and integration of information across different hazard domains. The first integration step in the committee's framework (Step 7) involves an initial identification of safer alternatives based on information compiled in previous framework steps. It is important to note that the decision-making process taking place during Step 7 is not expected to yield a single alternative among a set of possible alternatives. Other factors will be considered in a later integration step (Step 10). These, too, will be important, and may ultimately eliminate what appears to be a preferable alternative from a human health and ecotoxicity perspective on the basis of other valued considerations, such as its broader environmental impact, performance, cost, or technological feasibility.

Nonetheless, Step 7 represents a key transition. Most of the steps up to this point constitute activities that are traditionally considered to be aspects of risk assessment. Integrating evidence, however, also includes the application of explicit or implicit value judgments. The choices of which health end points are most important, how choices are made in the presence of uncertainty, and the relative importance of health and ecosystem end points bring societal value judgments into the alternative selection process. This suite of choices is generally considered to fall within the domain of risk *management*, as opposed to risk *assessment* or risk *characterization*. This chapter begins with a general overview of the data streams that will be integrated in Step 7, as illustrated in Figure 9-1. It then discusses strategies that can be used to address trade-offs and uncertainty.

The hypothetical data in Table 9-1 illustrate a number of challenges that may be expected while conducting an alternatives assessment: a) the presence of trade-offs, where one alternative is preferable with respect to one or more end points, but is less preferable for one or more other end points *within* a domain, such as human health or ecotoxicity; b) the presence of trade-offs *between* domains, where some alternatives are preferable from a human health perspective, while others are preferable from an ecotoxicity perspective; c) the presence of variable levels of uncertainty about the level of toxicity or exposure (as depicted by the colors); or d) complete absence of knowledge in that the level of toxicity cannot be determined even with caveats (as depicted by the gray entries). Depending primarily on the extent of the trade-offs and their degree of uncertainty, the task of determining the preferred alternative ranges from extremely simple to very challenging.

As noted earlier, Step 7 represents a key transition. Most of the steps up to this point are activities traditionally considered aspects of risk assessment. The step of integrating evidence, however, includes applying explicit or implicit value judgments. Therefore, the individual conducting the alternatives assessment may need additional guidance to complete this step. For example, if that individual is not considered the decision maker, then he or she will need to have the preferences of the decision maker made explicit in the form of decision rules or algorithms that can be applied in the face of trade-offs and uncertainty. The explicit consideration and documentation of those preferences were explained in Steps 2a and 2b, Scoping and Problem Formulation, respectively (see Chapter 4).

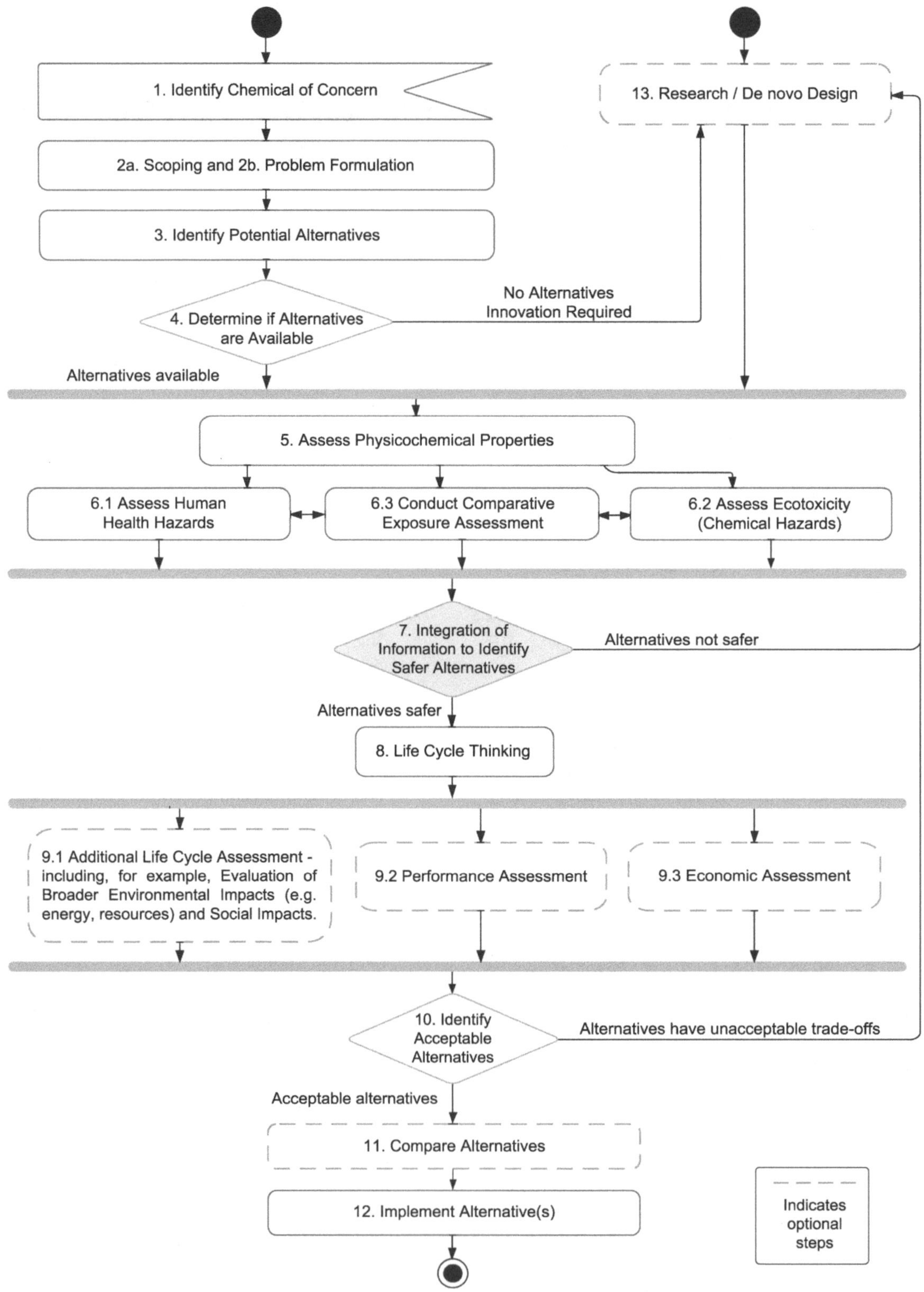

FIGURE 9-1 Committee's framework highlighting the integration step described in this chapter.

TABLE 9-1 Sample Results from Step 6 Providing Categorical (high, medium, low) Evaluations of Select Ecotoxicity and Human Health Impacts and Physicochemical Properties

Note: The relative uncertainty of each finding is depicted by colors (dark blue = known; light blue= limited certainty; pink=highly uncertain; gray = unknown). C= Chemical of concern.

	Human Health																		Ecotoxicity						Physicochemical					
Alternatives	Acute Toxicity			Carcinogenicity			Genotoxicity			Reproductive			Developmental			Neurotoxicity			Acute Aquatic Toxicity			Chronic Aquatic Toxicity			Persistence			Bioaccumulation		
	H	M	L	H	M	L	H	M	L	H	M	L	H	M	L	H	M	L	H	M	L	H	M	L	H	M	L	H	M	L
C	pink				pink				dark blue			light blue			dark blue			dark blue		light blue		gray	gray	gray		dark blue				dark blue
A			dark blue		light blue				dark blue	gray	gray	gray			dark blue	pink					light blue			dark blue	dark blue			light blue		
B			dark blue			dark blue			pink			pink			dark blue	gray	gray	gray	light blue				dark blue				dark blue			dark blue

TABLE 9-2 Sample Results of Comparative Exposure Assessment

Note: >> indicates that the alternative may involve substantially greater exposure than the chemical of concern, ≈ indicates that exposures may be considered substantially equivalent, and << indicates that the alternative may involve lower exposures due to intrinsic properties of the chemical or the specific functional use.

Alternatives	Human Health Exposure Routes												Eco Exposure								
	Oral			Dermal			Inhalation			Ocular			Water			Air			Soil		
	>	≈	<	>	≈	<	>	≈	<	>	≈	<	>	≈	<	>	≈	<	>	≈	<
A		X			X		X				X			X		X				X	
B		X				X		X			X				X		X			X	

INFORMATION NEEDED TO IMPLEMENT STEP 7 IN THE COMMITTEE'S FRAMEWORK

The information that the user will primarily rely on to complete Step 7 was evaluated and collated in Steps 5 (Assess Physicochemical Properties), 6.1 (Assess Human Health), 6.2 (Assess Ecotoxicity), and 6.3 (Conduct Comparative Exposure Assessment). The result of Step 6 is an assessment of human health and ecotoxicity hazards and an indication of how each alternative's exposure is expected to compare with that of the chemical of concern. In most cases, the alternatives will present different hazards both across domains (e.g., ecotoxicity vs. human health hazards) and within an evaluated domain (e.g., neurotoxicity vs. respiratory sensitization). Tables 9-1 and 9-2 present a summary of the types of evidence that may be gathered in Steps 5 and 6.

STRATEGIES TO ADDRESS TRADE-OFFS AND UNCERTAINTY

The two key underlying challenges (trade-offs and uncertainty) inherent in data integration can be viewed as separate but potentially overlapping. Figure 9-2 shows trade-offs and uncertainty in two dimensions, along with the decision-making strategies required as a result

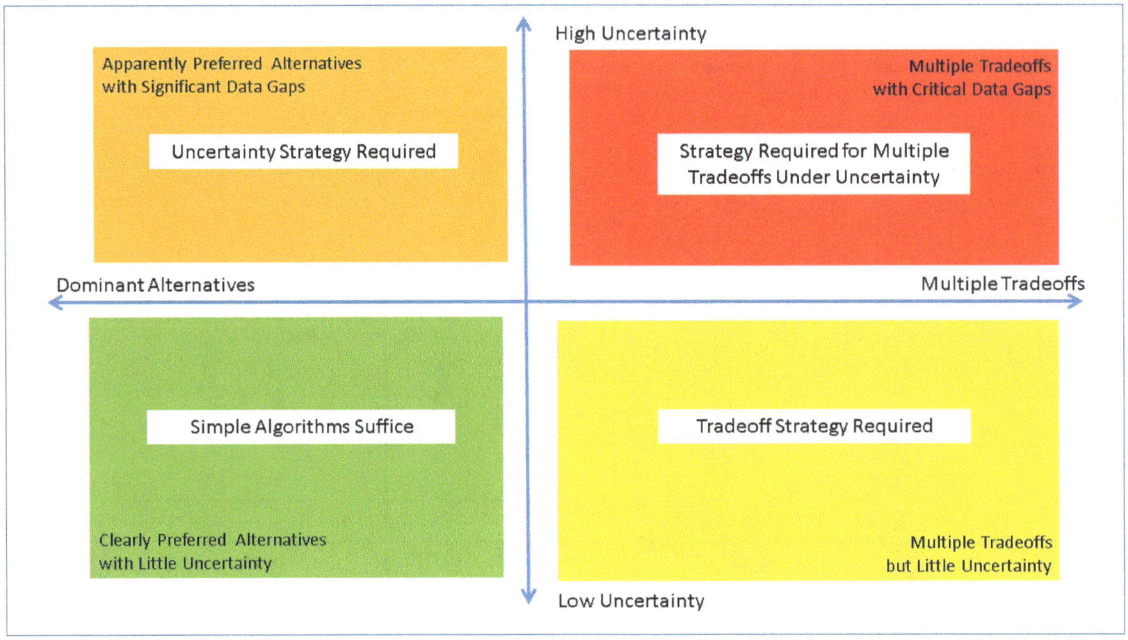

FIGURE 9-2 Strategies to address trade-offs and uncertainty in alternatives assessment. The pattern of results from Step 6 will present potential trade-offs among alternatives, as well as varying levels of certainty. In some cases, both trade-offs and uncertainty will be key challenges that will need to be addressed.

of the evidence gathered in Step 6. In the lower-left (green) quadrant, the choice among alternatives is made clear because sufficient information is available so that an alternative with no trade-offs within or among human and ecotoxicity domains can be chosen. In the upper-left (orange) quadrant, there appears to be a preferable alternative based on what is known and best estimates, but uncertainty about the findings remain, or there is a critical gap in the information available for the otherwise preferred alternatives. In the lower-right quadrant (yellow), there is adequate information available, but the pattern of findings is such that there are trade-offs within human health and ecotoxicity outcomes or between human health and ecotoxicity outcomes such that no alternatives are preferable for all end points or from all perspectives. In the upper-right quadrant (red), both challenges exist in that there are trade-offs among end points that are simultaneously affected by uncertainty.

The simplest case (green quadrant; no trade-offs, adequate and qualitatively equal levels of knowledge) can be addressed by simple algorithms that identify the preferred alternative. An example of such an algorithm is to identify alternatives that are preferable to a baseline alternative in at least one end point category, and not worse in any other.

Uncertainty Strategies (Orange Quadrant)

As discussed in Chapters 6-8, it is likely that there will be varying levels of uncertainty surrounding human or ecological toxicity and relative exposure. Box 9-1 describes types of uncertainty, and this section describes selected strategies for addressing them. With respect to variability (e.g., in exposures and in the responses of human or ecological receptors to those exposure), this is assumed to have been addressed to the extent possible in Steps 6.1, 6.2 and 6.3.

Consider the following example summarized in Table 9-3, which is a hypothetical output of Step 6. This example has been deliberately simplified such that all other end points are considered equal among the alternatives and are equally well known. Alternatives A and B are preferable (have lower or equivalent toxicity) to the chemical of concern given the best estimate available (the medium and low categories are based on the

BOX 9-1

TYPES OF UNCERTAINTY

There are numerous ways in which uncertainty may be defined and categorized. Taxonomies have been proposed that help differentiate uncertainty according to its many sources (Morgan and Henrion 1990; NRC 1994; Cullen and Frey 1999; and Krupnick et al. 2006). One key distinction that is often advocated is the conceptual separation of uncertainty and variability. The term *uncertainty* is most often used to describe limitations in knowledge. Uncertainty means that we do not know what the true situation is (the uncertainty can be qualitative or quantitative in nature). The term *variability* is used to describe real differences that exist in the world among individuals, behaviors and the natural world. When variability is described, it reflects the fact that there is no single true number that fully describes a phenomenon. In practice, it is often difficult to completely separate uncertainty and variability. This is particularly difficult when attempting to express *uncertainty* (or lack of knowledge) in the nature and extent of *variability*.

In describing uncertainty as a limitation of knowledge, scientists have found it useful to distinguish between two main types of uncertainty, *parameter uncertainty* and *model uncertainty*. When describing aspects of toxicity or exposure in quantitative terms, there will often be uncertainty with respect to specific values that need to be assigned. This type of uncertainty has often been called *parameter uncertainty*. Whenever there is an incomplete understanding of a causal nature (i.e., there are competing explanations for some observed phenomenon), there will necessarily be alternate mathematical models that might legitimately be used to make predictions about the level of risk. The existence of competing explanations, (and competing models), is often referred to as *model uncertainty*. When the model uncertainty is so great that it leads to questions about the very existence of causal relationships (as opposed to competing models based on the strength and exact nature of the relationship), it may be referred to as *fundamental causal uncertainty* (NRC 2009).

Examples of parameter uncertainties include the numerical thresholds for human or environmental exposure, below which no adverse effects are expected, unmeasured physicochemical properties, and estimates of the amount of exposure in a use scenario. Examples of model uncertainty include cases where the specific mode-of-action of carcinogenicity of a chemical may be unknown or debated, or where there are different interpretations of the evidence from in vitro experiments or in silico predictions, leading to alternate views of whether a specific form of toxicity should be assumed. Fundamental causal uncertainty may take the form of an unclear causal linkage between a chemical and a form of toxicity because of confounding factors and significant inconsistency in the toxicity database. Depending on the approach used to present evidence for human health and ecotoxicity, all of these types of uncertainty may need to be addressed in this step.

TABLE 9-3 Excerpt of an Evidence Table Demonstrating Differing Levels of Uncertainty among Alternatives, with no Apparent Trade-off among End Points based on Best Estimates

Note: Uncertainty in each toxicological finding is depicted by colors (dark blue = known; light blue= limited certainty; pink=highly uncertain; gray = unknown)

Alternatives	**Reproductive**			**Developmental**			**Neurotoxicity**		
	H	M	L	H	M	L	H	M	L
C									
A									
B									

best judgment available after considering multiple lines of reasoning). However, Alternative A does not have any evaluation available for neurotoxicity. Although Alternative B was given a low toxicity rating, toxicity for two end points was considered "highly uncertain."

There are a number of possible strategies for addressing the presence of uncertainty, assuming that each is presented as a separate concept (as shown in Table 9-3 or in some other qualitative or quantitative format). If the strategies were entirely embedded in the toxicological evaluation process, it is possible that toxicity would be deliberately overestimated to account for uncertainty. It is expected, however, that the toxicological evaluation process will have considered the concept of individual variability. The following section presents some strategies for addressing uncertainty.

Known best estimates basis: In this approach, only the best estimates are considered, and alternatives with unknown toxicity end points are excluded. In the example presented in Table 9-3, this strategy would prefer Alternative B to the chemical of concern. Alternative A would be excluded due to missing data. While being transparent about uncertainty, this approach does not directly apply the level of uncertainty in the algorithm, except to exclude alternatives with data gaps. According to this strategy, Alternative B would be preferred to the chemical of concern.

Uncertainty downgrade basis: In this strategy, the best-estimate toxicity value is, in some way, downgraded based on uncertainty (for example, downgraded by one level for moderate uncertainty or two levels for high uncertainty), and alternatives with missing data are excluded. According to this strategy, Alternative A would be considered unacceptable due to missing data, and Alternative B would be downgraded from {L,L,L} to {H,L,H} due to the high levels of uncertainty in the first and third end points. With this adjustment, Alternative B is no longer clearly preferred to the chemical of concern due to this uncertainty-based adjustment. This approach considers the uncertainty and "errs on the side of safety" by biasing the assessed toxicity to be greater where there is greater uncertainty in the finding. This strategy essentially "punishes" alternatives for uncertainty. The committee notes that this approach to addressing uncertainty (or any approach that excludes alternatives with limited or missing evidence on toxicity) can undermine the selection of safer alternatives and "erring on the side of safety" may be counterproductive. Depending on how it is implemented, this approach could lead to less safe alternatives consistently being preferred just by virtue of having been subject to many studies (and therefore having less uncertainty with respect to toxicity).

Quantitative uncertainty analysis: Uncertainty in toxicity values could be expressed quantitatively or illustrated graphically. This could take the form of a relatively simple expression of a range, or could be more elaborately expressed as a probability distribution, among other options. The benefit of this approach is transparency. It becomes easier to confidently conclude that one alternative is preferable to another if it is shown in a clear illustration. Therefore, this approach may be useful when the uncertainty is large enough that it presents a challenge in deciding on the preference ordering of alternatives. This approach can provide considerable insight, as illustrated by an example of a quantitative expression of uncertainty in comparing alternative chemicals (Finkel 1995).

Remaining neutral about uncertainty and missing data: A variation on the above strategies would be to note the presence of uncertainty and missing data but not exclude the alternative or otherwise demote it at this point in the selection process. The basis of temporarily treating an alternative neutrally with respect to uncertainty or missing data is to avoid prematurely removing potentially safer alternatives from the evaluation process. It may be assumed that while other assessments (economic, performance, etc.), are being conducted at later stages, the missing data can be replaced with direct or indirect evidence relating to the one or more end points for which data were missing. This approach should only be taken if the alternative appears to have sufficient merit on other grounds (e.g., safer with respect to some key end points) to warrant the effort of gathering more data. **The committee considers this approach as being most compatible with the multi-step consideration of alternatives recommended in the committee's framework.**

TABLE 9-4 Example of an Evidence Table Demonstrating Trade-offs with Alternatives with Adequate Information Indicated by Equal and Low Levels of Uncertainty for all End Points)
Note: Uncertainty of each finding depicted by colors (dark blue = known)

Alternatives	Developmental Toxicity			Neurotoxicity			Aquatic Toxicity		
	H	M	L	H	M	L	H	M	L
C		■				■			■
A			■	■			■		
B			■			■	■		

Trade-off Strategies (Yellow Quadrant)

Even under conditions of complete knowledge, an alternatives assessment may require the consideration of trade-offs among end points within a domain, or between health and ecotoxicity domains. This is further complicated in later stages by the consideration of other factors, such as broader environmental impacts, cost, performance, and social outcomes, which may require further consideration of trade-offs.

Table 9-4 includes a potential challenge faced in an alternatives assessment when there is no clearly preferred alternative. Alternative B is preferable from a human health perspective, but it is undesirable from the ecotoxicity perspective. While both Alternatives A and B have relatively lower ecotoxicity, Alternative B is not as safe as Alternative A from a human health perspective. And from a human health perspective, there is no clear preference between the chemical of concern and Alternative A, since it is not apparent that a rating of "medium" for developmental toxicity would necessarily be preferable to a rating of "high" for neurotoxicity. The reason that this becomes a difficult call is that the actual health consequences associated with either one could vary by many orders of magnitude in terms of severity or duration of the adverse health effect(s).

Similar to the situation for addressing uncertainty, there are a number of possible strategies for addressing the presence of trade-offs, (assuming for the moment that the health end points are known with equivalent certainty).

Improvement on key end point: In some contexts, the alternatives assessment may be motivated by the need to improve the safety of the product with respect to a specific end point (such as the original impetus driving the alternatives assessment, another regulatory requirement, or a commercial requirement of customers). In this case, it may be appropriate to remove any alternatives from consideration that do not improve upon the toxicity with respect to the specific end point of interest. The committee acknowledges that this criterion may be appropriate or even necessary from a practical perspective, while also recognizing that this approach may prematurely eliminate an option that does not improve the toxicity with respect to the original impetus of concern, but is potentially much safer when considering many other end points. That said, it is important to note that a focus on a key end point does not eliminate the need for an appropriate level of attention to the full range of human health hazard end points and ecotoxicity or for applying broader Life Cycle Thinking. By not taking these considerations into account, the assessor runs the risk of an unacceptable transfer of risks (i.e., burden shifting) or other types of regrettable substitution.

Strict ordering of end points: In this strategy, end points are strictly ranked such that the highest-ranked end point governs the overall preference ordering. In this case, if developmental toxicity was the higher-ranked end point, then Alternative A would be preferred to the chemical of concern. If the two had been equivalent for developmental toxicity, the ranking would be based on the next highest-rated end point. This approach requires a strict ordering of the importance of end points, which may not be justifiable on public health grounds and is not likely to be supported by all stakeholders.

Equal weighting of end points: In this strategy, each end point is considered to have equivalent importance, and the trade-off is resolved by assigning a relative weight to the high, medium, and low categories and then adding up the score. The total would indicate the preference ordering of alternatives. But this approach also has its limitations. Just as a strict ordering of end points is not necessarily appropriate, it is not necessarily preferable to treat all end points equally.

Weighted scoring of end points: In this strategy, end points are given an unequal weight, and the relative score is determined by summing up the weighted scores across the end points. This approach would also require placing a relative weight on the high, medium, and low categories or on the raw toxicity values. Weighted scoring of end points is one of the most common approaches in the discipline of Multi-Criteria Decision Analysis (MCDA). MCDA is directly applicable to the analysis of trade-offs in general. This discipline provides a diverse array of tools to use to arrive at a preference ordering of alternatives when considering multiple criteria involving trade-offs. As a general method, it can be applied either within the health and ecological considerations in this step, or later, during the final integration, when other factors ranging from costs to social impacts are considered. Or MCDA can be applied at both points. In addition to adding transparency and formality to the process of integration, MCDA tools often are implemented with software that allows for visualization of the weighting process, facilitating sensitivity analysis associated with the weights assigned to the different objectives. For example, a visualization tool like ToxPi (see Appendix C) could be used with MCDA tools to provide both transparent and formal weighting of end points in the tradeoff process. The ToxPi tool also has the benefit of helping visualize the assembled evidence and reducing the evidence to a unit-less score to support expert-driven decision making.

Rule-based ranking: Rather than using weights and arithmetic to indicate preferences among alternatives, the preferences can be ordered by a series of logical statements. The GreenScreen® algorithm uses such an approach by explicitly specifying the preference ordering of all possible combinations of toxicity findings. For rule-based systems, the underlying logic represents an unequal weighting of the importance of human and ecotoxicity end points, but the weighting process may or may not be explicitly described. While this appears to avoid the challenge of assigning "weights" explicitly, an implicit relative weighting is essentially embedded in the rule-based algorithm. The basis for implicit or explicit weighting should be carefully considered before applying a rule-based system to ensure that the organization's values with respect to the different health outcomes are appropriately represented. A key benefit associated with rule-based ranking is that the organization's value system, once codified in the form of these rules, can be consistently applied to make the alternatives assessment process less prone to idiosyncratic judgments or manipulation of the weighting schemes toward otherwise preferred outcomes.

Eliminate the "high" rating: In this strategy, the alternative is eliminated if it scores "high" on any toxicity end point. In the example shown on Table 9-4, this approach would eliminate both Alternative A (neurotoxicity) and Alternative B (aquatic toxicity).

Exposure weighting: In this strategy, the extent of exposure that may be associated with the various toxicity end points can be included to assign weight to those endpoints. For example, if the developmental toxicity was associated only with oral exposure, while the neurotoxicity was associated only with inhalation exposure, and oral exposure was expected to be much higher or more frequent given the specific functional use for the chemicals in question, then developmental toxicity could be considered more important. This would yield a preference for Alternative A over the chemical of concern. The inclusion of exposure considerations is addressed further in the next section.

Exposure tie-breaking: If a substantial difference in exposure potential was identified among the alternatives, then the framework's comparative exposure assessment (Step 6.3) may be used to provide a preference ordering for alternatives when they would otherwise be considered equivalent. For example, adverse exposure potential could be used to downgrade the toxicological finding, and inherently preferred exposure potential could be used to upgrade the toxicological finding.

Relative risk assessment with disease burden estimation: This strategy involves conducting a *relative* risk assessment and estimating the relative frequency with which the implicated health end points (those involved in the trade-off) might occur. If it is not possible to estimate the frequency of implicated health outcomes (e.g., due to the lack of a known dose-response relationship), surrogates for risk, such as a hazard index or margin of exposure, can be used to identify where risks to human or ecological endpoints appear to be more likely given expected exposure levels. The assessment can be further nuanced by considering the relative severity of the expected outcomes, if known, using comparative measures of burden of disease, such as Health-Adjusted Life Years (IOM 2006). To be consistent with the intent of many alternatives frameworks to avoid reliance on extrinsic exposure controls, the unmitigated exposure could be the basis of the evaluation. When considering unmitigated exposure, these relative risk estimations (or surrogate indicators) could then form the basis of focusing the attention (and weight in scoring approaches) on alternatives that appear to have reduced potential for harm. The approach to risk estimation (or at least, a more risk-based consideration of inherent toxicity) need only be done with the level of accuracy required to differentiate among the alternatives. It does not require the effort associated with a full risk assessment and health economic analysis.

Expert-manager judgment: This strategy relies on the application of expert judgment to replace all of the above algorithmic or scoring-based methods with selection by a group of presumably appropriate experts. The term "expert-manager" is used here because the expert is required to make explicit or implicit societal value judgments (e.g., the relative importance of human health and ecotoxicity among end points within each domain) in addition to applying their expertise. This approach has the benefit of using more information than is provided by the outputs of Step 6, including uncertainty and relative exposure considerations. However, this benefit comes at the cost of lower levels of transparency and idiosyncratic variability among experts, who are required by the process to impose value judgments, some of which may not be shared by the organization implementing the alternatives assessment.

List-based preference ordering:[43] In response to some regulatory, commercial, or other reasons, an organization may want to, or may be required to, apply the preference ordering based on an external organization's (for example, an important customer or an important regulatory jurisdiction) apparent preference ordering of health or ecotoxicity end points. This may be as simple as removing alternatives that appear on a list that has been designated as being "of concern." The choices of alternatives on this basis may lead to safer alternatives, but this essentially "outsources" the value judgments to an external organization, rather than eliminating them.

These strategies are just examples of a variety of possible means to address trade-offs where available alternatives present unavoidable applications of value judgments to determine their preference ordering. The strategies range in complexity, from simple decision rules to relative risk assessment. They are not all mutually exclusive; for example, simple decision rules could be used to eliminate a few alternatives and then a more complex weighting procedure could be applied to the remaining alternatives. Key considerations in choosing the means to implement trade-off decisions include the question of who is appropriately empowered to make societal value judgments, and whether these judgments are developed in advance of the implementation of alternatives assessment or are developed during the alternatives assessment. If the latter is true, the judgments may be more likely to be adjusted in a biased fashion toward a preferred or status quo alternative.

Strategies for Multiple Trade-offs under Uncertainty (Red Quadrant)

In some cases, the alternatives assessment process may be challenged by a combination of both value-based trade-offs as well as uncertainty about one or more end points. Up

43 This strategy may have been applied in Step 2, if the number of alternatives made an initial screening necessary.

TABLE 9-5 Example of an Evidence Table Demonstrating Both Trade-offs and Differences In the Level of Uncertainty in Toxicological Evaluations

Note: Uncertainty of each finding depicted by colors (dark blue = known; light blue= limited certainty; pink=highly uncertain)

Alternatives	Developmental Toxicity			Neurotoxicity			Aquatic Toxicity		
	H	M	L	H	M	L	H	M	L
C		dark blue				pink			dark blue
A			dark blue	dark blue					light blue
B			pink			dark blue	dark blue		

to this point in the discussion of integration, uncertainty and value judgments have been considered separately. In Table 9-3, trade-offs were not apparent, leaving only uncertainty. Conversely, in Table 9-4, uncertainty was eliminated as a consideration, but trade-offs were apparent.

An example of the combination of uncertainty and value-based trade-offs is shown in Table 9-5. Alternative A is preferable to the chemical of concern with respect to developmental toxicity, but appears to be less desirable from a neurotoxicity perspective. However, the neurotoxicity of the chemical of concern is highly uncertain, yielding an ambiguous preference ordering dependent upon the user's approach to addressing the uncertainty in the neurotoxicity of the chemical of concern. Similarly, Alternative B appears to be preferable from a human health perspective; however, there remains a high level of uncertainty in the one end point that is the basis for the health-based preference, and it is clearly not preferred with respect to ecotoxicity.

This section focuses on how to consider both trade-offs and uncertainty. In cases where there are high uncertainty and apparent trade-offs, a greater focus on de novo design to create safer options is warranted (Step 13). For analyzing the existing options, it may be useful to note that despite the separation of alternatives assessment from risk assessment, alternatives assessment does have similar goals to comparative risk assessment (supporting decisions on relative safety among decision-making options). The field of comparative risk assessment is generally associated with comparing very different risks (and therefore dealing with value-laden trade-offs), including established and emerging risks and their associated levels of uncertainty (Finkel and Golding 1995; Davies 1996; Florig et al. 2001; Morgan et al. 2001; Willis et al. 2004; Linkov et al. 2006).

The research on comparative risk assessment may provide an appropriate basis for deciding which approach to use when dealing with complex comparisons with considerable uncertainty. Approaches to comparative risk assessment, such as those studied and reported by Florig and colleagues (2001), may be appropriate for the more challenging applications of alternatives assessment, including those situations that involve both uncertainty and value-based trade-offs, because they were designed with such challenges in mind. A key component of these approaches is the parallel use of both quantitative and semi-quantitative schemes and expert consensus-based approaches to ranking risks. Quantitative (including both scoring-based and rule-based) schemes can provide more objective treatment of the evidence, and provide a degree of transparency in their conclusions by having a direct and consistent link between evidence and conclusions. Expert consensus-based approaches allow for more complete consideration of aspects of the evidence base that involve difficult and unquantifiable evaluations, such as conflicting data or conflicting valuations of outcomes.

The expert-consensus method can be augmented by the preparation of a structured summary document containing the evidence for all alternatives, including some narrative discussion and the quantitative inputs used in the scoring approach, but leaving the final rankings aside. These parallel approaches can then be merged to consider the differences in the rankings from each process and to determine a final ranking based on consideration of the two parallel methods of ranking. This can be done by adjusting one ranking result in light of what was learned in the parallel approach.

CONCLUSIONS

The overall outcome of Step 7 is the identification of alternatives that are acceptable because they meet the criteria of being safer, with respect to health and ecotoxicity outcomes. Step 7 may result in some alternatives being eliminated from further consideration. Given that considerations from later steps in the framework may also eliminate some alternatives, it may be appropriate to avoid eliminating too many alternatives early-on, unless they are unambiguously unfavorable from a health or ecotoxicity perspective. When the alternatives assessment is motivated by the need to improve the safety of a specific end point (because, for example, the chemical is on a carcinogen list), the alternative chemical will, for pragmatic reasons, need to be an improvement of, or no worse than, the original chemical of concern in the domain that initiated the alternatives assessment. If several chemicals meet this minimum requirement, then the practitioners should consider whether the alternatives would lead to a reduced overall impact on human health, ecotoxicity, and/or the environment. Ultimately, the approach chosen to integrate the evidence must take into account organizational resources available for conducting the alternatives assessment. More elaborate approaches to alternatives assessment may be appropriate for major decisions. The analysis should be proportionate to the importance of the decision (e.g., the risk associated with the status quo, or to the potential benefit of finding a safer alternative given current levels of exposure to the product). Just as simple approaches are appropriate for small-scale decisions, complex and rigorous treatment is appropriate for major decisions that impact large populations and have large environmental footprints (e.g., fuel additives, energy use, common household products, products used by children or found in most homes, infrastructure and building materials, and food and agricultural uses).

10

Life Cycle, Performance, and Economic Considerations

At this stage in the framework (after completing Step 7), a list of possible alternatives has been developed after considering physicochemical properties, comparative exposure assessment, human health, and ecotoxicity. The next steps in the committee's framework (Steps 8 and 9, Figure 10-1) consider trade-offs between these domains and other factors, such as product efficacy, economics, process safety, and resource use.

Estimating the materials and energy consumed and substances emitted by a product over part or all of its life cycle, and the human, environmental, and social impacts associated with those flows, are topics beyond the human health and ecological impacts evaluated in earlier analyses. Thus, additional steps to consider whether a life cycle analysis[44] is required, and to provide guidance on selection of an appropriate life cycle approach when needed, are included in the committee's framework. Step 8 is a required element that uses Life Cycle Thinking (LCT) and other screening methods to determine if additional detail and quantitation are required. The need to complete subsequent analyses (optional Step 9.1) is based on the output of this initial analysis. Additional consideration of broad environmental impacts, such as greenhouse gas emissions and energy resources, and social impacts, such as labor practices and human rights concerns, also occurs during Step 9.

Box 10-1 provides the elements of the committee's suggested approach to Steps 8 and 9.1. These steps should be performed in accord with Step 2 (problem formulation) of the committee's framework. They may also include other life cycle concerns identified as important by the assessor while progressing through the alternatives assessment. Step 8, which is required under the framework, asks the assessor to determine if significant differences exist between the chemical of concern and the possible alternatives over their respective life cycles. Box 10-2 provides definitions for the terms used in this chapter.

BOX 10-1

ELEMENTS OF LIFE CYCLE ANALYSIS IN THE COMMITTEE'S FRAMEWORK

1. Use Life Cycle Thinking (LCT) to qualitatively determine if differences in material or energy flow or synthetic history exist between the original chemical and the potential alternatives. These may be assessed across a number of risks, including those to human health, the environment, or society. This analysis should determine if those risks exist at a place or time other than the subject application.
2. If the Life Cycle Thinking identifies a significant difference in these areas when the life cycle of the original chemical is compared to that of the life cycle of an alternative, then a Life Cycle Inventory or "screening LCA" or Life Cycle Impact Assessment should be performed to provide quantitative information. If these analyses reveal that additional, quantitative information is required to support decision-making, then the assessor may wish to proceed to Step 9.1 and perform a Life Cycle Impact Analysis (see Box 10-2).

Before accepting a chemical as an alternative, it must be determined that the chemical can perform adequately in the intended application(s) identified early in the alternatives assessment process (Step 2, see Chapter 4). This early problem formulation step should have identified performance and economic criteria. To follow up on the findings from Step 2, the committee also includes optional performance (Step 9.2) and economic (Step 9.3) assessments in its framework. These steps are considered optional because the entity performing the assessment may

[44] As used in this chapter, the term "life cycle analysis," written in lower case, refers collectively to the family of methodologies that use a systems approach to compile and evaluate the inputs, outputs, and potential environmental impacts of a product system throughout its life cycle. Specific methods, such as Life Cycle Thinking (LCT), Life Cycle Inventory (LCI) and Life Cycle Impact Assessment (LCIA), will be capitalized or represented by their initials.

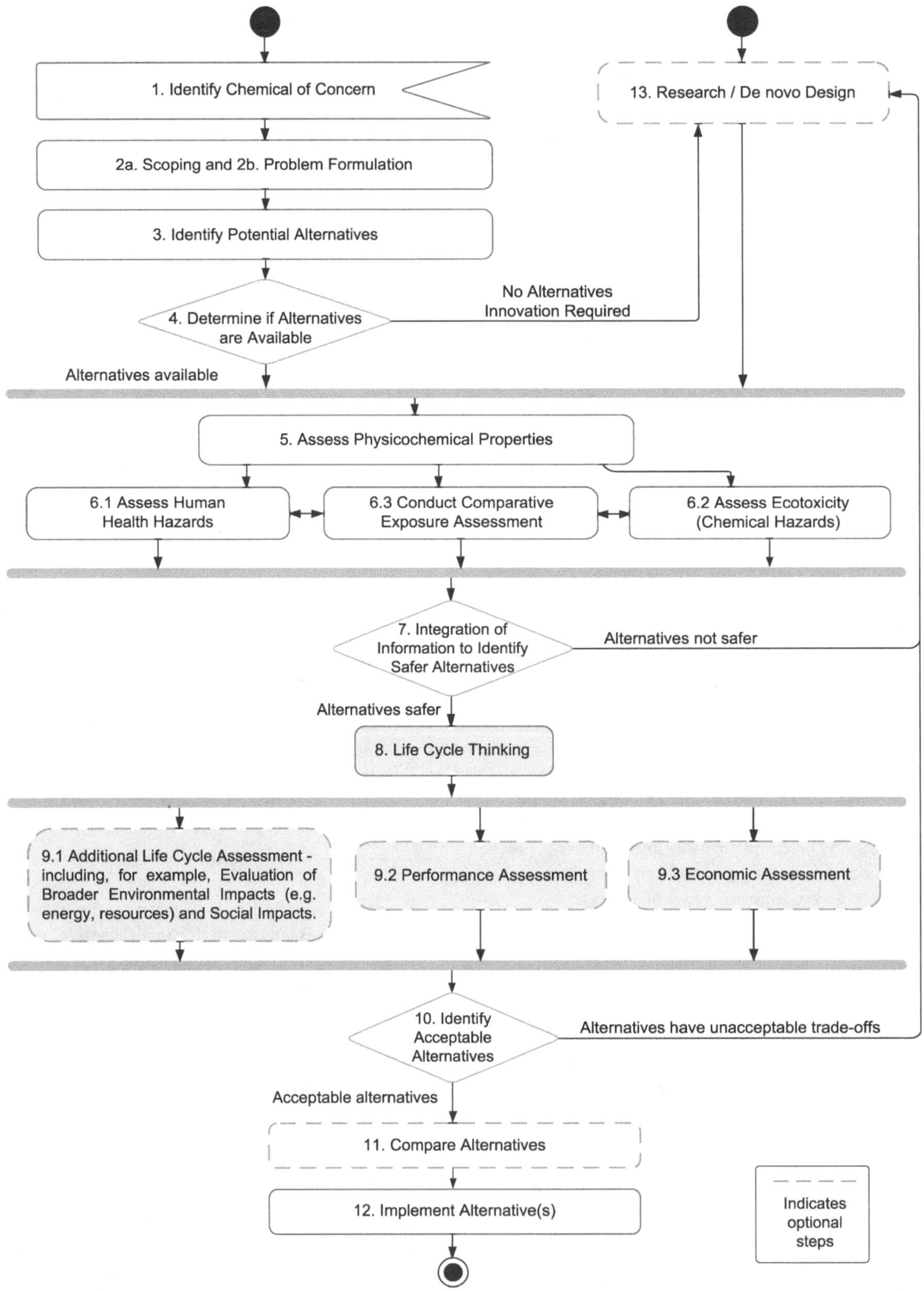

FIGURE 10-1 Excerpt of committee's framework highlighting the performance and life cycle assessments.

not be a business, and thus would not be in a position to evaluate performance and economics as a business would. The converse is obviously true; a business would be critically interested in establishing performance and economic performance criteria (Step 2 of the framework) and ensuring that any selected alternatives meet those criteria.

Note that it is beyond the scope of this committee to provide specific guidance on the best practices for performing the assessments in Steps 8 and 9. Instead, this chapter will provide an overview of these assessments and a brief discussion of how they might affect the final decision of which alternative chemical moves forward.

LIFE CYCLE, SOCIAL, PERFORMANCE, AND ECONOMIC CONSIDERATIONS IN OTHER FRAMEWORKS

Life Cycle

Three frameworks studied by the committee evaluate whether life cycle concerns indicate a need for a life cycle assessment, while three other frameworks suggest or require consideration of factors, such as greenhouse gas emissions, that would normally be addressed through a life cycle assessment. The six frameworks were IC2, BizNGO, the German Guide, CA SCP, REACH, and UCLA MCDA. LCT takes many different forms across these frameworks. Life cycle assessments come into play in three different ways: as a separate, specific element or module of an assessment, such as in BizNGO; as a requirement folded into many elements of the assessment, such as in the CA SCP assessment plan; or as a guiding principle or value of an overall analysis, such as in the German Guide. In IC2, the Life Cycle Module can be treated as a separate element, though life cycle effects are also noted as being relevant in the Cost and Availability, Social Impact, and Materials Management modules. In frameworks where considering life cycle of a chemical is called out specifically, it is described as a method to assist in distinguishing between potential alternatives by drawing attention to considerations outside of the area of technical feasibility. Recognizing the complexity of a full life cycle analysis, it is often left to the assessor to determine if it would be beneficial for the assessment to move beyond Life Cycle Thinking to a quantitative analysis.

BOX 10-2

TERMS

It is important that attention be given to language used when discussing life cycle considerations. For this reason, brief descriptions are provided here, and additional detail can be found later in this chapter.

- *Life cycle Assessment* (LCA) is a "compilation and evaluation of the inputs, outputs and the potential environmental impacts of a product system throughout its life cycle" (ISO 2006a).
- *Life Cycle Thinking* (LCT) as defined by Christiansen, is "a mostly qualitative discussion to identify stages of the life cycle and/or the potential environmental impacts of greatest significance e.g. for use in a design brief or in an introductory discussion of policy measures. The greatest benefit is that it helps focus consideration of the full life cycle of the product or system; data are typically qualitative (statements) or very general and available-by-heart quantitative data" (Christiansen et al. 1997).
- *Life Cycle Inventory* (LCI) is a "phase of life cycle assessment involving the compilation and quantification of inputs and outputs for a product throughout its life cycle" (ISO 2006a).
- *Life Cycle Impact Assessment* (LCIA) is a "phase of Life Cycle Assessment aimed at understanding and evaluating the magnitude and significance of the potential environmental impacts for a product system throughout the life cycle of the product" (ISO 2006a).

Social Impacts

Several frameworks (IC2, REACH, Lowell, the German Guide, UCLA MCDA, and UNEP support an option to consider social impacts beyond those already addressed in other steps. These frameworks consider whether there are worker issues, local community issues, or societal issues not addressed by other steps and whether differences between alternatives are expected to be significant. Two frameworks (IC2, REACH) assess potential social and socioeconomic impacts of each alternative across its life cycle.

Discussion of social impacts and how those assessments are performed also varies across the different frameworks. For example, in IC2, social impacts assessment is in a module that can be used if appropriate. Under REACH, the social impacts are contained in the socioeconomic analysis. In the Lowell framework, consideration of social impacts, including social justice performance, is described as important for future development of that framework. In the German Guide, social responsibility across the life cycle is a clear factor for assessing alternatives. The main themes across these different assessment approaches are corporate values about social responsibility, social justice as it relates to areas such as labor practices and human rights, and social impacts that affect communities and states with regard to management of chemicals during their manufacture, use, or disposal.

Performance Assessment

Assessment of performance is a critical element or module in every framework examined. Technical feasibility and performance is evaluated for each alternative, but for direct replacement chemicals, the performance of the chemical of concern is a starting point for evaluation. Thus, BizNGO notes that care should be taken to ensure that the performance requirements for existing products are not higher than necessary for the application so that screening out of potential alternatives is not done unnecessarily. Multiple frameworks note that if an alternative is in use in the commercial stream already, market information and assessments may provide useful technical and performance analyses that can be drawn upon for the new use. IC2 notes that feasible modifications of products or processes could be considered if an alternative falls outside the range of conditions required by the current chemical of concern.

Economic Analysis

Economic analysis generally falls into four categories across the reviewed frameworks:

- direct, business-relevant impact;
- market analysis, including potential changes to availability of the alternative, relevant regulations that might be affected, and competition from other vendors;
- costs to other entities, such as public agencies, stakeholders, and communities; and
- cost-benefit analyses.

The direct costs include positive and negative changes to revenue if an alternative is adopted. Both the market analysis and the cost-benefit analyses may entail some consideration of regulatory and social elements that can be easily quantified, such as cost of re-registration of approval of an end product or material, and those that may not be readily quantified, such as potential future liabilities in case of release, reduced risk of accidents during production, or potential changes to public perception of that product. In every assessment, the economic analysis is performed after the completion of the technical assessments. The complexity and detail of required or recommended analyses varies considerably across the various assessments; however, all recognize that there is a potential for no alternative to be viable due to cost concerns, and

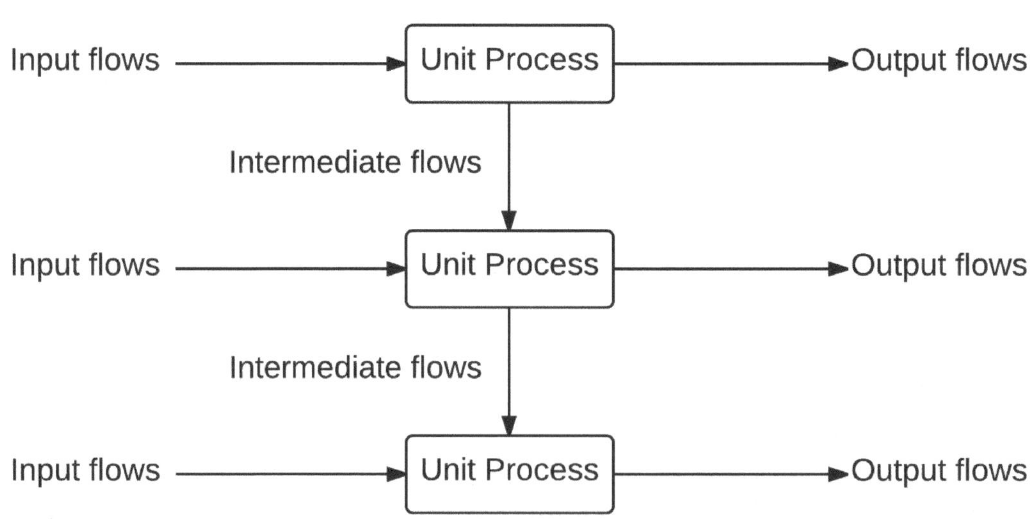

FIGURE 10-2 Unit processes within a product system (ISO 2006a). This excerpt is from ISO 14040:2006, Figure 2 on page 10, with the permission of ANSI on behalf of ISO. (c) ISO 2014 - All rights reserved.

this may result in additional considerations. For example, the CA SCP framework specifies that in cases where the requirement to identify a substitute chemical is initiated by regulatory schemes, if no financially viable alternative can be identified, then a clear description of end-of-life management plans for the chemical of concern must be presented as part of the assessment and cost comparisons.

LIFE CYCLE CONSIDERATIONS IN THE COMMITTEE'S FRAMEWORK

Considering Impacts beyond the Point of Chemical Use/Application

Up until this point in the committee's framework, all of the analyses have focused on the impacts of the chemical of concern and possible alternatives at the point of use. However, it is always the case that impacts to human health, the environment, and society may occur throughout a product's life cycle, not just at the point of application. Therefore, life-cycle analysis is appropriate for identifying and understanding the impacts posed by a chemical of concern and alternatives in a product's life cycle, from manufacture to disposal, and to determine if these impacts warrant preference for one possible alternative over another. In considering each chemical's role in the product's full life cycle, the assessor can identify where there may be "burden shifting"—eliminating an impact at one point in a product's life cycle with the consequence of an equal or greater impact appearing at another point in a product's life cycle.[45] The initial consideration of life cycle effects occurs in Step 8.

Step 8: Life Cycle Thinking

The committee framework includes qualitative LCT in Step 8. One purpose of LCT is for the assessor to thoughtfully consider potential upstream and downstream impacts. This section describes the components of such thinking. Step 8 often provides enough information from which to make a decision, and in these cases, a quantitative analysis may not provide additional value. LCT can therefore identify whether an additional, optional quantitative assessment would be useful. Fundamental to any life cycle analysis, including LCT, is mapping the *product system*. Each stage in the product system (raw material acquisition, etc.) can be viewed as a collection of one or more unit processes.[46] Product systems can be subdivided into a network of unit processes that are linked to each other by the flow

BOX 10-3

PRODUCT SYSTEM MAPPING: A PROCEDURE FOR IDENTIFYING LIFE CYCLE STAGES AND UNIT PROCESSES IN A PRODUCT SYSTEM

For the Substance of Concern:

Substep 1: At the unit process stage, identify all material and energy inputs to the unit process and all outputs (products, co-products, and by-products) and releases from the unit process.

Substep 2: For each material input, identify the unit process from which the material was an output. This is identified as *the present unit process*.

Substep 3: For the present unit process, identify all material and energy inputs to the unit process and all outputs (products, co-products, and by-products) and releases to the environment.

Substep 4: Repeat Steps 2 and 3 for all the inputs taken directly from Earth (minerals, agricultural products, forest products, water, air, etc.).

Substep 5: For each output identified in Step 1, identify the unit process to which the material is an input. This, too, is identified as the present unit process.

Substep 6: For the present unit process, identify all material and energy inputs to the unit process and all outputs (products, co-products, by-products, and releases) from the unit process.

Substep 7: Repeat Steps 5 and 6 until all the outputs are disposed (managed as waste, reused, or recycled).

The result of Substeps 1 through 7 will be a product life cycle map for the chemical of concern.

For Alternatives:

Substep 8: Repeat Steps 1 through 7 for each potential alternative.

The result of this exercise will be a product life cycle map for each potential alternative.

---

45 For example, introducing a biofuel may decrease the risk of harm to the environment by reducing emissions of greenhouse gases while increasing the risk of harm to the environment by increasing runoff of nutrients to waterways with concomitant eutrophication.

46 ISO 14040 (ISO 2006a) defines a "unit process" as "the smallest element considered in the life cycle inventory analysis for which input and output data are quantified."

BOX 10-4

SYNTHETIC HISTORY

The sequence of unit operations that proceed from acquisition of raw materials to production of chemical intermediates to production of the chemical of concern (or possible alternative) is of particular interest. This process is known as the "synthetic history" of a chemical. Examination of the synthetic history can quickly reveal unit processes that present impacts to human health or the environment (for example, building block chemicals or by-products of a production unit process). It is also possible to look at the synthetic history of a chemical and, without using LCT, screen for possible hazards. Using this approach could provide the basis for preferring one alternative to another without the rigor of mapping a product system.

The potential replacement of a dialkyl phthalate with its cyclohexyl analog as a polyvinyl chloride plasticizer serves as a useful illustration of this point. If we examine the life cycle (as noted in the description of LCT above), we see that the process and raw material history of the cyclohexyl alternative maps completely onto that of phthalate except for the final step, where the phenyl group is hydrogenated to form the presumably safer cyclohexyl product. In this case, the initial top-level LCT analysis clearly suggests that (assuming the cyclohexyl alternative is safer in its application) an LCIA does not need be performed, because the only difference in the synthetic history of the compounds is an extra hydrogenation step for the alternative. Conversely, if we were to propose an alternative for a given compound that exhibits a dramatically different life cycle (revealed in the LCT step), where clear "red flags" appear at some point during the compound's synthetic history, then an LCIA would still be unnecessary, because these "red flags" suggest that the proposed alternative would be a regrettable substitution. An example might be the proposed substitution of N-vinyl formamide for acrylamide. While each is a monomer for a high molecular weight water-soluble polymer, acrylamide is a potent neurotoxin, N-vinyl formamide is a safer alternative. However, acrylamide is derived in a single step from acrylonitrile via enzymatic hydrolysis, while N-vinyl formamide is manufactured in a multi-step process, where toxic hydrogen cyanide is a key raw material. The solution here may be to seek an alternative synthetic pathway to N-vinyl formamide.

of intermediate products, releases to the environment, and waste (ISO 2006b). A process for constructing a *product system map* is outlined in Figure 10-2 and Box 10-3. Note that the procedure is intended to be illustrative, not prescriptive. Other procedures for developing a product system map are available (e.g., EPA 2006; ISO 2006a).

Dividing a product system into its component unit processes facilitates identification of the inputs and outputs of the product system. Inputs from the environment into the unit operations of the product system are resources consumed (such as chemicals and energy). Useful outputs from the product system are products and co-products. Releases to air, water, and land are the environmental emissions of the product system. These mass flows are the basis for subsequent life cycle assessments.

After constructing the product system map, the next step is to compare the map of the chemical of concern system with the map of each potential alternative system. Unit operations that are unique to either system should be identified, and the inputs and releases to the environment noted and qualitatively assessed. If no unit process unique to an alternative presents a greater risk of harm to human health, the environment, or society than the chemical of concern in its subject application, then the alternative remains viable.

If a potential alternative has a unique unit operation containing a significant hazard not present in the product system of the original chemical, then a determination should be made as to whether the hazard is easily mitigated. For example, if the hazard is in a controlled workplace where engineering controls or effective personal protective equipment (PPE) are readily installed and occupational health protections in place, then the alternative may remain viable. Consideration of the "synthetic history" of the chemical subject to the alternatives assessment is also a useful exercise at this point (see Box 10-4). In cases where an alternative includes an "upstream" chemical hazard, another possibility is to perform an alternatives assessment to determine if safer alternatives to that upstream chemical exist and if not whether the alternative subject to the original assessment remains viable. Consideration of the "synthetic history" of the chemical subject to the alternatives assessment is also a useful exercise at this point.

STEP 9: OPTIONAL ASSESSMENTS

Step 9 contains three optional steps (See Box 10-5 for additional information):

9.1: Additional Life Cycle Assessment

9.2: Performance Assessment

9.3: Economic Assessment

Whether these optional steps are performed will be largely dependent on the problem formulation defined in Step 2 of the assessment. There may be cases where new concerns arise during Steps 3-8 that trigger inclusion of these assessments, but this is likely to be a rare occurrence. All of the optional assessments in this step should be considered comparative in nature. They can be used to assist in decision making by allowing the assessor to compare the original chemical and a given alternative or the original chemical and the potential alternatives to identify a best fit for the purpose. For businesses, these may be particularly useful steps for assessing the market viability and potential effect on costs within the company.

Additional Life Cycle Assessment (Step 9.1)

While Step 8 is required in the committee's framework and will often provide an adequate level of detail, assessors and decision makers may find that they require additional information to inform their decision-making process (as defined through the problem formulation step), and will continue to Step 9.1. Figure 10-3, provides a useful conceptual structure for identifying the stages of a product life cycle. Stages are composed of the "unit processes" identified in Step 8. Though the life cycle stages may be considered individually to identify process-specific hazards, it is important to remember that when a change is made to one life cycle stage, it may also result in changes to other life cycle stages.

> **BOX 10-5**
>
> **ELEMENTS OF STEP 9 IN THE COMMITTEE'S FRAMEWORK**
>
> - (9.1) Use the information provided from Step 8 and perform an LCI, "screening LCA", or LCIA for the chemical of concern and each alternative to determine if unique impacts to human health, the environment, society, or other areas identified during the problem formulation step exist for the chemical of concern or its alternatives.
> - (9.2) Consider the performance criteria for a given chemical to meet the functional use requirements for the product. Determine if the potential alternatives are favorable for the desired application and meet the performance requirements.
> - (9.3) Use tools and standards common to the field, such as cost of materials, cost of the product—including, for example, production costs, energy costs, equipment costs, and direct costs—and net present value calculations to evaluate the economic impact of each alternative.

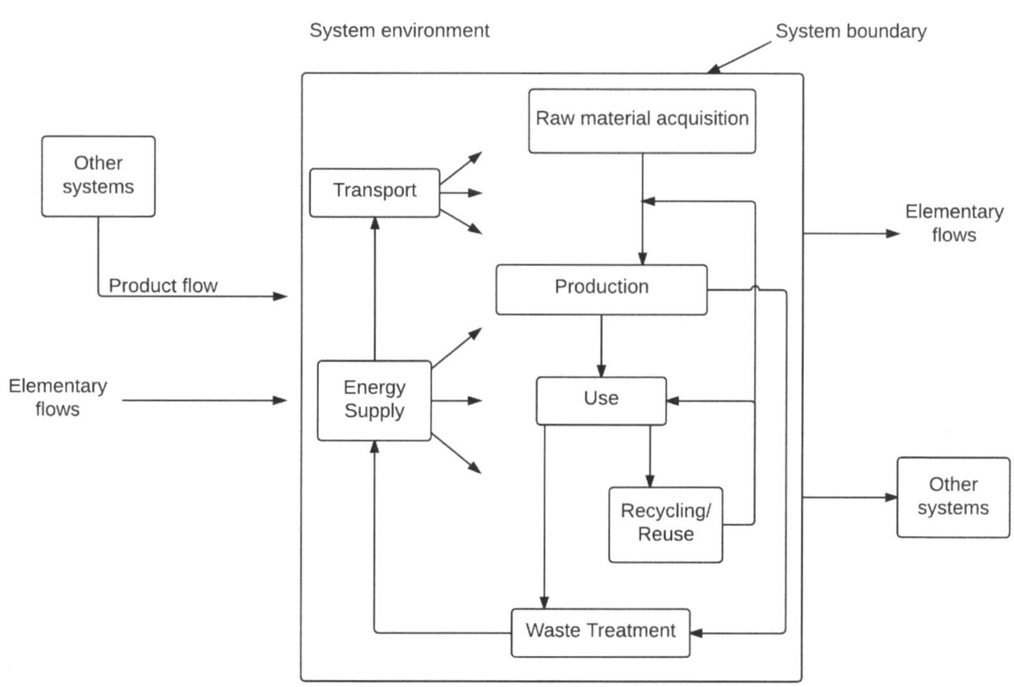

FIGURE 10-3 Example of a product system for life cycle assessment (ISO 2006a). This excerpt is from ISO 14040:2006, Figure 3 on page 10, with the permission of ANSI on behalf of ISO. (c) ISO 2014 - All rights reserved.

TABLE 10-1 Output of a Life Cycle Inventory (LCI)
Note: This is an abbreviated LCI output. This example only shows data for substances whose names begin with A. A complete LCI resource and release table typically has hundreds of entries.

Substance	Compartment	Unit	Total
Admium, 0.30% in sulfide, Cd 0.18%, Pb, Zn, Ag, In, in ground	Raw	μg	651.9
Barite, 15% in crude ore, in ground	Raw	mg	326.13
Basalt, in ground	Raw	mg	32.022
Borax, in ground	Raw	μg	1.4415
Acenaphthene	Air	pg	93.16118
Acetaldehyde	Air	μg	992.203
Acetic acid	Air	μg	491.599
Acenaphthene	Water	ng	28.0235
Acenaphthylene	Water	ng	1.75263
Acetaldehyde	Water	μg	1.2376
Acetic acid	Water	μg	24.109
Acetone	Water	pg	341.96
Acidity, unspecified	Water	μg	3.725295
Acrylate, ion	Water	ng	246.43
Actinides, radioactive, unspecified	Water	mBq	3.8371
Aluminum	Water	mg	66.79575
1,4-Butanediol	Water	pg	861.11
Aclonifen	Soil	mg	18.691
Aldrin	Soil	ng	2.6778
Aluminum	Soil	mg	1.952555
Antimony	Soil	pg	355.65
Arsenic	Soil	ng	787.777
Atrazine	Soil	pg	702.49
Barium	Soil	μg	948.0311

Screening Life Cycle Analysis

Depending on the problem formulation defined in Step 2 of the assessment or the surfacing of a material change in product systems identified in Step 8, a more quantitative comparison of the inputs and releases to the environment may be necessary to adequately evaluate the impact of a chemical substitution. For example, changing from a plastic to a metal housing for a computer may eliminate the need for an added flame retardant, but may also result in increased environmental and social impacts from mining. In these cases, a preliminary quantitative assessment, such as a screening LCA, may be performed as part of Step 9.1.

As shown in the description of LCT associated with Step 8, dividing a product system into its component unit processes facilitates identification of the inputs and outputs of the product system. When the data from the mass flows are summed across all unit operations (all resources consumed, all releases to air, water, and land) the result is a Life Cycle Inventory, or LCI (Table 10-1).

An obvious disadvantage of a system-specific, ISO-compliant LCI is that collecting the resource, output, and release data for each unit process in a product system is an enormous undertaking, and the resulting list of several hundred resources used combined with the list of several hundred releases to the environment may be difficult to interpret. Fortunately, databases and software tools have been developed to perform LCI analyses. These software tools use data that are not necessarily specific to the product system under consideration. For example, they may use industry average data or data from an unrelated facility making a similar product.

Life cycle inventories conducted with such data and software are often referred to as "screening LCAs" to differentiate them from life cycle studies, which use system-specific data. Additionally, screening LCAs often do not include peer review or fully meet the other requirements of ISO 14040 and ISO 14044 (ISO 2006a,b). Despite these limitations, screening LCAs can be used to estimate the materials and energy flows needed to conduct Life

TABLE 10-2 Commonly Used Life Cycle Environmental and Human Health Impact Categories

Impact Category	Common Possible Characterization Factor	Description of Characterization Factor
Global warming	Global warming potential	Converts LCI data to carbon dioxide (CO_2) equivalents Note: Global warming potentials can be 50, 100, or 500 year potentials.
Stratospheric ozone depletion	Ozone depleting potential	Converts LCI data to trichlorofluoromethane (CFC-11) equivalents.
Acidification	Acidification potential	Converts LCI data to hydrogen (H+) ion equivalents.
Eutrophication	Eutrophication potential	Converts LCI data to phosphate (PO_4) equivalents.
Photochemical smog	Photochemical oxidant creation potential	Converts LCI data to ethane (C_2H_6) equivalents.
Terrestrial toxicity	LC_{50}	Converts LC_{50} data to equivalents; uses multi- media modeling, exposure pathways.
Aquatic toxicity	LC_{50}	Converts LC_{50} data to equivalents; uses multi- media modeling, exposure pathways.
Human health	LC_{50}	Converts LC_{50} data to equivalents; uses multi- media modeling, exposure pathways.
Resource depletion	Resource depletion potential	Converts LCI data to a ratio of quantity of resource used vs. quantity of resource left in reserve.
Land use	Land availability	Converts mass of solid waste into volume using an estimated density.
Water use	Water shortage potential	Converts LCI data to a ratio of quantity of water used vs. quantity of resource left

SOURCE: Adapted from EPA 2006

Cycle Impact Analyses, which are discussed in the next section. Such analyses may assist in determining whether there is value in moving forward with a system-specific, ISO-compliant life cycle analysis.

Life Cycle Impact Analysis (LCIA)

Environmental and Human Health Impacts

An LCIA is a quantitative evaluation of potential human health, environmental, and social impacts of the material flows (resources acquired from the environment and releases to the environment) identified during the Life Cycle Inventory. That is, an LCIA attempts to establish a relationship between a product system and risk of harm to human health, the environment, and society. Other risks and impacts may be included if identified during problem formulation. The LCIA approaches this role by looking at each resource acquired and each release to the environment and assessing its impact relative to a "standard" material.

This is best illustrated by considering the impact of a product system on global warming. Carbon dioxide is the primary gas contributing to global warming. Methane also contributes to global warming and is approximately 22 times more potent than carbon dioxide. That is, 1 kilogram (kg) of methane has the same global warming impact as 22 kg of carbon dioxide. A product system might add carbon dioxide and methane to the atmosphere through incomplete combustion of natural gas. By converting the mass of methane released to carbon dioxide equivalents (multiplying the mass by 22) and adding the mass of carbon dioxide released, a global warming potential (GWP) equivalent to so many kg of carbon dioxide released can be calculated. This approach can be taken for all resources acquired and releases to the environment for a basic set of impacts. Table 10-2 summarizes some commonly used impact indicators.

Note that the aquatic toxicity and human health characterizations used here are not equivalent to the assessments made in Chapters 7 and 8. LCIA aggregates the total mass of hazardous substances released to the environment without consideration of exposure pathways available at each point of release. Indeed, due to the spatial scales of LCA datasets and the number of chemicals being assessed simultaneously, the mass data are often divorced from any location and concentration data, so no assessment of risk is possible. This may change in coming years as spatial representations of both LCI and LCIA data and methods for developing these representations are improving, particularly for air and water emissions. These improvements are due to efforts such as ImpactWorld method or the USEtox fate-exposure-effect model. Until these become commonplace, however, as reported in the LCIA, the human health and aquatic toxicity characterizations are directional indicators of the mass of hazardous materials released to the environment; in no way do they consider the actual risk of harm from the releases. An extension of this point is that in comparing two product systems, releases with a local effect, such as human or aquatic toxicity, are best handled by LCT and evaluations of risk to human and aquatic health as described in Chapters 7 and 8.

More generally, some of the releases identified using LCIA, such as greenhouse gases (GHGs), will have global impacts. Others, such as oxides of sulfur or nitrogen, will have regional impacts. Still others, such as inherently toxic chemicals, will have local impacts. Each field of impact (global, regional, or local) needs to be evaluated differently. Releases with global impacts, such as GHGs, may be aggregated over a product's life cycle because it is the global atmospheric concentration of GHGs that is of concern, not the concentration at the point of origin. In contrast, releases with only a local impact should be identified using LCT and the relative risk of harm assessed using the methods described in Chapters 7 and 8.

Finally, the choice to proceed from an LCT to an LCIA would likely only be warranted if additional information is required to resolve trade-offs to reach a substitution decision. If screening LCAs or LCT can provide sufficient insight to inform trade-off resolution as part of the substitution decision, it may not be necessary to conduct system-specific, ISO-compliant Life Cycle Impact Analyses.

Social Impacts

The committee acknowledges that an alternatives assessment may consider social impacts of a chemical choice. In contrast to other frameworks that considered social impacts separately from other life cycle impacts, the committee considers social impacts as part of the life cycle assessment because LCT and LCIA methods increasingly integrate social impacts (Jorgensen 2008).

Many factors leading to production and disposal may differ between the chemical of concern and the potential alternatives, including the routes and methods for acquiring the raw materials needed for production, the sites and methods of manufacture, and the availability of disposal methods. These differences may result in differential social impacts, and a company may wish to compare the effect of choosing a given chemical on, for example, workers' rights and safety, community rights, and rights of indigenous peoples. These issues are typically associated with developing economies, but areas of concern are found in developed countries as well. Because social impacts may occur at any point in the life cycle of a product, identifying the possible occurrence of social impacts requires a life cycle approach similar to that used when assessing possible risks to human health and the environment that occur at a time or place beyond the point of use or application. For this reason, the committee advises considering environmental life cycle impacts and social life cycle impacts concurrently rather than separately.

The committee does not recommend a specific set of social impacts to be considered. Rather, those impacts should be decided between the entity authorizing the alternatives assessment and its stakeholders during problem formulation early in the assessment process (Step 2). Table 10-3 summarizes social impact categories and possible characterization factors that may be considered.

Identifying and Managing Consequential Impacts

Life cycle considerations are, by their nature, complex. LCIs produce a large number of outputs, and it is rare for one product system to show advantage over another product system for every impact indicator. This reality strongly argues for the entity authorizing the alternatives assessment and affiliated stakeholders to identify, prioritize, and document life cycle considerations during problem formulation (Step 2) of this framework. It may also be necessary to use an integration approach similar to that described in Chapter 9 to determine whether one alternative is preferred over another.

The committee also notes that life cycle differences are primarily relevant if they are inherent

TABLE 10-3 Typical Social Impact Categories and Possible Characterization Factors

Social Impact Categories	Possible Characterization Factors
Human rights	Non-discrimination, including indicators on diversity, such as composition of employees on all levels according to gender, age group, disabled, part-time workers, and other measures of diversity Freedom of association and collective bargaining Child labor, including hazardous child labor Forced and compulsory labor
Labor practices and decent work conditions	Wages, including equal remuneration on diverse groups, regular payment, length and seasonality of work, and minimum wages Benefits, including family support for basic commodities and workforce facilities Physical working conditions, including rates of injury and fatalities, nuisances, and distance to workplace Psychological and organizational working conditions, such as maximum work hours, harassments, vertical, two-way communication channels, health and safety committees, job satisfaction, and worker contracts Training and education of employees
Society	Corruption, including incidents/press reports concerning fraud, corruption and illegal price-fixing, and violation of property rights Development support and positive actions toward society, including job creation, support of local suppliers, general support of developing countries, investments in research and development, infrastructure, and local community education programs Local community acceptance, such as complaints from society and presence of communication channels Ensuring commitment to sustainability issues from and toward business partners
Product responsibility	Integration of customer health and safety concerns about the product, such as content of contaminants/nutrients, other threats/benefits to human health (including special groups) due to product use, and complaint handling system Information about the product to users, such as labeling, information about ingredients, origin, use, potential dangers, and side effects Marketing communications, such as ethical guidelines for advertisements

SOURCE: Adapted from Jørgensen et al. 2008.

to, or otherwise directly associated with, the specific alternatives. For example, if generic databases are used as sources for the global warming potential associated with producing certain alternatives, and those data show an apparent difference, care should be taken to understand if the differences are based on factors inherent to the manufacturing process (such as a process that requires an elevated temperature) or due to where the substance may have been made at the time the data was collected (e.g., a country with coal-generated electricity vs. a country with wind-powered electricity). Differences that are not inherent or directly linked to a particular alternative may be of limited value in differentiating between alternatives, especially if those differences are the primary or only differences between alternatives. Fortunately, most life cycle analysts are familiar with these concerns and should be able to identify meaningful differences for the purpose of an alternatives assessment.

Conclusions about Including Life Cycle Considerations

Clearly, performing an LCIA adds significant effort, time, and cost to an alternatives assessment. Therefore, the decision to proceed with such an assessment should be based on a clear need. Need, or lack thereof, can be demonstrated by LCT and identification of significant differences between product systems.

There are no hard and fast rules that prescribe when such an assessment should proceed and when it can be avoided. The scope of the alternatives assessment, as defined by stakeholders during the problem formulation step, should ultimately determine this choice. Regardless of the decision, the basis for including or excluding an LCIA should be clearly documented.

Performance and Economic Factors in the Committee's Framework

The performance and economics of alternatives are primary considerations in substitution decisions. A substance will often be considered a possible alternative because it has already been used to provide the needed function, but if it is not known whether performance and economic criteria are met, then additional analyses will often be desirable. See Chapter 4 for a discussion of these concerns. Chapter 11 has a discussion of pilot testing as a means to evaluate unintended performance and health and safety impacts during the implementation phase of an alternative.

The elements of performance and economics are specific to the substance being evaluated and to its application. For example, when considering a chemical substitution for a flame retardant used in polymeric electronics housings, the final product must meet flame retardant requirements for each jurisdiction in which the product is sold. Typically, a range of acceptable performance and economic requirements will exist for products performing the same function. For example, some products are available in a "premium" format that offers higher performance at an increased price, and an "economy" format that offers lesser performance at a lower price. The range of cost-performance options that need to be considered is often based on internal and external stakeholder input and assurances that a range of customer needs are being met. Engaging direct customers or downstream users may be necessary to understand the critical functions or functionality and economics of a product.

Performance Assessment (Step 9.2)

A product provides specific functionality under a defined set of conditions. Customers for a product expect and often require that alternatives are favorable for the desired application and that they meet certain performance requirements. Often, customers expect a "drop in replacement," or a functionally identical product when considering an alternative. This expectation is often hard to achieve and may require additional discussion and deeper understanding of the customers' needs and expectations. There also may be additional specifications that the product must meet before it can be approved or used. Most companies understand the need to test their products before commercialization using internal testing regimens or consensus standards and methods, such as those published by ASTM International, ANSI, ISO, and others.

Economics Assessment (Step 9.3)

Although the statement of task did not require the committee to directly address economic factors in its framework, understanding the potential financial impacts of alternatives is important in most substitution decisions. It should be noted that economic assessments are not a requirement of the committee's framework since there may be situations in which financial analyses cannot be completed. For this reason, economic analyses are considered an optional step in the framework. In cases where an economic assessment is required by regulators, as with CA SCP or REACH legislation, then obviously this option must be exercised. However, there may be times when the user conducting the alternatives assessment is different from the entity that will be executing the substitution, so there may be insufficient financial information for a thorough evaluation at this stage in the assessment. This situation could arise when an alternatives assessment is being conducted by a regulator, a consortium, or a public-private partnership. In these cases, or any time financial information is not immediately needed or available, economic analyses may be deferred to later stages of the assessment or delegated to users of the final report.

Chemical substitution in a product is expected to have an economic impact, since most supply chains have been optimized to minimize cost. Thus, the most likely economic impact of a chemical substitution will be an increase in the cost of materials or retooling of manufacturing equipment to accommodate the alternative. The cost of materials is one of several factors contributing to the cost of a product (cost of goods sold, or COGS). Direct labor costs, direct energy costs, equipment costs, and other direct costs also contribute to the total cost. Any price increase in COGS for the final product will be the cost differential between the cost of the alternative and the cost of the chemical of concern. This is an important consideration because the economic viability of a product is typically measured in margin percent, the price minus the COGS divided by the price, times 100. Thus, if an ingredient represents 10% of the cost of a product and an alternative costs double that amount, the product cost will increase by 10%; it will not double.

Other production costs, such as increased processing time and energy, may also factor into the economics of the substitution. For example, a less reactive monomer may have a longer cure time in a reactor. This would reduce the productivity of the reactor (less product per hour) and increase the product cost. However, these costs can only be known after prototype products are made and evaluated, which is beyond the scope of this committee's charge.

This simple analysis reflects the comparative costs of materials for a given substitute, assuming that it is a one-for-one "drop-in" replacement, where no other changes in the final formulated product are required. For consumer products, drugs, materials, plastics, and other items of commerce, which are highly formulated, the cost and time required for reformulation to accommodate the substitute may be considerable. While the simple analysis is a useful illustration of the concept, a total economic analysis would be needed to include the costs and time to re-formulate a final product and, depending on the product, any reregistration costs that may be required. This broader analysis could also include consideration of indirect costs, such as those of waste and end-of-life management and potential medical costs. As described in the summary of other frameworks earlier in this chapter, in some cases, these analyses might be required as part of local or state regulatory requirements.

The committee acknowledges that some manufacturers consider an increased cost of goods as an impediment to substitution. In contrast, these same economic considerations may also stimulate development of novel innovations by other entities (see Chapter 13). Most companies, however, manage increased material costs by looking at their product holistically, and adjusting other costs, margin expectations, and price to offset the cost increases (and concomitant benefits) of a chemical substitution. In addition, over time, an initially more expensive chemical or material may become more cost competitive as the supply chain adjusts.

Another approach companies use to calculate the worth of a product innovation is Net Present Value (NPV). NPV is based on cash flow to the company over time (based, for example, on sales of a product), and calculates the equivalent amount of capital needed to produce that same cash flow at an assumed internal rate of return (IRR). If the investment to bring the product to market is less than the NPV, then the product is economically desirable. An obvious disadvantage of the NPV approach is that no consideration is given to the loss of value caused by harm to human health, the environment, or society, nor is consideration given to liabilities associated with managing restricted hazardous substances.

An example of a cost-effective substitution that may not have occurred if an NPV analysis had been conducted is one company's substitution of a surfactant in laundry and dish products to eliminate a carcinogenic byproduct. The company's product contained sodium lauryl ether sulfate (SLES), an anionic surfactant used in some laundry and dishwashing products, as well as for other applications. During production of SLES, a by-product, 1,4-dioxane, is formed. The World Health Organization and the NTP have categorized 1,4-dioxane as a possible human carcinogen. In this scenario, the company chose to eliminate 1,4-dioxane in its products by replacing SLES with sodium lauryl sulfate (SLS), which does not contain 1,4-dioxane. At considerable investment, the company successfully formulated a higher-performance product that could be produced at a lower cost than the original formulation.

Subsequently, intense pressure from consumer and environmental advocacy groups, and a law-suit by the State of California, forced conventional companies to limit the presence of 1,4-dioxane in their consumer products. Thus, though there was considerable initial outlay of funds to develop the alternative formulation, ultimately the substitution avoided liability, improved performance, and lowered the COGS for the company. A simple NPV analysis at the outset of the process may not have identified these potential future financial benefits to the company.

Conclusions on Performance and Economic Considerations

The committee's framework does not require a performance assessment to support a substitution decision because the entity requiring the alternatives analysis may not be a commercial entity, and therefore may not have the ability to prototype and test alternatives. However, it is likely that the substitution decision will eventually affect a commercial entity, which will conduct performance tests to ensure that its products meet user needs, industry standards, and regulatory requirements. Companies routinely perform such tests when innovating new products, and the committee expects they will do so when implementing a chemical substitution.

Similarly, the committee's framework also does not require an economic assessment to support a substitution decision because the entity authorizing the alternatives analysis may not be a commercial entity, and therefore may not have access to the information necessary to support an economic analysis. However, it is likely that the substitution decision will eventually affect a commercial entity, which will conduct economic analyses to ensure that its products meet user needs, industry standards, and regulatory requirements at a commercially viable price. As with a performance evaluation, companies routinely perform such economic analyses when innovating new products, and the committee expects they will do so when implementing a chemical substitution.

11

Identifying, Comparing, and Implementing Alternatives

The final steps (Steps 10, 11, and 12) of the framework integrate information from previous evaluations in order to identify acceptable alternatives, compare alternatives to make a decision (an optional step), and implement selected alternatives. By Step 10, assessors should have sufficient information to determine which, if any, of the potential alternatives have a lower overall negative impact to human health, ecotoxicity and other considerations, as well as meet other requirements established in Step 2. Figure 11-1 shows where these steps fall in the framework, and Figure 11-2 provides more information about what is involved in Steps 10-12.

IDENTIFYING ACCEPTABLE ALTERNATIVES WITHIN EXISTING FRAMEWORKS

All of the reviewed frameworks integrate information across different domains to identify acceptable alternatives, but they give varying levels of guidance on how to do this. The CA SCP and REACH frameworks set acceptability criteria at the beginning of the chemical alternatives assessment process and then measure alternatives against those criteria. The UCLA MCDA framework provides a structure for integrating information from different domains, but focuses on ranking alternatives rather than determining alternatives' acceptability. Both TURI and UNEP do not give specific guidance on determining acceptability, but they both demonstrate within their case studies how to organize disparate data in matrices, as well as how to use simple markers (e.g., +, -, or =) to denote better or worse performance against the chemical of interest in each criteria (Table 11-1).

The EPA's DfE framework relies on the stakeholders participating in an assessment to evaluate certain aspects of the alternatives, and leaves the integration of disparate data to the individual companies implementing the alternatives. The BizNGO, German Guide, and Lowell frameworks provide little or no guidance on how to integrate the disparate information from different steps in the alternatives assessment to determine the acceptability of alternatives.

STEP 10: IDENTIFYING ACCEPTABLE ALTERNATIVES IN THE COMMITTEE'S FRAMEWORK

Consistent with the reviewed frameworks, a step to identify acceptable alternatives based on information from different domains has also been included in the committee's framework. Inclusion of this step is also aligned with the committee's Statement of Task, which states that the framework should be able to consider the full range of benefits and shortcomings of substitutes, including balancing factors such as product functionality, product efficacy, process safety, and resource use. It is beneficial to retain a dedicated step for determining basic acceptability, without forcing a ranking or further narrowing the list of alternatives, because having more than one acceptable alternative may be desirable under certain circumstances. For example, if the entity performing the alternatives assessment is a regulator considering taking action on the chemical of interest, offering a range of alternatives to replace the chemical allows complex industries and supply chains the flexibility to select the option that best suits each company's needs. Also, alternatives assessments that identify multiple acceptable alternatives can spur innovation if alternatives that have minor shortcomings in certain areas in the initial assessment can be further developed so that they become preferred replacements. And finally, if irresolvable issues are encountered during the implementation of a selected alternative, it may be useful to have other alternatives that have been identified as acceptable to consider.

While Step 7 has the goal of integrating information about the potential human health and ecotoxicity impacts of the alternatives to determine if alternatives meet the definition of *safer*, Step 10 has the goal of integrating the additional disparate data from Steps 8 and 9 to determine which

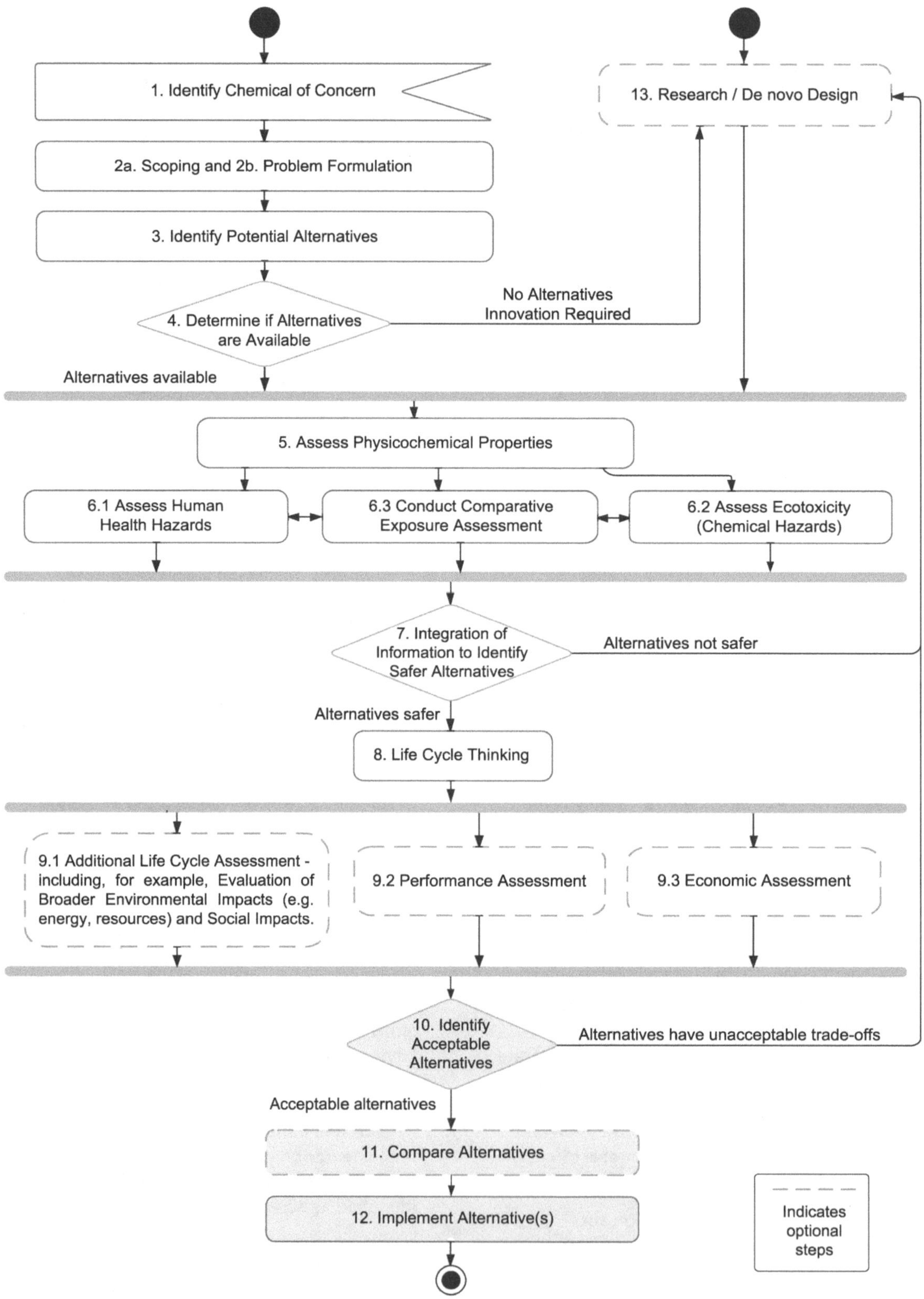

FIGURE 11-1 Committee's framework highlighting steps to identify, compare, and implement alternatives.

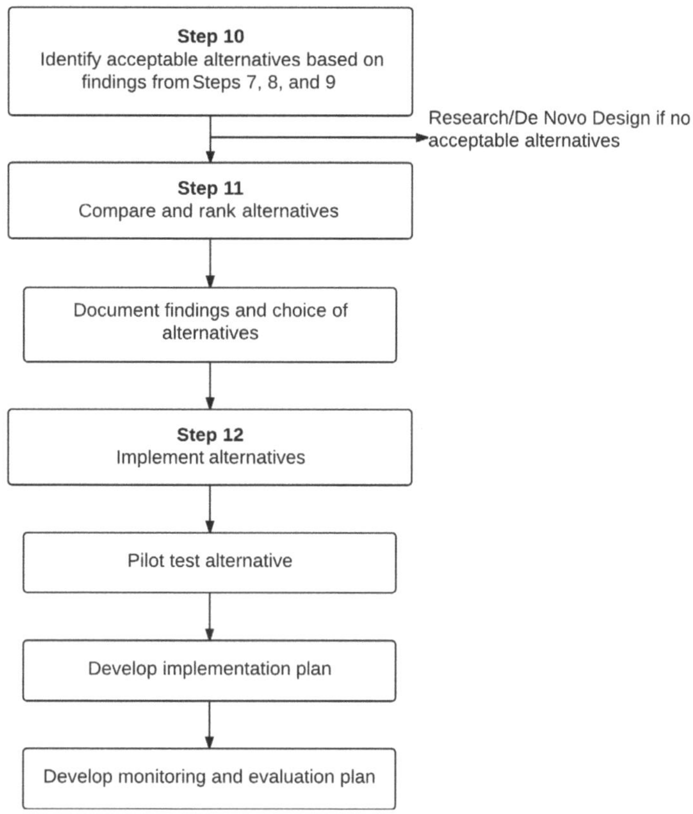

FIGURE 11-2 Additional detail about Steps 10-12.

alternatives, if any, are *acceptable*. For the purposes of this framework, an alternative is considered *acceptable* if it meets the requirements established in Step 2, and does not have undesirable aspects or trade-offs so that it no longer has a lower overall negative impact to human health and/or the environment. This definition of acceptability depends on the requirements set in Step 2 as opposed to factors that entities may simply have a *preference* for, because this step is focused on identifying *acceptable*, not preferred or optimal, alternatives. For example, companies will generally have a preference for lower cost alternatives, but unless a clear requirement is set (such as a maximum price that the entity will consider), the preference for lower cost should be addressed as a part of comparing alternatives in Step 11.

By Step 10, assessors should have sufficient information to determine which, if any, of the potential alternatives can be considered acceptable. Coming out of Step 7, each alternative will have been assessed to determine how it compares to the chemical of concern in the original domain of concern and environmental and human health hazards, as well as exposure. Results from Steps 8 and 9.1 may provide additional information about the broader potential environmental impacts of the alternatives, as well as their constituents and breakdown products. This information should be used to determine if the remaining alternatives continue to meet the requirement to have lower overall negative impact to human health and/or the environment.

Another important aspect of Step 10 is that it is a critical point for documenting the findings of all of the analyses that have been performed throughout the assessment, as well as documenting any monitoring or other measures that may be required to make particular alternatives acceptable. As noted in Chapter 3, thorough documentation of findings allows for more effective critical evaluation of alternatives assessment results and comparability across assessments. The organization of the reports and documentation is left to the discretion of the

TABLE 11-1 Example of a Summary Matrix for Multiple Alternatives across Several Criteria in a Case Study based on the TURI Framework

Assessment Criteria		Lead (Referenced)	Comparison Relative to Lead: Bismuth	Ceramic	Steel	Tin	Tungsten
Technical and Performance Criteria	Density	11.34 g/cm³	-	-	-	-	+
	Hardness (desirable for "feel" and noise)	Soft Mohrs: 1.5	+	+	+	= (pure) + (alloy)	+
	Malleability (split-shot application)	Yes	-	-	-	=	-
	Low melting point (for home production)	622°F	+	-	-	+	-
	Corrosion resistant	Yes	=	=	-	=	=
Environmental Criteria	Highly toxic to waterfowl	Yes	+	?	+	+	+
	Toxic to aquatic species	Yes	+	?	+	+	+
	Primary drinking water standards (MCL Action Level)	15 µg/L	?	?	+ (iron)	+ (FL & MN)	?
Human Health Criteria	Carcinogenicity	EPA B2 IARC 2B	+	+	+	+	+
	Developmental toxicity	Yes (Prop 65)	+	+	+	+	+
	Occupational exposure: REL (8-hour TWA)	0.050 mg/m³	?	+	+	+	+
Cost	Retail price	Low	-	-	-/=/+	-	-
	Availability of end product	Excellent	-	-	-	-	-

Note: + Better = Similar - Worse ? Unknown
SOURCE: Adapted from TURI (2006).

assessor, but summary tables or other graphic methods should be used to compile and present results for multiple alternatives against multiple criteria.[47]

If no alternatives are determined to be acceptable at the conclusion of Step 10, research can be initiated to develop new alternatives and/or improve existing ones, a process informed by observations about how each alternative failed to meet the requirements established in Step 2 or the expected negative impacts to human health and the environment that were considered unacceptable to the entity conducting the alternatives assessment.

STEP 11: COMPARING ALTERNATIVES IN THE COMMITTEE'S FRAMEWORK

If a single alternative must be selected for implementation, or if it is necessary to identify

47 Similar requirements are also found in the CA SCP, REACH, TURI, and UNEP frameworks.

preferred alternatives, ranking or other comparative methods may be applied to the alternatives identified in Step 10. Additional information about this optional step is provided in the next section.

Comparing Alternatives within Existing Frameworks

Four frameworks (IC2, CA SCP, Lowell, and UCLA MCDA) use information from different domains to evaluate alternatives so that they can be ranked, categorized, or narrowed to a single choice for implementation. The IC2, CA SCP, and Lowell frameworks allow ranking, but give no guidance on specific methods on how this, as well as categorizing or narrowing the choice of alternatives, should be done. As a result, the choice of approach is left up to the discretion of assessor. The UCLA MCDA framework deals more comprehensively with ranking. The framework referred to as UCLA MCDA is actually a specialized form of the more general approach of *decision analysis*, which is a field that applies decision theory to real-world, complex problems. For this reason, it is well suited for integrating disparate information for each alternative and evaluating that information against multiple criteria (Siddall 1972; Keeney and Raiffa 1976; Triantaphyllou 2000; Wang 2002; Figueira et al. 2005; Hatamura 2006; Edwards et al. 2007). Applying MCDA methods requires the creation of a model that reflects the decision maker's preferences, value trade-offs, and goals (Belton and Stewart 2002). The UCLA report *Developing Regulatory Alternatives Analysis Methodologies for the California Green Chemistry Initiative* (Malloy et al. 2011) demonstrates how such an approach could be applied within a chemical alternatives assessment.

In the UCLA report, two case studies are presented in which an MCDA model created to compare a regulated hazardous substance and its alternatives is used to analyze alternatives to perchloroethylene (PCE) for dry cleaning and lead (Pb) solders in electronics. The variables for the model were first selected from the human health, environmental, resource usage, performance, and economic factors that must be evaluated under California Assembly Bill AB 1879, the enabling statute for CA SCP. For each major area of interest (upper-level criteria), sub-criteria with metrics against which alternatives could be scored were identified (*measurement sub-criteria*). Weights for the criteria within the model were based on averaged scores of expert and stakeholder ratings of the relative important of the different criteria.

The authors were able to rank the alternatives in both cases using two commonly used MCDA methods: multi-attribute utility theory (MAUT)[48] and outranking.[49] When the authors varied the assigned criteria weights, they found relatively small variations in the rank order using different stakeholder weighting levels. The authors were also able to run the model with different assumptions about missing data (such as assuming missing data were to receive the worst or best possible score for an end point) to see if these differences affected the rank order of alternatives. When they used different assumptions and policies for handling data gaps, they found that different assumptions could result in significant differences in the relative rank of alternatives. The authors also examined the impact of converting continuous data (such as LD_{50}) to categories (high, moderate, low), and found that the rank order of alternatives with respect to top performers was unchanged, but that the remaining alternatives were significantly reordered. Based on the successful application of MCDA methods in the case studies, the authors concluded that MCDA was a viable way to assist in the evaluation of complex data within a chemical alternatives assessment.

Step 11: Comparing Alternatives in the Committee's Framework

A step for comparing alternatives has been included as an option in the committee's framework to address the need to differentiate among acceptable alternatives in order to select a single alternative for implementation or to identify preferred alternatives from the list of acceptable ones.

The decision analysis methods used in the MCDA example are one way to integrate disparate information to rank or differentiate alternatives. Those methods may be most helpful when evaluating complex data across many criteria, for cases with many alternatives, or when the substitution decision is expected to have a high impact. Although MCDA methods may be useful in some cases, they may be more complicated than required for many

48 MAUT is an optimization approach that represents the decision-maker's preferences as utility functions, and attempts to maximize the decision-maker's overall utility.

49 "Outranking models compare the performance of two alternatives at a time, in terms of each criterion, to identify the extent to which one alternative out-performs the other, then aggregates that information for all possible pairings to rank the alternatives based on overall performance on all criteria" (Malloy et al. 2011).

assessments. There are other ways to rank, compare, and select alternatives, including simple matrix methods (such as the one shown in Table 11-1), as well as the decision rules described in Chapter 9.

Ultimately, the choice of integration method is beyond the scope of the committee and is left to the assessor. All assumptions, data, and methods should be documented regardless of the method used. The criteria and weighting used within these decision analysis methods are context-dependent and based on values, and therefore left to the discretion of the assessor or entity conducting the alternatives assessment.

STEP 12: IMPLEMENTING ALTERNATIVES

By the end of Step 11, the assessor will have either identified preferred alternatives or initiated research on de novo green chemistry alternatives. In those cases where acceptable alternatives are identified, the next step is implementation of the selected alternative(s) in particular applications.

Implementing Alternatives within Existing Frameworks

Implementation of alternatives is addressed only to a limited degree in the frameworks reviewed by the committee. Most of the frameworks end with the *selection* of a preferred alternative. CA SCP requires an implementation plan as well as confirmation that a substitution has occurred. Two frameworks, BizNGO and Lowell, contain steps entitled "Select and Implement Safer Alternative" and "Select and Implement/Review Selection," respectively. The Lowell framework states that the final step, Review Selection, reflects the fact that technologies are not perfect in terms of environment and social acceptability. Specific chemical selections will need to be re-visited and re-evaluated over time, based upon emerging science and changing social expectations. Alternatives assessment is an iterative process on the journey towards sustainable technologies (Rossi et al. 2006).

The most detailed attention to implementation is in REACH. In particular, the European Chemical Agency (ECHA 2011) document *Guidance on the Preparation of an Application for Authorization* states that entities seeking an authorization (noting that no feasible alternatives are available) must consider: "What research and development activities are needed and/or planned to develop an alternative substance(s) or technology(ies), or develop equipment or processes enabling the use of alternative(s); and (2) What testing must be done and what criteria need to be satisfied before an alternative can be used for a particular function" (ECHA 2011).

The guidance further documents some of the particular implementation challenges for an alternative that might substantiate a longer substitution transition period, including:

- "The transfer to the alternative requires investments that take considerable time (time needed to plan the necessary changes, to purchase the equipment needed, to build any constructions, to install, to train the personnel, etc.);
- The transfer to an alternative substance requires regulatory approval (e.g., production of aircraft or medical equipment), or change to an alternative technique requires a review of permit;
- The transfer to an alternative requires customer approval (e.g., for use in products that must be tested for technical performance over long time periods, or where the transfer to an alternative up in the supply chain may affect the quality of the end products and testing by several downstream user levels is required);
- An alternative substance is currently not produced in sufficient quantity; and
- Costs related to investment in new equipment/techniques may depend on other planned investments, age of the current equipment, etc." (ECHA 2011).

Under REACH, if an applicant for authorization identifies a suitable alternative, that entity must develop a substitution plan for the alternative, documenting timing, supply chain consultation, and how the transition will occur, including evaluating risk trade-offs. Figure 11-3 provides a graphic of the substitution planning steps under REACH.

Finally, several occupational health chemical substitution frameworks not considered in Chapter 2 include steps focused on implementation and evaluation of the consequences of the substitutions. For example, OSHA's framework, *Transitioning to Safer Chemicals,* has the steps, "Piloting the Alternative" and "Implementing and Evaluating the Alternative," with information on each step (OSHA 2014). The European Commission's Directorate General for Employment, Social Affairs and Equal

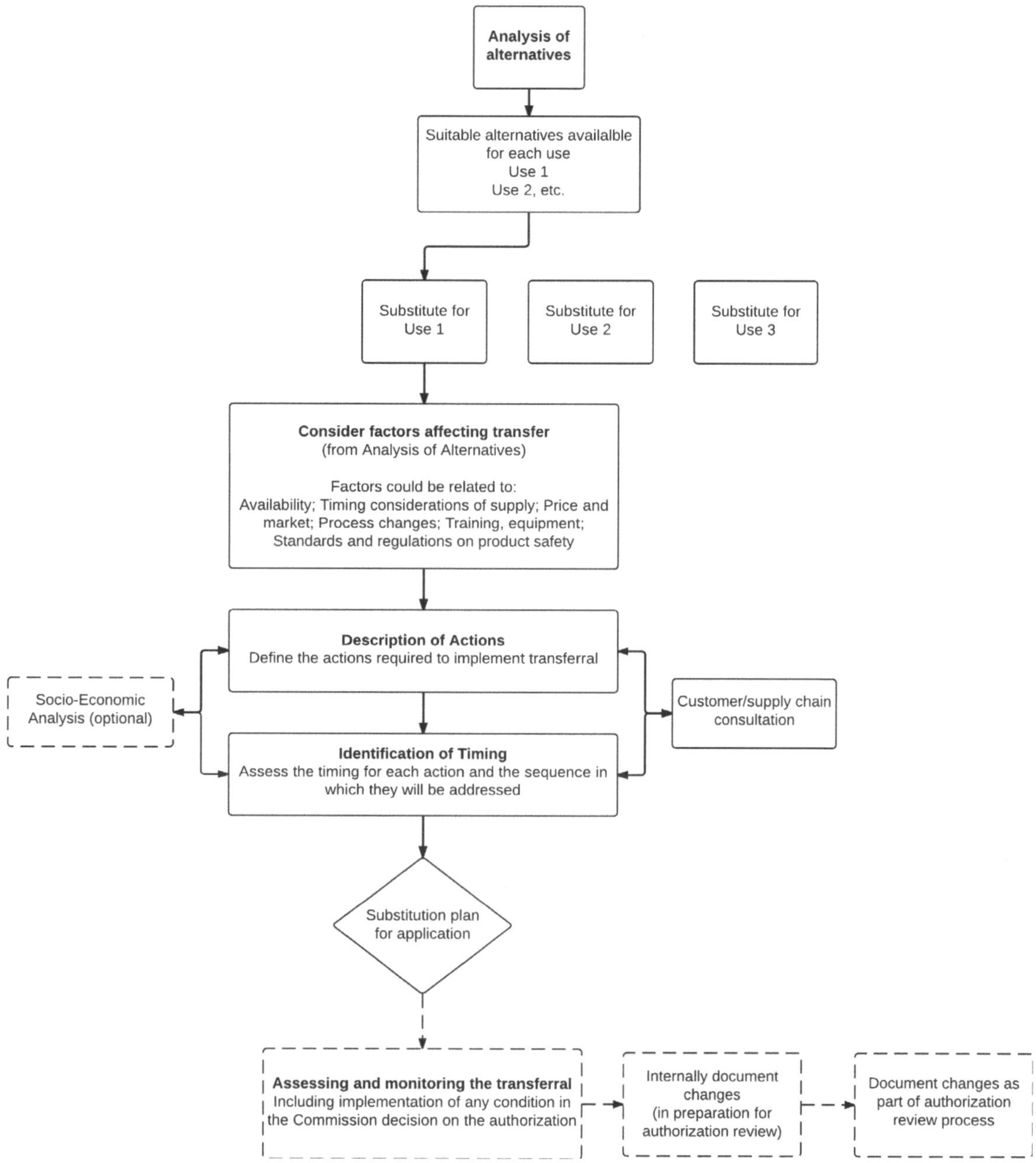

FIGURE 11-3 Steps to follow in a substitution planning process required for companies seeking authorization for a Substance of Very High Concern under REACH. SOURCE: Adapted from ECHA 2011.

Opportunities has published a substitution framework (developed by the Finnish consulting group GAIA) entitled "Guidance for minimizing chemical risk to workers' health and safety and the environment," which includes implementation guidance in a *Plan-Do-Check-Act* approach (Pessala et al. 2012).

Implementation is an underdeveloped topic within chemical alternatives assessments, but one that is critical for minimizing unintended health, environmental, and performance consequences, as well as ensuring continuous improvement in transitioning to safer chemicals and products. Many alternatives assessments only peripherally consider the actual adoption of alternatives and the challenges that might occur either up or downstream of the production process, or the potential unforeseen health and environmental hazards that may be created at this stage. Implementation is often the most challenging part of the substitution process and may require ongoing monitoring to identify and minimize potential trade-offs. It is important to give attention to implementation early in the chemical alternatives assessment, including involving stakeholders affected by a chemical substitution. Such attention can make informed substitution more successful and develop a culture of continuous improvement toward safer processes and products. It can also help reduce the potentially high costs of additional substitutions by identifying potential problems before full-scale implementation.

Implementing Alternatives in the Committee's Framework

An implementation step has been included in the committee's framework, consistent with the best practices in the existing frameworks. This step is intended to support action related to the implementation of safer substitutes by helping entities identify alternatives and mitigate expected or unintended consequences in the substitution process, and ensuring a more successful, informed substitution. Ultimately a chemical alternatives assessment is not worthwhile if the alternatives are not adopted.

An implementation step will prepare for the following challenges that may result:

- Identified acceptable alternatives may work in a specific range of applications but not in others that have specific processing or operating requirements. Or, alternatives may change product functionality. For example, lead free solders may not perform well in high pressure or low gravity applications (NASA 2009).
- Identified acceptable alternatives may require significant process design or formulation chemistry changes to achieve functionality that may not have been considered. These changes may affect product quality or may lead to increased or modified exposures or new hazards.
- Implementing alternatives may require work practice changes that can affect worker exposure pathways, increase potential hazards (toxicity and physical), and affect productivity if they do not work as well (Bartlett et al., 1999).
- Changes may make end of life collection and recycling more challenging or lead to unexpected end of life exposure concerns.
- New understanding about the toxicity of a chemical substitute or a chemical used alongside the substitute in a process or product. New understandings about environmental fate or life cycle may require adjustment of earlier assumptions.
- The large amount of information collected in the evaluation phase, including potential conflicts in information and inertia within a firm or sector to make changes, might lead to paralysis that inhibits action on alternatives adoption.

While some of these challenges will have been addressed in earlier technical and environmental and health and safety evaluations, some particular changes may not have been foreseen and thus encountered for the first time during implementation.

Goal and Objectives of the Implementation Step

The overall goal of this step is to enhance the implementation of safer alternatives while avoiding unintended consequences of substitutions. Planning for implementation supports the transition to safer chemicals, processes, and products and allows for continuous improvement, updating understanding as scientific knowledge evolves on hazards and exposures, and minimizing or avoiding adverse health and ecosystem impacts that might be identified in the application phase of an alternative.

The objectives of the implementation step are to:

a. Document final choices of preferred alternatives, including the rationale and potential information gaps that need to be filled;

b. Identify potential unintended consequences that might occur at the application phase of a substitute and implement modifications to minimize these; and

c. Develop evaluation and continuous improvement plans, including a plan for updating and modifying assumptions and data used in the assessment if substantial new, unanticipated information arises that could affect the evaluation and choice of alternatives.

While the implementation and evaluation step of the committee's framework is primarily focused on avoiding unintended consequences in the application phase of substitutes, there are some overlaps with adoption support,[50] particularly in the areas of pilot trials and greater integration of alternatives assessment processes as a precursor to adoption efforts.

The Implementation Process

Implementation generally consists of the following series of steps, with a strong emphasis on stakeholder engagement:

1. Pilot testing, or small-scale testing of a substitute to identify (a) issues related to performance of alternatives, including process or product modifications that are needed to make the alternative function to specifications; and (b) changes in product or process chemistries or work practices (both in product manufacture or use) that might affect worker or consumer health.
2. Developing an implementation plan, including outlining and documenting the processes and actions needed to implement the substitution, including research and mitigation needs.
3. Monitoring and evaluation, which are essential to the early identification of potential unintended consequences of substitutions and to the documentation of the beneficial impacts of substitutions and potential improvements. Monitoring needs are context dependent and could involve simple measures, such as air and water monitoring or waste audits, as well as workplace industrial hygiene evaluations. It could also include more complex and formal adverse events post-market monitoring, such as formal adverse reporting systems.

These steps can be completed using a pilot testing/supply chain partnership. In this model, trade organizations and/or government or academic research centers (such as the National Institute for Standards and Technology [NIST] or the Massachusetts Toxics Use Reduction Institute [TURI]) work with a sector in a pre-competitive manner or with a particular firm to evaluate the functionality (and, in some cases, the health and safety implications) of alternatives for a chemical of concern. This type of testing is designed to both share the costs of evaluating the concrete application of a substitute to ensure adequate performance in situ, as well as identify process or formulation conditions that might have to change to ensure functionality, such as the use of new solvents that might present health and safety or environmental concerns. This information can then be fed back into a revised chemical alternatives assessment, if necessary.

Another model of implementation is Intervention Research, an occupational health prevention strategy, reflected in the P2OSH framework (Quinn et al. 2006). The P2OSH framework has an iterative series of steps that involve piloting and then implementing alternatives, and exploring how the adoption process might result in changes to materials used, health and safety of workers, direct costs of adoption, and changes to performance. With this information, the company or organization can determine whether full-scale implementation of an alternative should move forward, or whether design, process, or product modifications should be instituted to minimize potential unintended consequences of a substitution.

Ultimately, these steps will not only support action related to implementation of substitutes, but also identify and mitigate expected or unintended health and safety, ecosystem, performance, or economic consequences during the substitution process.

---

[50] The implementation step within the committee's framework is distinct from the concept of "adoption support," which includes the policies (restrictive, purchasing, or other), incentives, technical assistance, and other support provided to businesses to increase the rate of adoption of safer alternatives.

12
Case Studies

To illustrate how the committee's framework can be applied, two case studies are presented in this chapter. The case studies represent different users in contrasting decision contexts with diverse priorities. Case Study 1 was written from the perspective of a fictitious manufacturing company with limited expertise. Case Study 2 is intended to demonstrate how new types of data can be used by a company with sufficient scientific resources.

CASE STUDY 1: CHEMICAL SUBSTITUTION OF A RESTRICTED SUBSTANCE (decaBDE)

In Case Study 1, we present a scenario where the use of a substance, the flame retardant decabromodiphenyl ether (decaBDE), is restricted through regulation, and an alternative must be selected from available chemical and material options that have a range of trade-offs. This case study was written from the perspective of a fictitious company—KayDisplay, a small U.S. manufacturer of specialty displays for retail kiosks. In this scenario, the company wants to expand its market by selling products in the European Union (EU), but its current products contain a substance (decaBDE) that is restricted in the EU and is being phased out in the United States (EPA 2012g). This case study illustrates how a chemical alternatives assessment was conducted by a single company as part of an internal feasibility study to determine whether there are alternatives to using materials with decaBDE in order to be able to sell their products in the EU.

While considering this case study, it is important to note that:

- KayDisplay is a fictitious corporate entity, and has been envisioned as a small company headquartered in Washington State, with limited in-house expertise in chemistry, material sciences, and toxicology.
- The chemical alternatives assessment reflects the internal effort of a single company, and not the more extensive assessments that might be expected of regulators facilitating a multi-stakeholder review of a substance prior to regulatory action.
- Conducting a meaningful chemical alternatives assessment and implementing an informed substitution at a smaller company, like KayDisplay, can only be successful when published information is available. In this particular case, KayDisplay has access to recent multistakeholder and regulator-created alternatives assessments from which to draw.
- The use of tools or modules in this case study should not be interpreted as committee endorsement. Instead, these tools should be viewed as plausible options for an entity to use in this situation.
- The committee's framework will be applied through Step 7 (comparative chemical hazard assessment) and context-dependent steps (Step 8 and beyond) will be described narratively.
- Alternatives to decaBDE have been studied extensively, so this scenario offers a relatively data-rich case through which to demonstrate the committee's framework.

Steps 1- 4 of the Committee's Framework

Step 1: Identify Chemical of Concern

The substance of interest for this assessment is the brominated flame retardant decabromodiphenyl ether (decaBDE). EU legislation restricts the use of certain hazardous substances in electrical and electronic equipment (EC 2003), including decaBDE, and KayDisplay's kiosk displays would be regulated under Restriction of Hazardous Substances (RoHS, Directive 2002/95/EC), if the company were to place these products on the market in the EU.

Step 2: Scoping and Problem Formulation

Electronic hardware put on the market in the EU cannot contain decaBDE or other polybrominated biphenyl ethers (PBDEs) at levels in excess of 1000 ppm in any homogenous material found in the product. As designed, the KayDisplay enclosure is made of a low- gloss blend of polyphenylene ether and high-impact polystyrene (PPE/HIPS), with 15%wt decaBDE added to meet UL V-0 flammability rating requirements.

Step 2a: Scoping

Identify Stakeholders and Determine Their Role

The IC2 includes a "Stakeholder Involvement Module," which KayDisplay will use to consider potential stakeholders. As a small firm, KayDisplay is unable to directly contact regulators, governments, or nongovernment organizations, but will consult with key executives and technical experts within the company, relevant suppliers, and customers. Initial input from stakeholders includes:

- *Company representatives:* Senior leadership and executives support eliminating decaBDE to expand the company's market to the EU. They support selecting alternatives that are not expected to be restricted in the future as long as they are technically and economically feasible. They do not need to be involved in technical or context-dependent assessments, but must approve the final decision.
- *Technical experts*: The primary person responsible for conducting this assessment is the mechanical designer of the enclosure because she is responsible for selecting the material for the parts. Other internal stakeholders will be consulted, including the product managers, procurement engineers, manufacturing engineers, regulatory compliance experts, and product marketing. These inputs will be noted when relevant.
- *Supply chain:* The direct supplier of the plastic enclosure will be consulted to identify potential alternatives and to provide input on performance and economic issues. The supplier does not want to lose KayDisplay as a customer, but the supplier is sensitive to cost and therefore not willing to acquire new capital equipment to support a change.
- *Customers:* KayDisplay's products are sold to companies that assemble kiosks for retail sales (business to business). Key customers in the U.S. were consulted, along with potential EU customers. U.S. customers were most interested in maintaining fire safety and avoiding cost increases. Potential EU customers expect safe, RoHS-compliant products containing no decaBDE, and would prefer that the product qualify for an ecolabel. One ecolabel of interest to KayDisplay's potential customers is the Total Cost of Ownership (TCO), a European sustainability certification for information technology products, including displays. Products must meet several requirements to be TCO certified, including a requirement that plastic parts weighing more than 25 grams must not contain flame retardants or plasticizers with organically bound bromine or chlorine (TCO Development AB 2012).

Goals, Principles, Decision Rules and Constraints

As a small company in a competitive market, KayDisplay is under significant cost pressure, so it must minimize cost increases. However, the company understands that the current solution is highly cost-optimized, so it may not be possible to bring in a new material or design at cost parity. If there must be a material or process cost increase to meet the new requirement, the company will favor alternatives that offer a performance or aesthetic improvement, which could potentially be used to market the product at a higher price point to compensate. KayDisplay would prefer to use the same design for both the U.S. and EU markets to minimize costs and to increase inventory flexibility.

Based on EU customers' heightened interest in health and environmental issues, as well as executive support for reducing the risk of future regulations, the product team will attempt to include options that could meet the criteria to earn TCO Display 6.0 certification. However, if cost targets cannot be met within the ecolabel requirements, RoHS-compliant halogenated alternatives may also be considered.

KayDisplay has not conducted a formal alternatives assessment before and has no established principles or policies to guide the assessment. Through an internet search, it was able to locate several sets of principles from which to choose. The product team found a set that aligned with company values and included reducing hazard, minimizing exposure, using the best available information, requiring disclosure and transparency, resolving trade-offs, and taking action. The company will use a "missing data neutral" approach and not assume missing data would receive either the worst or best possible score for an end point or criterion.

As a small company, KayDisplay relies on guidance from outside experts to complete some of

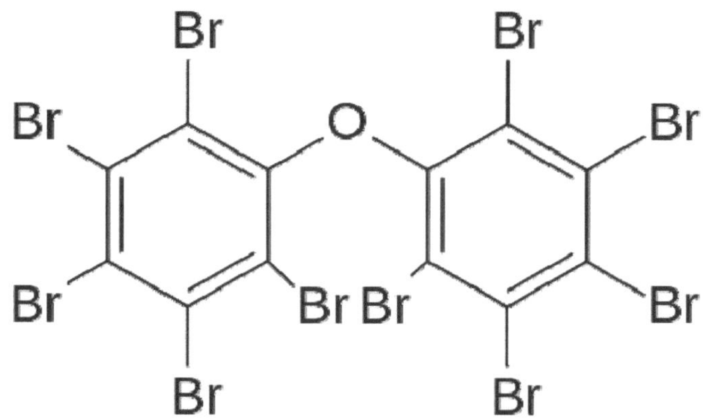

FIGURE 12-1 Chemical structure of decabromodiphenyl ether (decaBDE), CAS number 1163-19-5.

the analyses in the chemical alternatives assessment because it does not have experts on certain tools or methods on staff.

Step 2b: Problem Formulation

Gather Information on Chemical of Interest

Since KayDisplay does not have a chemist or toxicologist on staff, the company is dependent upon published information to gather information about the substance of interest. Fortunately, decaBDE has been studied extensively. The team was able to gather the following information about decaBDE:

Identifying the Chemical. DecaBDE has been identified and described in previous publications. According to Lassen et al. (2006):

- "DecaBDE is a polybrominated diphenyl ether (PBDEs), a group of aromatic brominated compounds in which one to ten hydrogens in the diphenyl oxide structure are replaced by bromine."
- "Decabromodiphenyl ether, or Deca-BDE, as indicated by the name, has ten bromine atoms attached to the diphenyl oxide structure and a bromine content of 82%-83%. It is used as a flame retardant" (Figure 12-1).
- "The CAS No (chemical identification number) of decabromodiphenyl ether is 1163-19-5. The substance is also known as decabromodiphenyl oxide (DBDO) or bis(pentabromophenyl) ether."
- "Three different PBDEs have been commonly commercially available. They are referred to as penta-, octa- , and decabromodiphenyl ether, but each product is, in fact, a mixture of brominated diphenyl ethers."
- The commercial product decaBDE may contain up to 3% of other PBDEs, mostly nonabromodiphenyl ether.

Function and Application and Performance Requirements. DecaBDE is an additive flame retardant:

- Flammability rating: In the U.S., V-0 grade plastics are required for display enclosures. Although the EU has less stringent requirements, the same products will be sold in both markets, so the flammability rating for the alternative materials must be V-0 at 1/16 inch thickness (Lassen et al. 2006).
- Mechanical properties: The alternative must meet or exceed current mechanical properties and performance as listed in the datasheet for the PPE/HIPS resin (Table 12-1).
- Manufacturing: The plastic enclosure parts are injection-molded. Significantly changing the material or using another resin might require new molds. The injection molding supplier would charge KayDisplay for any significant process changes, as well as the non-recurring engineering (NRE) expense of the new molds. Information about the costs associated with mold and process changes are important and would be used for economic analysis. Table 12-2 presents characteristics of the current injection mold process.

TABLE 12-1 Mechanical Properties for the PPE/HIPS Resin Used in KayDisplay's Kiosks.

Mechanical	Value	Unit
Tensile Stress, yld, Type I, 50 mm/min	540	kgf/cm^2
Tensile Stress, brk, Type I, 50 mm/min	490	kgf/cm^2
Tensile Strain, yld, Type I, 50 mm/min	5.1	%
Tensile Strain, brk, Type I, 50 mm/min	40	%
Tensile Modulus, 5 mm/min	24400	kgf/cm^2
Flexural Stress, yld, 1.3 mm/min, 50 mm span	860	kgf/cm^2
Flexural Modulus, 1.3 mm/min, 50 mm span	22400	kgf/cm^2
Tensile Stress, yield, 50 mm/min	51	MPa
Tensile Stress, break, 50 mm/min	48	MPa
Tensile Strain, yield, 50 mm/min	4.2	%
Tensile Strain, break, 50 mm/min	40	%
Tensile Modulus, 1 mm/min	2200	MPa
Flexural Stress, yield, 2 mm/min	77	MPa
Flexural Modulus, 2 mm/min	2200	MPa
Hardness, H358/30	95	MPa
Hardness, Rockwell R	116	-
IMPACT	**Value**	**Unit**
Izod Impact, notched, 23°C	16	cm-kgf/cm
Izod Impact, notched, -30°C	11	cm-kgf/cm
Instrumented Impact Total Energy, 23°C	428	cm-kgf
Izod Impact, notched 80*10*4 +23°C	11	kJ/m^2
Izod Impact, notched 80*10*4 -30°C	7	kJ/m^2
Charpy 23°C, V-notch Edgew 80*10*4 sp=62mm	14	kJ/m^2
Charpy -30°C, V-notch Edgew 80*10*4 sp=62mm	7	kJ/m^2
THERMAL	**Value**	**Unit**
Vicat Softening Temp, Rate B/50	140	°C
HDT, 1.82 MPa, 3.2mm, unannealed	117	°C
CTE, -40°C to 40°C, flow	9.2E-05	1/°C
CTE, -40°C to 40°C, xflow	9.5E-05	1/°C
CTE, -40°C to 40°C, flow	9.2E-05	1/°C
CTE, -40°C to 40°C, xflow	9.5E-05	1/°C
Ball Pressure Test, 125°C +/- 2°C	Passes	-
Vicat Softening Temp, Rate B/50	139	°C
Vicat Softening Temp, Rate B/120	142	°C
HDT/Bf, 0.45 MPa Flatw 80*10*4 sp=64mm	133	°C
HDT/Af, 1.8 MPa Flatw 80*10*4 sp=64mm	117	°C
PHYSICAL	**Value**	**Unit**
Specific Gravity	1.06	-
Density	1.06	g/cm^3
Water Absorption, (23°C/sat)	0.23	%
Moisture Absorption (23°C / 50% RH)	0.06	%
OPTICAL	**Value**	**Unit**
Gloss, untextured, 60 degrees	20	-

TABLE 12-2 Physical Properties for the Injection Mold Process Used by KayDisplay's Current Supplier

Mold Shrinkage, flow, 3.2 mm (5)	0.5 - 0.7	%
Melt Flow Rate, 280°C/5.0 kgf	8	g/10 min
Melt Volume Rate, MVR at 280°C/5.0 kg	8	cm^3/10 min
Drying Temperature	70 - 90	°C
Drying Time	2 - 3	hrs
Melt Temperature	265 - 285	°C
Nozzle Temperature	260 - 280	°C
Front - Zone 3 Temperature	260 - 285	°C
Middle - Zone 2 Temperature	240 - 260	°C
Rear - Zone 1 Temperature	200 - 220	°C
Hopper Temperature	60 - 80	°C
Mold Temperature	40 - 70	°C

<u>Human Health and Environmental Effects, Exposure Pathways, and Life Cycle Segments.</u>

- Hazards. The human health impacts, environmental impacts, and exposure pathways associated with PBDEs are well established. PBDEs are persistent, they bioaccumulate, and are of high concern to human health because they adversely affect the endocrine (e.g., thyroid) system and neurological development (de Wit 2002). Studies have demonstrated that decaBDE breaks down into more toxic PBDEs through photodegradation, microbial degradation, and metabolism (Rossi and Heine 2007). DecaBDE is an additive flame retardant (not reacted into the polymer molecule), so it can leave the material under certain conditions and enter the environment. People are exposed to PBDEs through inhalation, ingestion and dermal absorption of dust particles in the air where electronic products are installed and used (Johnson-Restrepo and Kannan 2009). Occupational exposure occurs through the same routes, but at higher concentrations at locations producing PBDEs or formulations containing PBDEs, plastic component manufacturing facilities (such as injection molders), and electronics waste recycling and disposal facilities.

- Regulations. Although this assessment is focused on decaBDE as the substance of interest, no other PBDEs can be considered as possible replacements because they are also restricted by the RoHS Directive.

Determining Assessment Methods

For this Case Study, Steps 1 through 7 will be completed in their entirety to demonstrate the framework. Actions planned for Steps 8 through 12 will only be described narratively.

- Step 3 (identify potential alternatives) will be completed through consultation with the current supplier of the plastic injection molded parts and online and offline literature searches.

- Step 5 (assess physicochemical properties) will be completed through literature searches, relying heavily on the EPA's 2014 DfE report entitled, *An Alternatives Assessment for the Flame Retardant Decabromodiphenyl Ether (DecaBDE)* and in accordance with guidance provided in Chapter 5.

- Step 6 (assess human health hazards, assess ecotoxicity, and conduct comparative exposure assessment) will be completed through literature searches, relying heavily on the DfE's DecaBDE alternatives assessment, as well as guidance presented in Chapters 6-8.

- Step 7 (identify safer alternatives) will be completed using the GreenScreen® for Safer Chemicals tool, with a preference for choosing alternatives that are Benchmark 2 or better. GreenScreen® assessments may be supplemented with additional investigations, if needed. Data gaps will be handled in accordance with the GreenScreen® guidelines.

Step 8 and beyond will not be executed as part of this case study, but to complete the exercise of fully planning the assessment, the following steps and tools will be selected:

- Step 8 (Life Cycle Thinking) would be completed as described in the "Life Cycle Module" of the IC2. Published life cycle assessments would be used to understand the contribution of the housings to the overall environmental impacts of display products. Findings from Step 8 could trigger additional life cycle investigations (Step 9.1) and/or exposure assessments (Sub-step 6 of Step 6.3).
- Step 9.2 (performance assessment) would be completed by screening materials based on properties on their respective datasheets, by prototyping enclosure parts in the alternative materials, and subjecting the prototype parts to standard inspection and qualification tests. Flammability ratings may be verified. The "Performance Module" of the IC2 may be consulted for additional considerations.
- Step 9.3 (economic assessment) would be completed to assess the internal costs and benefits of different options, including changes in material cost, manufacturing costs and NRE charges, costs of compliance for RoHS (such as analytical testing to prove compliance), costs of certification for the TCO ecolabel, and potential market benefits from improved environmental features (such as having ecolabel certification), performance, and aesthetics. Net present value may be used to evaluate the merits of the proposal to enter the EU market, which is the driving force for eliminating decaBDE. The payback period will be calculated. Externalized costs will not be considered. The "Cost and Availability Module" of the IC2 may be consulted for additional considerations.
- Step 10 (identify acceptable alternatives) would be completed by comparing results of Step 9 to the requirements established in Step 2, and by ensuring that the alternatives had lower overall impact to the environment based on any findings in Step 8 and/or 9.1 (Life Cycle Thinking and additional life cycle assessment). Assessment methods, assumptions, data, results, and conclusions would also be documented.
- Step 11 (comparing) would be accomplished using a comparison summary matrix and weighted ranking of the performance, economic, and environmental criteria for each alternative. The best solution would be selected based on the results of Step 11.
- Step 12 (implementation) would be completed by integrating the implementation plan for the alternative solution into the overall plan for KayDisplay's entry into the EU market. The list of stakeholders would be reviewed to determine if others needed to be consulted. The alternative would be piloted and then ramped up to volume production, addressing issues as they are identified. Finally, a milestone date would be set to review the implementation and to consider new potential alternatives prior to designing the next model.

Steps 3 and 4: Identify Potential Alternatives and Initial Screening

An extensive list of potential alternatives can be found in the literature, so the KayDisplay mechanical designer grouped the alternatives to narrow the assessment (Table 12-3).

Based on preliminary screening, KayDisplay will primarily consider PPE/HIPS with halogenated and non-halogenated flame retardants and a material change to PC/ABS with non-halogenated flame retardants.

After consulting with the injection molder and conducting online and offline literature searches, the KayDisplay mechanical designer identifies the following options:

- PPE/HIPS with a halogenated flame retardant,
- PPE/HIPS with a non-halogenated flame retardant, and
- PC/ABS with a non-halogenated flame retardant.

To identify potential halogenated and non-halogenated flame retardant alternatives, KayDisplay again refers to the DfE's *DecaBDE Alternatives Assessment (AA)*. KayDisplay is able to share the extended list of alternatives in the report with the injection molding supplier. After conferring with the supplier about available resins and comparing the properties in the resins' technical datasheets to those in Tables 12-1 and 12-2, the alternatives are narrowed to those listed in Table 12-4.

Therefore, the chemical alternatives to be evaluated in the assessment are:

- Decabromodiphenyl ethane [DBDPE],
- Antimony trioxide [ATO],
- Resorcinol bis-diphenylphosphate [RDP], and
- Triphenyl phosphate [TPP].

TABLE 12-3 Potential Alternatives

Class of alternative	Comments	Alternative to be Assessed
PPE/HIPS with no added flame retardant	- Cannot meet U.S. flammability requirements + Meets ecolabel criteria + Material cost of PPE/HIPS is low *Would require different products for U.S. and EU markets, so this option will not be considered.*	NO
PPE/HIPS with a halogenated flame retardant[a]	+ Meets U.S. flammability requirements - Does not meet ecolabel criteria + Material cost of PPE/HIPS is low *The advantage of low material cost (lower product cost) might offset not having the ecolabel in the EU market, so this option will be considered.*	**YES**
PPE/HIPS with a non-halogenated flame retardant	- Meeting U.S. flammability requirements with non-halogenated flame retardants in HIPS may be difficult (according to literature) + Meets ecolabel criteria + Material cost of PPE/HIPS is low *If the flammability and performance targets can be met, this option offers both lower material cost than PC/ABS or metal and also the market benefit of ecolabel listing, so this option will be considered.*	**YES**
PC/ABS with a halogenated flame retardant	+ Meets U.S. flammability requirements - Does not meet ecolabel criteria - Material cost of PC/ABS is significantly higher than PPE/HIPS + May get performance and aesthetic improvements *This option has the combination of higher material cost and lost ecolabel market opportunity, and will not be considered.*	NO
PC/ABS with a non-halogenated flame retardant	+ Meets U.S. flammability requirements + Meets ecolabel criteria - Material cost of PC/ABS is significantly higher than PPE/HIPS + May get performance and aesthetic improvements *Although the material cost will be higher, the combination of meeting both the U.S. flammability requirements and ecolabel requirements while also potentially gaining performance and aesthetic benefits make this a viable option, and it will be considered.*	**YES**
Metal (aluminum or magnesium)	+ Meets U.S. flammability requirements + Meets ecolabel criteria - Significant material cost increase - Would require changing suppliers - Would require significant design changes - Would require significant manufacturing changes *Having to change suppliers combined with significant material cost increases make this option an undesirable choice, and it will not be considered.*	NO

[a] The option of continuing to use decaBDE at levels below 1000ppm will not be considered because decaBDE is not effective as a flame retardant at that low level.

TABLE 12-4 Remaining Alternatives

Class of alternative	Alternative(s)	CAS Number(s)
PPE/HIPS with a halogenated flame retardant	Decabromodiphenyl ethane [DBDPE] (with 5% antimony trioxide synergist) [ATO][a]	84852-53-9 [DBDPE] 1309-64-4 [ATO]
PPE/HIPS with a non-halogenated flame retardant	Resorcinol bis-diphenylphosphate [RDP] (with 5% triphenyl phosphate contamination) [TPP]	125997-21-9; 57583-54-7 [RDP] 115-86-6 [TPP]
PC/ABS with a non-halogenated flame retardant	Resorcinol bis-diphenylphosphate [RDP] (with 5% triphenyl phosphate contamination) [TPP]	125997-21-9; 57583-54-7 [RDP] 115-86-6 [TPP]

[a] DecaBDE also requires the use of Antimony Trioxide (ATO).

TABLE 12-5 Physicochemical Properties of DecaBDE and Potential Alternatives

Property	DecaBDE	DBDPE	ATO	RDP	TPP
Structure	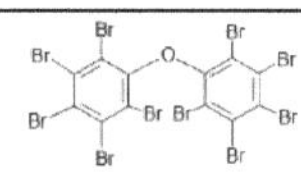	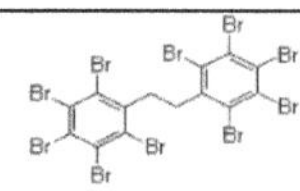	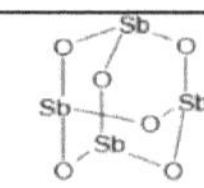	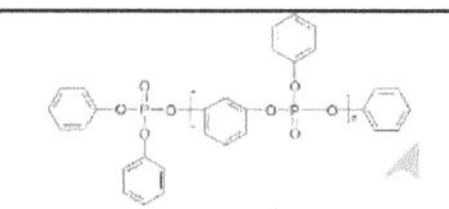	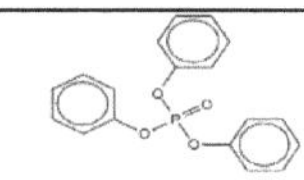
MW	959.2	971.2	291.5	574.46 (n=1) (57583-54-7) 822.64 (n=2) (98165-92-5)	326.29

Physical State of Chemical (ambient conditions)

Physical state indicates if a chemical substance is a solid, liquid, or gas under ambient conditions, and is determined from the melting and boiling points. Chemicals with a melting point more than 25°C are considered solid. Those with a melting point less than 25°C and a boiling point more than 25°C are considered liquid, and those with a boiling point less than 25°C are considered a gas.
Relevance to exposure: Physical state influences the potential for dermal and inhalation exposure. For solids, there is potential for the inhalation and ingestion of dust particles and dermal contact. For liquids, there is potential for direct dermal contact but not for direct inhalation of the liquid (except in operations that produce aerosols). In the case of these alternatives, all are solid at room temperature except for RDP, but once RDP is blended into a polymer, it has the same exposure potential as a solid, so the assessment will consider the inhalation and ingestion of dust particles and dermal contact in the solid form for all alternatives.

Property	DecaBDE	DBDPE	ATO	RDP	TPP
Physical Form at Ambient Conditions	Solid	Solid	Solid	Liquid	Solid
Melting Point (°C)	300-310	350	656	-12 to -16 (liquid at room temperature)	50.5
Boiling Point (°C)	> 320 (decomposes)	>350 (estimated)	1425	300 370 (decomposes)	245 at 11 mm Hg

Vapor Pressure

Relevance to exposure: Vapor pressure indicates the potential for a chemical to volatilize into the atmosphere. If a chemical has a vapor pressure leading to volatilization at room temperature or typical environmental conditions, then the chemical may evaporate and present the potential for inhalation of the gas or vapor. For a Design for the Environment (DfE) chemical alternatives assessment, inhalation exposure is assumed to occur if the vapor pressure is greater than 1 x 10^{-8} mm Hg. A default value of <10^{-8} was assigned for chemicals without data that are anticipated to be non-volatile this is based on EPA HPV assessment guidance (EPA 2011b).

Property	DecaBDE	DBDPE	ATO	RDP	TPP
Vapor Pressure (mm Hg)	3.5 x 10^{-8} at 21 °C	<7.5x10^{-7}	<10^{-8}	1.9 x 10^{-5} at 20°C	6.28 x 10^{-6}

Log K_{ow} (LogP), Water Solubility (mg/L), and dE (eV)

Relevance to bioavailability: *Log K_{ow} can be used to evaluate absorption and distribution in biological organisms, potential acute aquatic toxicity by narcosis, and potential general population exposure via ingestion.* Generally, chemicals with a log K_{ow} < 5 are orally bioavailable to mammals; chemicals with logK_{ow} < 4 are water soluble and available to aquatic species. LogK_{ow} is linearly related to bioaccumulation factor (BAF) up to log K_{ow} ~ 5, where lower water solubility levels off and bioavailability becomes asymptotic.
Relevance to aquatic toxicity: LogP "usually correlates well with acute aquatic toxicity. For non-ionic organic chemicals that are toxic through narcosis, acute and chronic toxicity increases exponentially with increases in logP up to a value of about 5-7" (Voutchkova et al. 2011). Chemicals with logP <2 have higher probability of having low acute and chronic aquatic toxicity (Voutchkova et al. 2011).
Relevance to environmental transport: *Chemicals with a high log K_{ow} also tend to bind strongly to soil and sediment.*
Log K_{ow} cannot be measured for inorganic substances, polymers, and other materials that are not soluble in either water or octanol. This is indicated in the table with "No data."
Water solubility indicates the potential of a chemical to dissolve in water and form an aqueous solution. Water soluble chemicals present a higher potential for human exposure through the ingestion of contaminated drinking water (including well water). In general, absorption after oral ingestion of a chemical with water solubility less than 10^{-3} mg/L is not expected. Water soluble chemicals are more likely to be transported into groundwater, absorbed through the gastrointestinal tract or lungs, partition to aquatic compartments, and undergo atmospheric removal by rain washout. A substance with water solubility at or below 10^{-3} mg/L is considered insoluble.
HOMO-LUMO gap (ΔE, eV): The energy separation between the highest occupied and lowest unoccupied molecular orbitals (HOMO–LUMO gap, ΔE) is related to broad chemical reactivity (Fukui et al. 1952). A molecule with a small ΔE is considered

TABLE 12-5 (Continued)

more chemically reactive for covalent bonding than one with a larger ΔE. Chemicals with ΔE > 6.5 eV (as calculated by DFT) are much less likely to be acutely or chronically toxic to aquatic species (Kostal et al. in press; Voutchkova-Kostal et al. 2012).
Conclusions:
Aquatic toxicity: DecaBDE and TPP have logP > 2 and ΔE < 6.5 eV, which puts them in the high risk category for high acute and/or chronic aquatic toxicity. DBDPE also has ΔE < 6.5 eV but its high logP value (14) suggests it is not very bioavailable to aquatic species, so is likely to be of low/moderate aquatic toxicity.
Bioaccumulation: DecaBDE and DBDPE is likely to have high tendency to bioaccumulate; TPP will likely have a lower bioaccumulation tendency due to its lower logP and higher water solubility; The likelihood of bioaccumulation for RDP will depend strongly on its dissociation to monomer units in the environment.
Environmental transport: Of the alternatives assessed, DBDPE is likely to bind most strongly to soil and sediment (highest logKow).

Log K_{ow} (LogP)	6.27	14 (estimated)	No data	4.93	4.59
Water Solubility (mg/L)	$< 1.00 \times 10^{-4}$ at 25 °C	7.2×10^{-4}	14 at 30°C	1.05 at 20°C	1.9
dE (eV)	5.0	5.3	No data	No data	5.0
Physical hazards					
Flammability (Flash Point)	Not flammable	Not flammable	Not combustible	302°C	220°C
Explosivity	Not expected to form explosive mixtures with air	Not expected to form explosive mixtures with air	Not expected	Not explosive	Not expected to form explosive mixtures with air
Metabolites, Degradates, Transformation Products					
Pyrolysis by-products are of particular importance to electronics due to improper and informal waste practices.					
Metabolites, Degradates, Transforma-tion Products	Photodegradation, anaerobic biodegradation, fish metabolism to lower brominated diphenyl ether (BDE) congeners; Pyrolysis – polybrominated dibenzofurans and polybrominated dibenzo-p-dioxins	Photodegradation —potential to form lower brominated congeners; Pyrolysis—possible polybrominated dibenzofurans and polybrominated dibenzo-p-dioxins	None	Metabolites: hydroxy-RDP, dihydroxy-RDP, resorcinol diphenyl phosphate, and hydroxyl-resorcinol diphenyl phosphate, resorcinol (108-46-3), resorcinol conjugates, resorcinyl glucuronide and resorcinyl sulfate. Environmental degradation of RDP has been demonstrated in experimental studies, but the degradates have not been identified. Degradation of RDP by sequential dephosphorylation could produce phenol, diphenyl phosphate, or resorcinol.	Diphenyl phosphate (CASRN 838-85-7) and phenol (CASRN 108-95-2)

NOTE: Most data and text in Table 12-5 are from the DfE DecaBDE AA. However, information in this section is simulated, and presented as if it had been obtained by environmental scientists and chemists at KayDisplay's resin formulator. All *italicized* text is taken from EPA 2014i.

Step 5: Assess Physicochemical Properties

The physicochemical properties of decaBDE, DBDPE, ATO, RDP, and TPP are compiled in DfE's *DecaBDE AA* and presented in Table 12-5. KayDisplay does not have chemists or toxicologists on staff, so they will rely on the EPA's DfE *DecaBDE* report data and conclusions.

Step 6.1: Assess Human Health (Chemical Hazards)

The human health effects of decaBDE, DBDPE, ATO, RDP, and TPP have been compiled in *DfE's DecaBDE AA*. Similar to Step 5, KayDisplay will rely on the determinations published in DfE's *DecaBDE AA* because the company does not have chemists or toxicologists on staff to complete comparable work (see Table 12-6).

It should be noted that this tabular format is only one way of presenting summary data. There are other approaches, such as ToxPi, which are illustrated in the second case study and in Appendix C.

Step 6.2: Assess Ecotoxicity Hazards

The ecotoxicity effects of decaBDE, DBDPE, ATO, RDP, and TPP have been compiled in DfE's *DecaBDE AA*. As in Step 5, KayDisplay will rely on the determinations in the EPA DfE report because the company does not have chemists or toxicologists on staff (see Table 12-7).

Although several of the alternatives under consideration (e.g., ATO, RDP) will be found primarily in sediment and soil, the DfE *DecaBDE AA* only evaluates aquatic toxicity because ecotoxicity data for terrestrial species was limited or completely absent for the chemicals assessed. Therefore, potential for impacts of the alternatives on high trophic level and terrestrial wildlife is unclear and could not be fully assessed.

Step 6.3: Conduct Comparative Exposure Assessment

Human and environmental exposures to decaBDE are described in Section 5.1.5 of DfE's *DecaBDE AA* and the EPA report, *An Exposure Assessment of Polybrominated Diphenyl Ethers* (EPA 2010b). Because the manufacturing process for the enclosure part, the product-use pattern, and end-of-life hardware disposal are expected to be the same for decaBDE and its alternatives, the exposure scenarios and routes will be considered the same for alternatives as for decaBDE, which is consistent with DfE practice (Lavoie et al. 2010).

- Human exposure (occupational) from EPA 2014: "According to the U.S. EPA's 2010 exposure assessment of polybrominated diphenyl ethers (PBDEs), individuals in occupations that would lead to higher exposures to specific congeners have higher concentrations of PBDE congeners in their blood than the general public (EPA 2010b). Workers involved in the manufacturing or recycling and disposal of products containing PBDE flame retardants have greater exposure to the chemical compared to the general population (Sjodin et al. 1999; Thomsen et al. 2001; Thuresson et al. 2006)."
- Human exposure (consumer/user) from EPA 2014: "Consumer exposure to decaBDE is possible given that it can be released from common home products and become a component in house dust (Stapleton et al. 2004; Takigamie et al. 2008). It is also possible that workers exposed to decaBDE may inadvertently carry particles containing the chemical home with them. This may lead to exposure to family members through household dust or direct contact, as has been proven with other hazardous chemicals such as pesticides and lead (Thompson et al. 2003; Minnesota Department of Health 2010). DecaBDE has been found in dust within automobiles (Lagalante et al. 2009) and automobile air (Mandalakis et al. 2008). The primary route of consumer exposure to decaBDE is through the ingestion of dust or, for infants, ingestion of breast milk, followed by food and water ingestion and dermal absorption (Lorber 2008; Petito Boyce et al. 2009; EPA 2010a). Inhalation may also be a relevant route of exposure (EPA 2010b). Children have higher levels of exposure to decaBDE than do adults (Petito Boyce et al. 2009), likely due to higher hand- to- mouth behavior." Information about exposure of decaBDE and alternatives is shown on Table 12-8 on toxicokinetics.
- Environmental exposures from EPA 2014i: "Environmental releases of decaBDE can occur during each stage of a product's life cycle, including chemical manufacturing, product manufacturing, product storage and use, and end-of-life handling (EPA 2009)". This is expected to be true for alternatives, as well. Tables 12-9, 12-10, and 12-11 list persistence, transport, and bioaccumulation levels for decaBDE and alternatives.

TABLE 12-6 Human Health Effects Data from Dfe's *DecBDE Alternatives Assessment*

		Human Health Effects										
Chemical	CASRN	Acute Toxicity	Carcinogenicity	Genotoxicity	Reproductive	Developmental	Neurological	Repeated Dose	Skin Sensitization	Respiratory Sensitization	Eye Irritation	Dermal Irritation
Decabromodiphenyl Ether	1163-19-5	L	M	L	L	H	*L*	M	L		L	L
Decabromodiphenyl Ethane	84852-53-9	L	*M*§	L	L	*H*§	*L*	L	L		VL	VL
Antimony Trioxide[a]	1309-64-4	L	M*	M	*M*	*L*	*L*	H	L		L	M
Resorcinol Bis-Diphenylphosphate; RDP	125997-21-9	L	*M*§	L	L	M	M	M	*L*		L	VL
Triphenyl Phosphate	115-86-6	L	*M*	L	L	L	L	H	L		L	VL

NOTE: VL = Very Low hazard L = Low hazard M = Moderate hazard H = High hazard VH = Very High hazard Endpoints (VL, L, M, H, and VH) were assigned based on empirical data. Endpoints in italics (*VL, L, M, H, and VH*) were assigned using values from predictive models and/or professional judgment.
§ Based on analogy to experimental data for a structurally similar compound.
* Ongoing studies may result in a change in this endpoint.
[a] This compound is included in the ongoing EPA Work Plan evaluation for Antimony Trioxide.
SOURCE: Adapted from EPA 2014i.

TABLE 12-7 Ecotoxicity Data from DfE's Alternatives Assessment

		Aquatic Toxicity**		Environmental Fate	
Chemical	CASRN	Acute	Chronic	Persistence	Bioaccumulation
Decabromodiphenyl Ether	1163-19-5	*L*	*L*	VH	*H*
Decabromodiphenyl Ethane	84852-53-9	L	*L*	VH	*H*
Antimony Trioxide[a]	1309-64-4	H	M	*H*[R]	*L*
Resorcinol Bis-Diphenylphosphate; RDP	125997-21-9	VH	VH	M	*H*‡
Triphenyl Phosphate	115-86-6	VH	VH	L	M

NOTE: VL = Very Low hazard L = Low hazard M = Moderate hazard H = High hazard VH = Very High hazard Endpoints (VL, L, M, H, and VH) were assigned based on empirical data. Endpoints in italics (*VL, L, M, H, and VH*) were assigned using values from predictive models and/or professional judgment.
‡ The highest hazard designation of any of the oligomers with MW <1,000.
[R] Recalcitrant: Substance is comprised of metallic species that will not degrade, but may change oxidation state or undergo complexation processes under environmental conditions.
**Aquatic toxicity: EPA/DfE criteria are based in large part upon water column exposures which may not be adequate for poorly soluble substances such as many flame retardants that may partition to sediment and particulates.
[a] This compound is included in the ongoing EPA Work Plan evaluation for Antimony Trioxide. SOURCE: Adapted from EPA 2014i.

TABLE 12-8 Toxicokinetic Data

Toxicokinetics	
DecaBDE	*Although experimental findings in human and animal studies suggest that decaBDE is poorly absorbed following oral and dermal administration, even low levels of decaBDE are physiologically relevant due to its chemical properties. 82.5-91.3% of decaBDE is eliminated from the body in the feces with ≤0.05% excreted in urine. DecaBDE is mainly excreted as unchanged parent compound but may also be excreted in the form of metabolites. Some conversion of parent compound may be mediated by intestinal epithelium or microflora. Monitoring studies in humans, with unknown levels of exposure, demonstrate that decaBDE can be absorbed, distributed to mammary tissue, and secreted in human breast milk during lactation.*
Alternative	**Expected Toxicokinetics**
DBDPE	*Decabromodiphenyl ethane, as a neat material, is estimated to not be absorbed through the skin and to have poor skin absorption when in solution. Decabromodiphenyl ethane is expected to have poor absorption via the lungs and gastrointestinal (GI) tract. Decabromodiphenyl ethane is poorly absorbed in the GI tract following oral exposure and is mainly excreted in the feces. If absorption does occur, decabromodiphenyl ethane is distributed to the serum, liver, kidney, and adipose tissues and undergoes biotransformation to form metabolites.*
ATO	*Antimony trioxide is expected to have no absorption through skin and has poor absorption through the lungs and gastrointestinal (GI) tract, according to experimental data. Following oral exposure, the majority of antimony trioxide is excreted in the feces. The compound accumulates in lungs with inhalation exposure due to slow absorption and clearance.*
RDP	*Resorcinol bis-diphenylphosphate was readily absorbed via the oral route and was absorbed to a lesser extent following dermal exposure. Metabolism was extensive with metabolites excreted in feces, urine, and in expired air as CO2.*
TPP	*Triphenyl phosphate is hydrolyzed in the liver to produce diphenyl phosphate as the primary metabolite. TPP can be detected in human breast milk.*

Note: Italicized text taken from EPA 2014i.

Step 7: Identify Safer Alternatives

The combined hazard table for decaBDE, DBDPE, ATO, RDP, and TPP from the DfE's *DecaBDE AA* report is shown in Table 12-12.

"Confidence in the categorization of endpoint hazard levels," in Section 4.2: Data Sources and Assessment Methodology of the DfE *DecaBDe AA*, deals with how data were collected, prioritized and reviewed for use in the development of hazard profiles. According to the report, "High-quality experimental studies lead to a thorough understanding of behavior and effects of the chemical in the environment and in living organisms. Analog approaches and SAR-based estimation methods [were] also useful tools and are discussed throughout this section" (EPA 2014i).

KayDisplay recognizes that there are varying levels of confidence (per Chapter 6) in the different end point categorizations (vH, H, M, L, vL), and the company understands that measured data are not necessarily higher confidence than models. However the company has insufficient expertise to differentiate the confidence levels, and therefore will assume approximately equal confidence levels for the categorizations of end points for the purpose of this assessment.

- *Relative hazards:* In reviewing the hazard summary table for the alternatives, KayDisplay finds that DBDPE/ATO shows improvements over decaBDE in repeated dose toxicity and irritation, but not in the original areas of concern (persistence, bioaccumulation, and neurodevelopmental toxicity), nor in transformation products. RDP/ATO shows improvements over decaBDE/ATO in the original areas of concern, but does not offer clear improvements in every impact area, and appears to have higher aquatic toxicity.

- *Trade-off resolution:* In order to help resolve this trade-off and make a decision, KayDisplay had originally considered applying a scoring scheme. However, the company found that constructing a robust scoring scheme, or chemical ranking and scoring (CRS) system, is difficult and can lead to incorrect conclusions (Davis et al. 1994; Swanson and Socha 1997). For example, if a scoring system assigned each chemical very high (vH) four points, each high (H) three points, each medium (M) two points, each low (L) one point, and each very low (vL) zero points, the results would indicate that a substance with all Ms (score 28) would appear worse than a PBT like decaBDE (score 23) if each end point were

TABLE 12-9 Persistence for DecaBDE and Alternatives

Persistence	
DecaBDE	*VERY HIGH: Empirical and predicted data indicate that all PBDEs (including decaBDE) are highly persistent in the environment (Environment Canada 2006), and decaBDE has been found in high and increasing concentrations in the sediment of lakes, rivers, streams and estuaries (Song, Li et al. 2005; Environment Canada 2006; Illinois Environmental Protection Agency 2006).* *The persistence potential for decaBDE is Very High; it is not expected to degrade rapidly under aerobic conditions. Slow degradation through debromination may occur under anaerobic conditions. The anaerobic experimental results are indicative of limited removal, but at very low rates that are possibly background level degradation under the test conditions. Experimental studies indicate no degradation after 2 weeks in a ready biodegradation test, but no data were located for soil or water. Results from biodegradation estimation models also suggest decaBDE is recalcitrant under aerobic conditions. Non-guideline experimental studies indicate decaBDE may be capable of undergoing limited anaerobic biodegradation; however the removal rate also suggests Very High persistence. The initially formed degradation products are also expected to be persistent. DecaBDE is not expected to hydrolyze in the environment based on experimental data. Experimental data indicate that decaBDE may undergo photolysis to debrominated transformation products. Data concerning the kinetics of these photolysis reactions were not located.*
Alternative	**Expected Persistence**
DBDPE	*VERY HIGH: Very high persistence of decabromodiphenyl ethane is expected based on experimental biodegradation data. Decabromodiphenyl ethane was determined to not be readily biodegradable in a 28-day MITI test, nor was it inherently degradable in a 90-day aerobic sewage/soil test using pre-exposed inoculum. Decabromodiphenyl ethane is not expected to undergo hydrolysis since it does not contain hydrolysable functional groups. The atmospheric half-life of decabromodiphenyl ethane is estimated to be 4.5 days, although it is expected to exist primarily in the particulate phase in air. Laboratory studies have demonstrated photolysis of decabromodiphenyl ethane, although the rate of this process under environmental conditions has not been established.*
ATO	*HIGH: Antimony trioxide is an inorganic substance containing metallic atoms that are likely to be found in the environment for more than 180 days after release, resulting in a very high persistence/recalcitrant hazard designation. Based on water solubility studies under a range of pH values, antimony trioxide is expected to slowly dissolve, resulting in the release of antimony ions and, depending on pH, be oxidized or reduced to other oxidation states. Additionally, results from a pure culture study using autotrophic bacterium indicate that antimony may be oxidized by bacteria. Antimony trioxide is not anticipated to undergo hydrolysis under environmental conditions. Antimony trioxide does not contain functional groups expected to absorb light at environmentally significant wavelengths, and therefore is not expected to photolyze. No degradation processes for antimony trioxide under typical environmental conditions were identified.*
RDP	*MODERATE: Moderate persistence is expected for resorcinol bis- diphenylphosphate based on experimental biodegradation studies that indicate the potential for biodegradation of the commercial polymeric mixture. The commercial mixture was determined to be inherently biodegradable using the guidelines of Directive 84/449/EEC, C.6 "Biotic degradation - the Closed Bottle test" test. After 28 days, 37% biodegradation occurred, and after 56 days, 66% biodegradation occurred. Resorcinol bis-diphenylphosphate oligomers (n=1 and n=2) do not contain chromophores that absorb at wavelengths >290 nm, and therefore, are not expected to be susceptible to direct photolysis by sunlight. The atmospheric half-life of resorcinol bis-diphenylphosphate oligomers are estimated to be 6.1 (n=1) and 4.1 (n=2) hours, although they are expected to exist primarily in the particulate phase in air. Enzymatic or basic hydrolysis leading to the production of phenol (CASRN 108-95-2), diphenyl phosphate (CASRN 838-85-7), and resorcinol (CASRN 108-46-3) through sequential dephosphorylation is theoretically possible but has not been demonstrated.*
TPP	*LOW: The persistence of triphenyl phosphate is based on experimental data. Under aerobic conditions in a Japanese MITI ready biodegradability test (OECD Test Guidelines (TG) 301C), 90% biodegradation of triphenyl phosphate occurred after 28 days, and 93.8% triphenyl phosphate removal as dissolved organic carbon (DOC) occurred over 20 days in an OECD 303A guideline study. TPP does not meet the criteria for very low persistence because the percent removal in the criteria does not occur within a 10-day window. In loamy sand, a half-life of 37 days was observed under aerobic conditions. Triphenyl phosphate was determined to be inherently biodegradable in a river die-away test, after degrading 100% over 3 days in river water. Triphenyl phosphate may degrade under anaerobic conditions, with primary degradation of 31.1% after 3 days (89.7% after 40 days) in river sediment. However, removal under anaerobic conditions is not anticipated to be an important fate process. Triphenyl phosphate will undergo hydrolysis under alkaline conditions, with half-lives of 3 days at pH 9; it is relatively stable to hydrolysis under neutral and acidic conditions, with half-lives of 28 days at pH 5 and 19 days at pH 7. Triphenyl phosphate is not expected to be susceptible to direct photolysis by sunlight, since it does not absorb light at wavelengths >290 nm. The atmospheric half-live of vapor-phase triphenyl phosphate is estimated to be 12 hours.*

Note: Italicized text taken from EPA 2014i.

TABLE 12-10 Transport for DecaBDE and Alternatives

Transport	
DecaBDE	*DecaBDE has also been measured in ambient atmospheric particulates (Illinois Environmental Protection Agency 2006) and in the Arctic environment, providing evidence that it is subject to long-range transport (Environment Canada 2006).* *The transport evaluation for decaBDE is based on both estimated and experimental physical and chemical properties. Based on the Level III fugacity models incorporating the located experimental property data, decaBDE is expected to partition primarily to soil. It is not expected to dissociate at environmentally-relevant pHs. DecaBDE is expected to have low mobility in soil based on its estimated Koc. Therefore, leaching of decaBDE through soil to groundwater is not expected to be an important transport mechanism. Estimated volatilization half-lives for a model river indicate that it will have moderate potential to volatilize from surface water. Volatilization potential from a model lake is expected to be low. In the atmosphere, decaBDE is expected to exist primarily in the particulate phase. Particulate phase decaBDE will be removed from air by wet or dry deposition.*
Alternative	**Expected Transport**
DBDPE	*Based on the Level III fugacity models incorporating the located experimental property data, decabromodiphenyl ethane is expected to partition primarily to soil. Decabromodiphenyl ethane is expected to be immobile in soil based on its estimated Koc. Leaching of decabromodiphenyl ethane through soil to groundwater is not expected to be an important transport mechanism. Estimated volatilization half-lives indicate that it will be non-volatile from surface water. Volatilization from dry surface is also not expected based on its vapor pressure. In the atmosphere, decabromodiphenyl ethane is expected to exist solely in the particulate phase, based on its estimated vapor pressure. Particulates may be removed from air by wet or dry deposition.*
ATO	*The limited mobility observed under experimental conditions and the low vapor pressure indicates that antimony trioxide is anticipated to partition predominantly to soil and sediment. It will not volatilize from water. Soil mobility and sediment adsorption tests indicate that antimony trioxide will be immobile in soil, and therefore will not be expected to migrate into groundwater.*
RDP	*The environmental fate is described for the oligomer where n=1, which is the primary component of the commercial product. Based on the Level III fugacity models incorporating the located experimental property data, resorcinol bis-diphenylphosphate is expected to partition primarily to soil and sediment. Resorcinol bis-diphenylphosphate is expected to be immobile in soil based on its estimated Koc. Leaching of resorcinol bis-diphenylphosphate through soil to groundwater is not expected to be an important transport mechanism. Estimated volatilization half-lives indicate that it will be non-volatile from surface water. Volatilization from dry surface is also not expected based on its vapor pressure. In the atmosphere, resorcinol bis-diphenylphosphate is expected to exist solely in the particulate phase, based on its estimated vapor pressure. Particulates may be removed from air by wet or dry deposition. The higher MW components of the commercial product are anticipated to behave similarly to that described above.*
TPP	*Level III fugacity models incorporating available physical and chemical property data indicate that at steady state, TPP is expected to be found primarily in soil and, to a lesser extent, water. Triphenyl phosphate is expected to have moderate mobility in soil, based on measured Koc values in silty clay, loamy sand, and silt loam. Leaching through soil to groundwater may occur, though it is not expected to be an important transport mechanism. Triphenyl phosphate may volatilize from moist soil and water surfaces based on its Henry`s Law constant. Volatilization from dry surface is not expected based on its vapor pressure. In the atmosphere, triphenyl phosphate is expected to exist in both the vapor phase and particulate phase. Particulates may be removed from air by wet or dry deposition.*

Note: Italicized text taken from EPA 2014i.

equally weighted. A weighted scoring scheme could be an improvement, but as noted above, constructing a robust weighted scoring scheme is difficult and would be beyond the capabilities of KayDisplay.

Instead of creating its own system, KayDisplay referred to the "Hazard Assessment Module" of the IC2, which recommends using GreenScreen® for Safer Chemicals as a way of integrating information across human health and environmental topics (Clean Production Action 2014).

The GreenScreen® benchmark scoring system uses structured decision logic to assign a single integer score to each chemical being assessed. This scheme incorporates national and international precedents to weigh and prioritize combinations of hazard end points.

The GreenScreen® defines four hazard levels for substances:

- Benchmark 1 — "Avoid - Chemical of High Concern"

TABLE 12-11 Bioaccumulation for DecaBDE and Alternatives

Bioaccumulation	
DecaBDE	*HIGH: Laboratory studies demonstrate decaBDE's bioavailability and metabolism in fish (Illinois Environmental Protection Agency 2006). DecaBDE has been detected in some but not all species of fish studied (Dodder et al. 2002; European Chemicals Bureau 2002; Johnson-Restrepo et al. 2005; Environment Canada 2009; Roberts et al. 2011). Also, decaBDE has been measured in birds and their eggs (Lindberg et al. 2004; Vorkamp et al. 2005) and in mammals, including polar bears, seals, marmots, and foxes (Christensen et al. 2005; Illinois Environmental Protection Agency 2006; Voorspoels et al. 2006; Environment Canada 2009). Further, terrestrial species tend to have higher levels of decaBDE than aquatic species for both birds (Jaspers et al. 2006) and mammals (Christensen et al. 2005). These observations indicate bioavailability of decaBDE to wildlife and human food sources, with potential for bioaccumulation and biomagnification of decaBDE and/or its degradation products.* *Based on estimated BAF values suggesting that the potential for bioaccumulation is high and located monitoring data indicating that decaBDE has been detected in higher trophic level organisms. DecaBDE degradation, transformation, and metabolism products also contribute to the high bioaccumulation hazard designation. These compounds are lower brominated congeners and also have been detected in monitoring studies (ATSDR 2004).*
Alternative	**Expected Bioaccumulation**
DBDPE	*HIGH: The bioaccumulation hazard designation is estimated based on decabromodiphenyl ethane monitoring data reporting detections in many different species, including those higher on the food chain. Although the estimated bioaccumulation factor is low, the persistence of decabromodiphenyl ethane and its detection in many species from different habitats and trophic levels indicates high potential for bioaccumulation hazard in aquatic or terrestrial species.*
ATO	LOW: Antimony trioxide is an inorganic compound and is not expected to bioaccumulate.
RDP	*HIGH: The estimated BCF value for the n=1 component has high potential for bioaccumulation. The higher MW oligomers that may be found in this mixture (n=2, 3, 4...) are expected to have moderate or low potential for bioaccumulation based on their large size and low solubility according to the polymer assessment literature (Boethling and Nabholz 1997).*
TPP	MODERATE: There is moderate potential for bioaccumulation based on experimental BCF values.

Note: Italicized text taken from EPA 2014i.

- Benchmark 2 — "Use but Search for Safer Substitutes"
- Benchmark 3 — "Use but Still Opportunity for Improvement"
- Benchmark 4 — "Prefer - Safer Chemical"

"Each benchmark includes a set of criteria that a chemical, along with its known and predicted transformation products, must pass" (Rossi and Heine 2007). For example, if a chemical met any of the following criteria, it would be classified as "Benchmark 1:

a. PBT = High P + High B + [very High T (Ecotoxicity or Group II Human) or High T (Group I or II* Human)]

b. vPvB = very High P + very High B

c. vPT = very High P + [very High T (Ecotoxicity or Group II Human) or High T (Group I or II* Human)]

d. vBT = very High B + [very High T (Ecotoxicity or Group II Human) or High T (Group I or II* Human)]

e. High T (Group I Human)" (Clean Production Action 2011)

The criteria for each benchmark become progressively more demanding, with Benchmark 4 representing the most preferred (least hazardous) chemicals.

GreenScreen® attempts to use all available data, including analogs, models, and expert judgment, to assess end points. It has a hierarchy of data adequacy to establish whether the hazard data were of sufficient quality to meet the requirements of the assessment process. End points with insufficient information to assess the hazard are assigned a data gap (DG). There are also minimum datasets which, if not met, will either lower the score or result in the chemical receiving a rating of "U," denoting that there is insufficient data to enable evaluation. This is consistent with KayDisplay's choice in Step 2 to be labeled, "missing data neutral."

As noted above, KayDisplay does not have chemists or toxicologists on staff, and therefore cannot complete GreenScreen® in-house. However, GreenScreen® is aligned with the DfE hazard criteria, and the Clean Production Action has published draft benchmark scores for many of the substances in the DfE DecaBDE AA (see Table 12-11).

Based on the GreenScreen® scores, RDP (Benchmark 2) with TPP (Benchmark 2) appears safer than DecaBDE (Benchmark 1) or DBDPE (Benchmark 1) with ATO (Benchmark 1). However, KayDisplay headquarters are located in Washington State, where water issues are of the highest priority, so the company will further investigate the potential aquatic toxicity of RDP/TPP.

KayDisplay was able to contact the chemical supplier of RDP, and the team learned that commercial formulations of RDP, which contain TPP contamination (<5%), have been subjected to acute ecotoxicity testing, and that the commercial mixture shows no toxicity at the maximum water solubility level, using what is called the Water Accommodated Fraction (WAF) methodology in accordance with OECD guidance. Although RDP/TPP will most likely sequester in sediments, tests using aquatic organisms as surrogates indicate that concerns with water issues are minimal and, for this application it appears to be acceptable.

Conclusion

Based on these analyses, KayDisplay concludes that alternatives based on RDP/TPP meet the requirement of being safer than those based on the original DecaBDE/ATO, so RDP/TPP alternatives will be evaluated further. Alternatives based on DBDPE/ATO (Benchmark 1) will not be evaluated further because DBDPE/ATO is only minimally safer than the original DecaBDE/ATO and does not meet the goal of being Benchmark 2 or better.

Steps 8-13

Once alternatives based on DBDPE/ATO have been eliminated, the remaining alternatives are:

- PPE/HIPS with RDP/TPP
- PC/ABS with RDP/TPP

Both alternatives meet the ecolabel requirement. However, the PPE/HIPS option with RDP/TPP offers a lower cost, but may not meet flammability and performance targets. In contrast, the PC/ABS option with RDP/TPP costs more, but is likely to meet flammability requirements and offer performance and aesthetic benefits. It is clear that additional assessments must be completed to select and implement a single alternative.

As noted earlier, Steps 8- 13 will not be completed as part of this case study.

TABLE 12-12 Combined Hazard Table from DfE Alternatives Analysis

		Human Health Effects											Aquatic Toxicity**		Environmental Fate	
Chemical	CASRN	Acute Toxicity	Carcinogenicity	Genotoxicity	Reproductive	Developmental	Neurological	Repeated Dose	Skin Sensitization	Respiratory Sensitization	Eye Irritation	Dermal Irritation	Acute	Chronic	Persistence	Bioaccumulation
Decabromodiphenyl Ether	1163-19-5	L	M	L	L	H	*L*	M	L		L	L	*L*	*L*	VH	*H*
Decabromodiphenyl Ethane	84852-53-9	L	*M*§	L	L	*H* §	*L*	L	L		VL	VL	L	*L*	VH	*H*
Antimony Trioxide[I]	1309-64-4	L	M*	M	*M*	*L*	*L*	H	L		L	M	H	M	*H*[R]	*L*
Resorcinol Bis-Diphenylphosphate; RDP	125997-21-9	L	*M*§	L	L	M	M	M	*L*		L	VL	VH	VH	M	*H*‡
Triphenyl Phosphate	115-86-6	L	*M*	L	L	L	L	H	L		L	VL	VH	VH	L	M

NOTE: VL = Very Low hazard L = Low hazard M = Moderate hazard H = High hazard VH = Very High hazard Endpoints (VL, L, M, H, and VH) were assigned based on empirical data. Endpoints in italics (*VL, L, M, H, and VH*) were assigned using values from predictive models and/or professional judgment.

§ Based on analogy to experimental data for a structurally similar compound. * This alternative may contain impurities. These impurities have hazard designations that differ from the flame retardant alternative, Brominated poly(phenylether), as follows, based on experimental data: HIGH for human health, HIGH for aquatic toxicity, VERY HIGH for bioaccumulation, and VERY HIGH for persistence. This chemical is subject to testing in an EPA consent order for this endpoint.

* Ongoing studies may result in a change in this endpoint.

‡ The highest hazard designation of any of the oligomers with MW <1,000.

[R] Recalcitrant: Substance is comprised of metallic species that will not degrade, but may change oxidation state or undergo complexation processes under environmental conditions.

**Aquatic toxicity: EPA/DfE criteria are based in large part upon water column exposures which may not be adequate for poorly soluble substances such as many flame retardants that may partition to sediment and particulates.

[I] This compound is included in the ongoing EPA Work Plan evaluation for Antimony Trioxide.

SOURCE: EPA 2014i.

TABLE 12-13 Clean Production Action Draft Benchmark Scores

Substance	**Draft Benchmark score**	**Basis of Benchmark Score**
DecaBDE	Benchmark 1	Very high persistence; high bioaccumulation; high developmental toxicity (1a, 1c, 1e).
DBDPE	Benchmark 1	Very high persistence; high bioaccumulation; high developmental toxicity (1a, 1c, 1e).
ATO	Benchmark 1	High systemic repeat dose toxicity and very high persistence (1c).
RDP	Benchmark 2	Very high ecotoxicity (2f); moderate Group I human toxicity end points (carcinogenicity) (2e); and high bioaccumulation and moderate toxicity (2d).
TPP	Benchmark 2	Moderate Group I human toxicity end points (carcinogenicity and endocrine activity) (2e); high Group II human toxicity end points (repeat dose systemic) and very high ecotoxicity end points (acute and chronic aquatic toxicity) (2f).

CASE STUDY 2: CHEMICAL SUBSTITUTION OF A HAZARDOUS BIOLOGICALLY ACTIVE COMPOUND (GLITAZONE)

In this case study, an alternatives assessment will be performed on three chemicals that were originally developed as pharmaceutical agents. The rationale for choosing this example was driven in part by the committee's statement of task requiring examples demonstrating "how high throughput and high content data streams could inform assessment of potentially safer substitutes early in the chemical development process" (see Chapter 1). This case study was specifically intended to illustrate how in silico and in vitro high throughput screening (HTS) data, animal toxicity data, and human health outcome data can be used to assess potential hazards associated with a chemical substitution.

When considering this case study, it is important to note the following:

- This case study represents a hypothetical situation where there is a need to find a substitution for a biologically active ingredient that has been identified to cause severe liver injury. This was the result of accidental ingestion by humans during or after the use of the product containing this active ingredient.
- Although based on a real-life historical problem, the presentation of data has been adapted to illustrate the use of the committee's framework. The approach shown is for illustration purposes only and is not intended as a commentary on any drug development or regulatory process.
- Many of the comparisons made here are based on data and knowledge that were not available at the time of regulatory approval for these drugs. The human health observations associated withthese chemicals drove much of the scientific investigation that led to the development of some of the key in vitro assays and their implications for safety that are discussed in this case study.
- This case study is not intended to imply that all chemical alternatives should be held to the same level of stringency (e.g., as commonly used in the development of pharmaceuticals).
- Publicly available data have been used throughout this case study. For example, the mammalian safety assessments for all three chemicals are taken from the original Summary Basis of Approval documents that are publicly available from the FDA through the Freedom of Information Act. These studies were conducted according to Good Laboratory Practice (GLP) guidelines and formed the basis for regulatory approval.

Steps 1- 4 of the Framework

Step 1: Identify Chemical of Concern

Concerns for human health have been identified with the primary biologically active ingredient, (*RS*)-5-(4-[(6-hydroxy-2,5,7,8-tetramethylchroman-2-yl)methoxy]benzyl)thiazolidine-2,4-dione, in a product that is widely used across the world. This ingredient (Figure 12-2) is commonly referred to by its abbreviated trade name, Glitazone-T, and is the chemical of concern in this scenario. Numerous reports of severe liver injury, sometimes fatal, in people exposed to products containing Glitazone-T have come to light, so there is a desire to reduce human exposure, eliminate Glitazone-T from the product, or find an alternative chemical substitute for this active ingredient.

Step 2: Scoping and Problem Formulation

Glitazone-T is the primary biologically-active ingredient in the products in which it is used. The exact mechanism of action of Glitazone-T has not been clearly established, although its stimulatory effect on the peroxisomal proliferator activated receptor gamma (PPARγ) is well known and thought to play a key role in its biological effectiveness. In vitro experiments with Glitazone-T showed that the activity of PPARγ increased by 50% at a concentration of 0.72 μM when tested in transfected HepG2 cells. In 3T3-L1 adipocytes, it was shown to reduce the uptake of 2-deoxyglucose by 50% at a concentration of 2 μM.[51]

Regulatory authorities have identified Glitazone-T as having potential adverse effects on human health. Products containing this active ingredient have been linked to numerous cases of severe liver injury, and in some cases, these effects result in fatalities (Watkins and Whitcomb 1998). The bioavailability of Glitazone-T is approximately 58%. Product effectiveness requires relatively high concentrations in the final formulation. As a

[51] Data available from FDA Summary Basis of Approval by FOIA request.

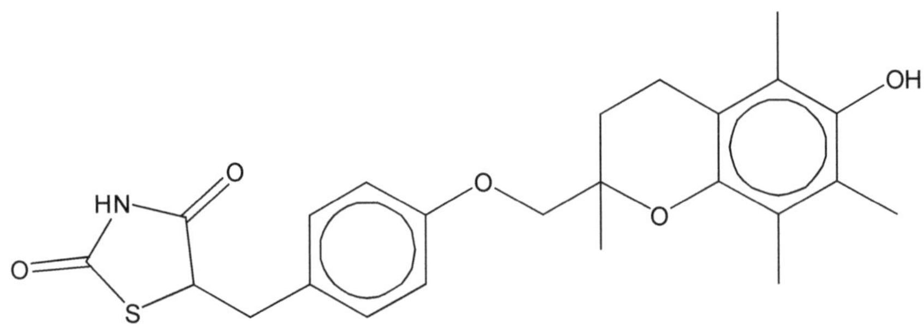

FIGURE 12-2 Chemical structure of Glitazone-T, CAS # 97322-87-7.

consequence, it is estimated that the maximum adult human daily exposure to the active ingredient is approximately 400 mg through the normal use of products containing Glitazone-T. Any proposed alternative must satisfy government bodies and product consumers that it has a substantially improved safety profile for human health.

Other considerations in Step 2 include identification of the following:

1. *Stakeholders*: Relevant internal stakeholder groups include safety experts, chemists, and pharmacologists. External stakeholders include relevant advocacy groups and regulatory agencies. These groups may have differing views on the relative importance of the various aspects of an alternatives assessment, such as the relative weight given to functional performance vs. environmental or human health concerns for any proposed alternative.
2. *Guiding assumptions and values implicit in the assessment*: Avoid persistent, bioaccumulative, and toxic (PBT) chemicals. Whenever possible, the GreenScreen® for Safer Chemicals® classification system will be used to assign health and ecological hazard ratings.
3. *Function and performance requirements for the substance of concern and alternatives:* Complete removal of Glitazone-T would eliminate any functional use of those products where this active ingredient is included, rendering the product nonviable from an economic perspective. Any alternative must be able to replace the biological activity of Glitazone-T, including activation of PPARγ, which is thought to be critical to the beneficial effects observed from using this class of product.
4. *Hazards of concern and potential exposure trade-offs that should be evaluated in the assessment:* Alternatives to Glitazone-T must have a lower potential for causing human hepatotoxicity. Ecotoxicity must also be considered, since release of Glitazone-T and its alternatives to wastewater can occur. Because of the beneficial aspects of the product, human health considerations are considered a primary motivation.
5. *Assessment Steps to be completed*: Steps 1-8 and 10 should be completed. Because product use is anticipated to be similar, a comparative Exposure Assessment (Step 6.3) and Life Cycle Thinking (Step 8) should be adequate and the optional Step 9 not needed.
6. *Identify safer alternatives:* In Step 7, assessments of in vivo data will be completed using the GreenScreen® tool. GreenScreen® assessments may be supplemented with additional data sources, such in vitro and in silico investigations, if needed. Remaining data gaps will be handled in accordance with the GreenScreen® guidelines. End points with insufficient information to assess the hazard are assigned a data gap (DG). For illustration purposes, the uncertainty of each in vivo finding will also be considered.[52] Factors used to evaluate parameter uncertainty will include robustness of the data (e.g., multiple studies, multiple species, adequacy of the reporting of the results), and model uncertainty (e.g., relevance of an assay end point to a human health end point of concern). A neutral approach to uncertainty and missing data will be used in this example (see Chapter 9 for more details).
7. Life Cycle Thinking (Step 8) will qualitatively determine if there are differences in material or energy flow or synthetic history exist between

52 Strategies for handling uncertainty in other endpoints could also be developed.

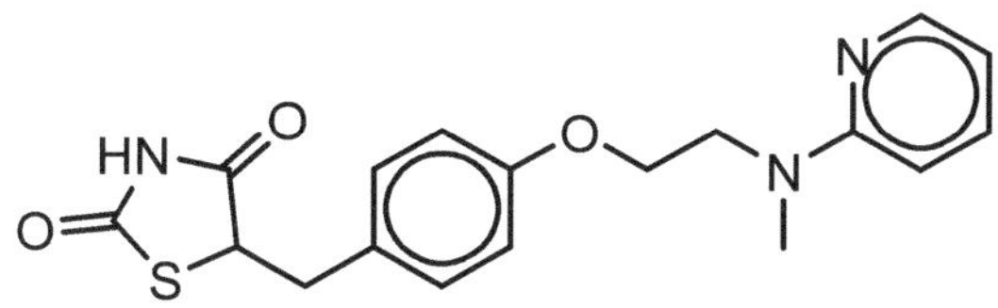

R-ThZD

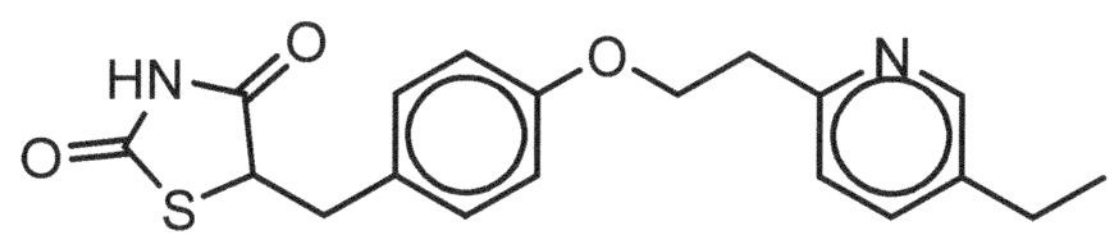

P-ThZD

FIGURE 12-3 Chemical structures of R-ThZD and P-ThZD.

the original chemical (Glitazone-T) and the potential alternatives.

Step 3: Identify Potential Alternatives

Numerous structural analogs to Glitazone-T are available, but for the most part these were deemed to have either lower potency against the PPARγ receptor or had physicochemical properties, such as solubility or bioavailability, that would reduce their effectiveness as a replacement for Glitazone-T. However, two viable alternatives have been identified: 5-(4-{2-[Methyl(2-pyridinyl)amino]ethoxy}benzyl)-1,3-thiazolidine-2,4-dione, commonly referred to as R-ThZD and 5-{4-[2-(5-Ethyl-2-pyridinyl)ethoxy]benzyl}-1,3-thiazolidine-2,4-dione, also known as P-ThZD. Structures for these alternatives are shown in Figure 12-3.

In in vitro experiments, R-ThZD and P-ThZD were shown to increase the activity of PPARγ by 50% at concentrations of 0.082 μM and 0.81 μM, respectively, when tested in transfected HepG2 cells. In in vitro 3T3-L1 adipocytes R-ThZD and P-ThZD were shown to reduce the uptake of 2-deoxyglucose by 50% at concentrations of 50 nM and 3 μM, respectively.

Step 5: Assess Physicochemical Properties

General assessment of physicochemical properties indicates that both alternatives have similar physical characteristics in terms of their melting point, boiling point, and vapor pressure (see Table 12-14). However, computational assessments of the aqueous solubility of both R-ThZD and P-ThZD suggest that these chemicals are significantly more water soluble than Glitazone-T.

Assessment of Ecological Impact Based on Physicochemical Properties

Comparison of the physicochemical properties of Glitazone-T with the other two Glitazone alternatives show the same thiazolidinone ring structure, but Glitazone-T has a phenolic functional group as well as a prospectively liable, if masked, carbonyl group (Weltman et al 2011). The pKa (base) value is also orders of magnitude different between these chemicals. Hence the environmental fate and impact of Glitazone-T, its metabolites, or degradation products are uncertain.

Assessment of the ecological impact of a chemical and its degradation or metabolic products is best based on direct data. For P-ThZD, it has been experimentally determined that it and its major metabolites do not significantly bioaccumulate, persist in the aquatic environment, show toxicity to aquatic organisms, or become absorbed by sewage solids (Drug Bank, 2013a). An evaluation of R-ThZD can be carried out by comparison of physicochemical properties of P-ThZD and R-ThZD. Both P-ThZD and R-ThZD have similar chemical structures, functional groups, molecular weights, and logPs, as well as calculated pK_as and polar surface areas (psa). It is reasonable to assume that environmental binding, persistence, degradation, and transformation of R-ThZD is well modeled by P-ThZD (Drug Bank, 2013b). In terms of chemical structure, the only difference is in the substitution of pyridine rings, which would have a minor effect on the reactivity.

Assessment of Human Health Impacts Based on Physicochemical Properties

In comparing the physicochemical properties of R-ThZD and P-ThZD to Glitazone-T, it can be hypothesized that the lower Log*P* values for R-ThZD and P-ThZD and higher predicted aqueous solubility (see Table 12-12) will increase their relative bioavailability when compared to Glitazone-T. Given that the in vitro potency of R-ThzD is superior to

TABLE 12-14: Physicochemical Properties for Glitazone-T, P-ThZD, and R-ThZD

Property	**Glitazone-T** (EC_{50} = 0.72μM)	**P-ThZD** (EC_{50} = 0.81μM)	**R-ThZD** (EC_{50} = 0.082μM)
MW	441.5	356.4	357.4
cLogP[a]	5.585	3.533	3.02
Polar surface area	110.16	93.59	96.83
LogD (shake flask pH 7.4)	3.65	2.45	1.93
Aqueous solubility (*pred.*)	0.04 mg/ml	46.8 mg/ml	1033 mg/ml
Rule of 5 violations	1	0	0
Acid pKa	6.27	6.27	6.27
Melting point	184°C (exp.)	271°C (pred.)	153°C (exp.)
Boiling point @ 760 mmHg (*pred.*)	657 ± 55 °C	575 ± 45°C	585 ± 35°C
Vapor pressure at 25°C (*pred.*)	0.0 ± 2.1 mmHg	0.0 ± 1.6 mmHg	0.0 ± 1.6 mmHg

[a]Values for cLogP in this table were determined using the Biobyte software package.

SOURCE: ChemSpider 2014a, b,c.

that for Glitazone-T against the PPARγ receptor, and the in vitro potency of P-ThzD is comparable to that for Glitazone-T, then higher bioavailability of these alternatives will lead to a decrease in their relative concentrations in the end products. A direct result will be a reduction in the level of human exposure to these biological active ingredients, assuming that similar product usage patterns are equivalent.

Step 6.1: Assess Chemical Hazards for Human Health

This section examines the various data streams available for hazard assessment by looking at in silico, in vitro, and in vivo data.

Computational Assessment of Safety

In silico predictions for a variety of different properties were obtained for Glitazone-T, P-ThZD, and R-ThZD using some available quantitative structure activity relationship (QSAR) models. Model outputs include predictions of cytotoxicity to cells; inhibition of the human Ether-a-go-go Related Gene (hERG) ion channel that is associated with prolonged cardiac QT interval; volume of distribution; free fraction in human plasma; and other end points (Table 12-15). The rationale for choosing these predicted properties is explained in more detail in Chapter 8.

- *Cytotoxicity:* Compounds that cause cytotoxicity at lower in vitro concentrations will generally have a higher probability of causing toxicity in vivo at lower plasma concentrations (Greene et al. 2010a). The in silico predictions suggest that P-ThZD and R-ThZD will have a higher LC_{50} values for cytotoxicity in cells compared to Glitazone-T. Thus, cytotoxicity associated with these chemicals likely occurs at higher in vivo (plasma) concentrations.
- *hERG inhibition:* hERG channel inhibition has been shown to cause QT interval prolongation in humans. This alteration of the cardiac electrical cycle has been implicated in the onset of ventricular tachyarrhythmias like torsades de pointes, which can result in sudden death. In silico predictions suggest that there is no increased risk of hERG Inhibition with either R-ThZD or P-ThZD when compared to Gitazone-T.
- *Volume of distribution, free fraction, and passive permeability:* Volume of distribution (Vd_{ss}) has

TABLE 12-15 Various Predicted Properties for Glitazone-T, P-ThZD, and R-ThZD

Predicted Property	Glitazone-T	P-ThZD	R-ThZD
Cytotoxicity LC_{50}	79.7 μM	254 μM	259 μM
hERG (human Ether-a-go-go Related Gene) IC_{50}	19.2 μM	13.5 μM	19.4 μM
VDss (L/kg) Volume of distribution	0.615	0.538	0.3214
Fu (%) (Free fraction in human plasma)	0.00154	0.00976	0.00369
RRCK (Russ Ralph Canine Kidney) (x 10^{-6} cm/sec) Passive permeability	5.16 (Moderate)	28.5 (High)	26.2 (High)
MDR (Multidrug Resistance) efflux Pgp (P-glycoprotein) active efflux	2.04 (Low)	0.967 (Low)	0.914 (Low)
Structural alerts	Yes Thiazolidinedione	Yes Thiazolidinedione	Yes Thiazolidinedione
Mitochondrial dysfunction	High	Medium	Medium
BSEP (Bile Salt Export Pump) Inhibition @100 μM	75%	83%	80%

Note: Data for this table were generated by a committee member using unpublished Pfizer data.

been shown to correlate with the lowest observable adverse effect level (LOAEL) in preclinical studies, where higher Vd_{ss} values lead to lower LOAEL concentrations (Sutherland et al. 2012). The Vd_{ss} for all three compounds is predicted to be low, suggesting that there would be no substantial increased safety concern from either R-ThZD or P-ThZD when compared with Glitazone-T. Passive permeability is linked to bioavailability, where highly permeable compounds have high bioavailability. Moderate passive permeability was predicted for Glitazone-T, whereas R-ThZD and P-ThZD are expected to be higher, indicating that R-ThZD and P-ThZD would have better bioavailability. In addition, since the pharmacological action of a compound is generally driven by the unbound fraction in vivo, then a higher free fraction indicates that lower total drug doses would be needed to elicit the desired effect of the compound. The free fraction for both R-ThZD and P-ThZD is predicted to be ~9-fold and ~3-fold higher than Glitazone-T, suggesting that the overall exposure required to achieve the intended effect would be lower.[53]

- *Thiazolidinedione structural alert:* The thiazolidinedione substructure has been identified as a structural alert associated with hepatotoxicity resulting in liver failure and/or cholestatic hepatitis (Greene et al. 2010b). Cyp3A4 enzyme induction has also been observed with compounds containing this structural group. The mechanism of toxicity is thought to be via CYP mediated oxidation of the activated methylene to give a reactive quinoid intermediate (see Figure 12-4), which can be trapped with glutathione (GSH) in a reactive metabolite assay. All three compounds contain this structural alert, so it cannot be determined if the two alternatives, R-ThZD and P-ThZD, would have an improved safety profile when compared to the hepatotoxic Glitazone-T compound.
- *Mitochondrial dysfunction and BSEP inhibition:* Many eukaryotic cells derive the majority of their energy needs from the mitochondrial

[53] This observation could also affect wastewater concentrations of these compounds.

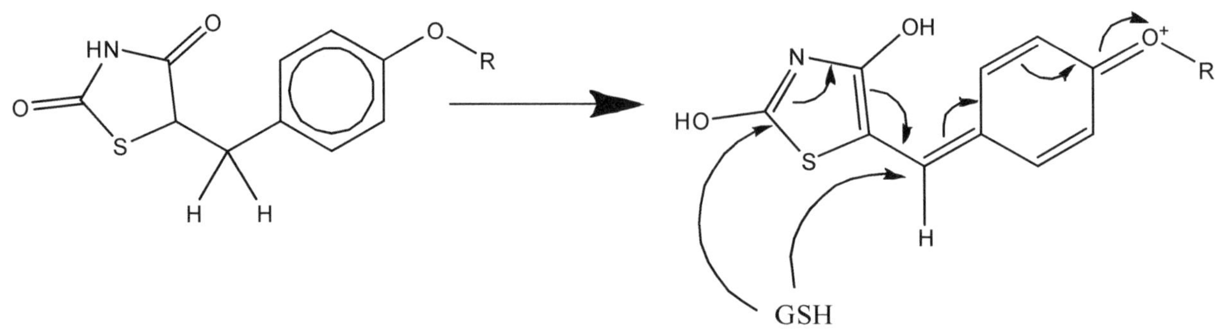

FIGURE 12-4 Formation of a reactive quinoid intermediate.

production of adenosine triphosphate (ATP). Interfering with mitochondrial production of ATP will deplete cellular energy stores, and may result in cellular stress and cell death. Glitazone-T is predicted to have a high likelihood of having an adverse impact on mitochondrial function, whereas R-ThZD and P-ThZD are predicted to have only a moderate likelihood of having an effect on mitochondrial function. Therefore, it might be expected that both R-ThZD and P-ThZD would have a lower likelihood of having adverse safety effects. As a co-factor to mitochondrial dysfunction, inhibition of the Bile Salt Extraction Pump (BSEP), an energy-dependent transporter, has been linked to causing cholestasis and hepatic injury. All three compounds are predicted to have similar inhibitory effects on BSEP. This information doesn't allow for differentiating between these chemicals on the basis of this potential mechanism of liver injury.

Based on the in silico analysis, R-ThZD and P-ThZD offer a slightly more favorable hazard profile than Glitazone-T due to a lower predicted potential for causing cytotoxicity, better predicted bioavailability, and lower plasma protein binding.

Using In Vitro Data to Assess Safety Hazards

- *In vitro absorption, distribution, metabolism, and excretion (ADME) assessments:* When comparing in vitro ADME data for all three compounds, R-ThZD was shown to have moderate metabolic stability in human liver microsomes and hepatocytes, whereas Glitazone-T had poor detection sensitivity in the experimental conditions. R-ThZD also had good passive permeability compared to Glitazone-T. No data were available for P-ThZD, but based on its close structural similarity and similar physicochemical properties to R-ThZD, along with similar in silico predictions for passive permeability, Vd_{ss} and protein binding, it might be expected that these two compounds would show similar profiles in the in vitro systems. Despite some observed differences in their interactions with specific biological pathways or proteins, metabolic stability and permeability have been shown to be strongly correlated with physicochemical characteristics, such as lipophilicity and pKa.
- *In vitro safety assays:* Glitazone-T was shown to cause cytotoxicity at lower concentrations (LC_{50}=78μM) compared to R-ThZD (259μM) and P-ThZD (263μM) in immortalized human liver epithelial (THLE) cells, but no significant difference was noted in a human liver carcinoma (HepG2) cell line. This might be due to the increased sensitivity of THLE to compounds that affect mitochondrial function when compared with HepG2 cells. This relative difference in mitochondrial dysfunction was confirmed in vitro, where Glitazone-T shows both uncoupling and inhibitory effects on isolated mitochondria at significantly lower concentrations than that observed for either R-ThZD or P-ThZD (see Table 12-16). However, it is worth noting that P-ThZD had a significantly greater inhibitory effect on the BSEP transporter than either Glitazone-T or R-ThZD.

When comparing these compounds for their effects on endoplasmic reticulum stress, which has been linked to a number diseases, R-ThZD showed an measurable increase in the nuclear translocation of XBP1, part of the endoplasmic reticulum stress pathway, at much lower concentrations than Glitazone-T and P-ThZD, which may indicate a slightly higher concern for adverse effects with R-ThZD.

TABLE 12-16 In Vitro Safety Data.

In Vitro Safety Assay	**Glitazone-T** (EC_{50} = 0.72 μM)	**P-ThZD** (EC_{50} = 0.81 μM)	**R-ThZD** (EC_{50} = 0.082 μM)
Cytotoxicity in THLE cells	78 μM	263 μM	259 μM
Cytotoxicity in HepG2 cells	242 μM	>300 μM	>276 μM
XBP1 Reporter Assay (ER Stress)	144 μM	279 μM	46 μM
BSEP Inhibition (Bile Salt Extraction Pump)	9.1 μM	0.15 μM	5.2 μM
Mitochondrial Uncoupling (Isolated mitochondria)	22.9 nmol/mg	>100 nmol/mg	88.7 nmol/mg
Mitochondrial Inhibition (Isolated mitochondria)	55 nmol/mg	>100 nmol/mg	>100 nmol/mg
Off-target pharmacology (%inhib@10μM >50%)	Dopamine Transporter, Norepinephrine Transporter, Thyroid Hormone Receptor, GABA A, CYP3A4, H3	None	None

NOTE: Data for this table were generated by a committee member using unpublished Pfizer data.

Finally, R-ThZD and P-ThZD had fewer off-target effects when compared to Glitazone-T in a panel of biochemical binding assays. Greater target promiscuity has been linked to a higher likelihood of observing toxicity at lower exposures (see Chapter 8 for more details).

- *ToxCast data:* Glitazone-T and P-ThZD have been profiled in numerous in vitro assays as part of the ToxCast initiative. Figure 12-5 shows the in vitro profile in the Apredica high content assays, where data for P-ThZD Glitazone-T are presented. This figure illustrates that Glitazone-T has increased effects on p53 and mitochondrial membrane potential when compared to P-ThZD, which suggests that P-ThZD may have a better safety profile.

Similarly, when comparing the profiles for these two compounds in the Attagene nuclear hormone receptor panels in Figure 12-6, it can be seen that aside from the intended biological activity of these molecules, Glitazone-T is having an effect on more of these receptors than P-ThZD. Based on these observations, it may be expected that P-ThZD would have a better safety profile than Glitazone-T.

This trend is also observed when comparing the in vitro profiles of the two chemicals in the BioSeek platform (Figure 12-7), where it can be observed that Glitazone-T has a much stronger response across almost all of the measured end points when compared to the profile for P-ThZD. Similar data were not available for R-ThZD, but based on the close structural similarity and similar physicochemical properties to R-ThZD, along with similar in silico predictions for passive permeability, Vd_{ss} and protein binding,it might be expected that these two compounds would show similar profiles in the in vitro systems, although differences could be present based on the observation that these two chemicals have different activities for XBP1 and BSEP.

Glitazone-T is more cytotoxic in THLE cells when compared to P-ThZD and R-ThZD, which probably reflects its greater impact on mitochondrial function. Similarly, there is a general lack of off-target activity for P-ThZD and R-ThZD when compared to Glitazone-T. Although P-ThZD is a more potent inhibitor of the BSEP transporter than either R-ThZD or Glitazone-T, this finding by itself may not translate into a direct biological effect in vivo. R-ThZD has a greater impact on inducing ER stress based on the XBP1 reporter assay, and so may be expected to show in vivo toxicity at lower plasma concentrations than P-ThZD. From the ToxCast profiles, P-ThZD has a "cleaner" profile across the three assay platforms when compared to Glitazone-T. Therefore, it might be expected to have a better in vivo safety profile aside from those effects related to the primary mechanism of action of these compounds. ToxCast data were not available for R-ThZD, and so comparisons between these two alternatives cannot be made. Based on the in vitro assessments of Glitazone-T, P-ThZD, and R-ThZD, it can be inferred that both P-ThZD and R-ThZD would have fewer effects on a biological system compared to Glitazone-T, making them potentially viable safer alternatives.

Mammalian Toxicity Assessment

Comparisons between Glitazone-T, R-ThZD, and P-ThZD were made based on the available data using the GreenScreen® classification system.[54]

Acute mammalian toxicity: Glitazone-T has an acute oral LD_{50} of greater than 2000 mg/kg in multiple species, and so it receives a hazard designation of Low. P-ThZD, however, has an acute oral LD_{50} = 181mg/kg in mice, which is considered to be Very High. Similarly, R-ThZD has a mouse LD_{50} = 300 mg/kg, so its acute mammalian toxicity is categorized as High.

Carcinogenicity: In mice, Glitazone-T showed an increased hemangiosarcoma incidence in females at 400 mg/kg and in males and females at 800 mg/kg. In mice, Glitazone-T showed an increased hepatocellular carcinoma incidence in females at 800 mg/kg (Herman et al. 2002). P-ThZD showed benign and/or malignant transitional cell neoplasms in rats at 4 mg/kg/day and an increased incidence of urinary bladder tumors at 63 mg/kg. R-ThZD showed a significant increase in benign adipose tissue tumors (lipomas) in rats at doses greater than or equal to 0.3 mg per kg (mg/kg/day) for 104 weeks. On the basis of this evidence, all three chemicals are categorized as Moderate.

Mutagenicity/genotoxicity: Glitazone-T was not mutagenic in bacteria at concentrations up to 10,000 μg/plate, with or without metabolic activation. In a Chinese hamster fibroblast assay, both aneuploid cells and giant cell forms were noted after exposure to 2.9 μg/ml without metabolic activation for 48 hours. With activation, the number of cells with endoreduplicated chromosome was increased with Glitazone-T at 58 and 64 μg/ml. Pronounced cytotoxicity and increased structural chromosome aberrations frequency were observed following 6 hours of exposure to Glitazone-T at 178 μg/ml without activation and at 163 μg/ml with activation. Results of the in vitro mouse lymphoma mutation assay at cytotoxicity-limited concentrations up to 30 μg/ml were mixed because minimal, but significant increases in mutation frequency were noted in two out of five trials without metabolic action and in two out of six trials with activation. The unscheduled DNA synthesis was observed in hepatocytes isolated 2 or 24 hours post-dose from rats given single oral doses of Glitazone-T at 1,000, 1,500, or 2,000 mg/kg. Thus, it is concluded that the Glitazone-T genotoxic potential should be categorized as Moderate.

P-ThZD showed no mutagenic or genotoxic potential in bacterial mutagenicity studies, in vitro mammalian tests, and in vivo micronucleus studies. Thus, it can be concluded that P-ThZD genotoxic potential should be categorized as Low.

The overall genotoxicity potential of R-ThZD appears to be equivocal since the tests of chromosomal aberration, unscheduled DNA, and in vivo mouse micronucleus were all negative, while the incidence of forward mutations at the thymidine kinase locus of mouse lymphoma LS 178Y cells was increased by R-ThZD in triplicate assays in the presence of S-9 mix. Thus, it can be concluded that R-ThZD genotoxic potential should be categorized as Moderate.

Reproductive & developmental toxicity: Pregnancy duration was slightly shorter in rats given Glitazone-T at 1000 mg/kg when compared with untreated controls. Growth rate of rat pups was reduced in both sexes following high dose (2000 mg/kg/day) Glitazone-T. This effect was particularly pronounced between postnatal days 29 to 57. Aside from these findings, Glitazone-T had little or no effect on fertility, teratology, and peri- and post-natal development in rodents and rabbits. Based on this information, the reproductive hazard categorization is Low and the developmental hazard categorization is Very Low.

In studies with P-ThZD (Takeda Canada 2012), rats exhibited delayed parturition, embryotoxicity, delayed development, and reduced fetal weights at oral doses > 40 mg/kg/day. In rabbits, embryotoxicity was observed at an oral dose of 160 mg/kg. Based on this information, both the reproductive and the developmental hazard categorizations are Moderate.

R-ThZD treatment of rabbits and rats was studied by GSK (GSK 2012). Treatment of rats during early pregnancy did not result in notable implantation or embryo impacts. However, treatment of both rats and rabbits during mid-late pregnancy was associated with growth retardation and fetal death. Teratogenicity was not observed. Placental pathology was observed with R-ThZD treatment of rats (>3 mg/kg/day) but not in rabbits (100 mg/kg/day). When rats were treated during pregnancy and lactation with R-ThZD, reductions in

[54] Alternative (e.g., GHS) classification schemes could be used.

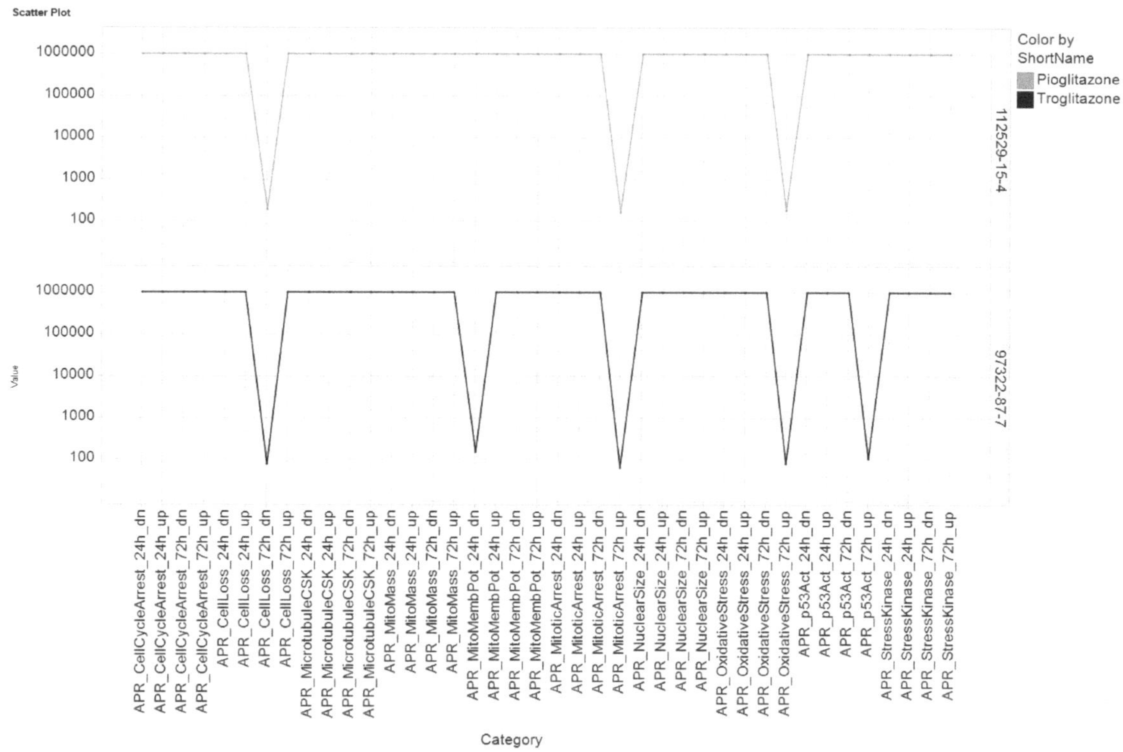

FIGURE 12-5 Apredica assay profiles for Glitazone-T (Trogliatazone) and P-ThZD (Pioglitazone). NOTE: Data in figure are from EPA ToxCast Initiative; figure generated using Spotfire.

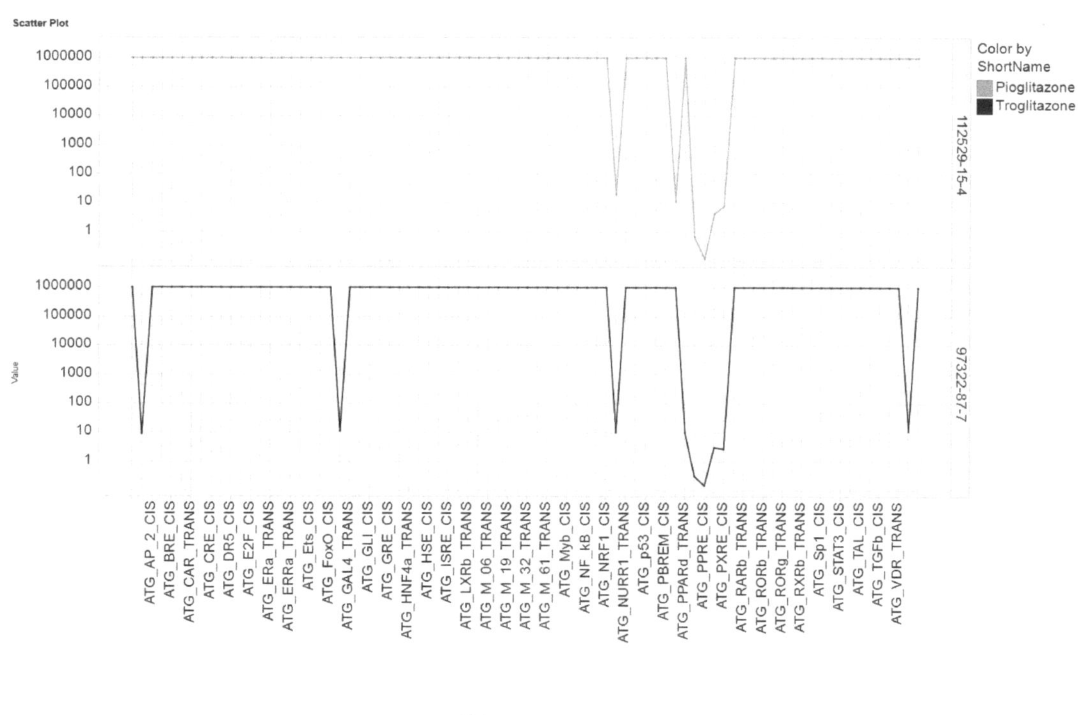

FIGURE 12-6 Attagene Nuclear Hormone Receptor panel assay profiles for Glitazone-T (Trogliatazone) and P-ThZD (Pioglitazone). NOTE: Data in figure are from EPA ToxCast Initiative; figure generated using Spotfire.

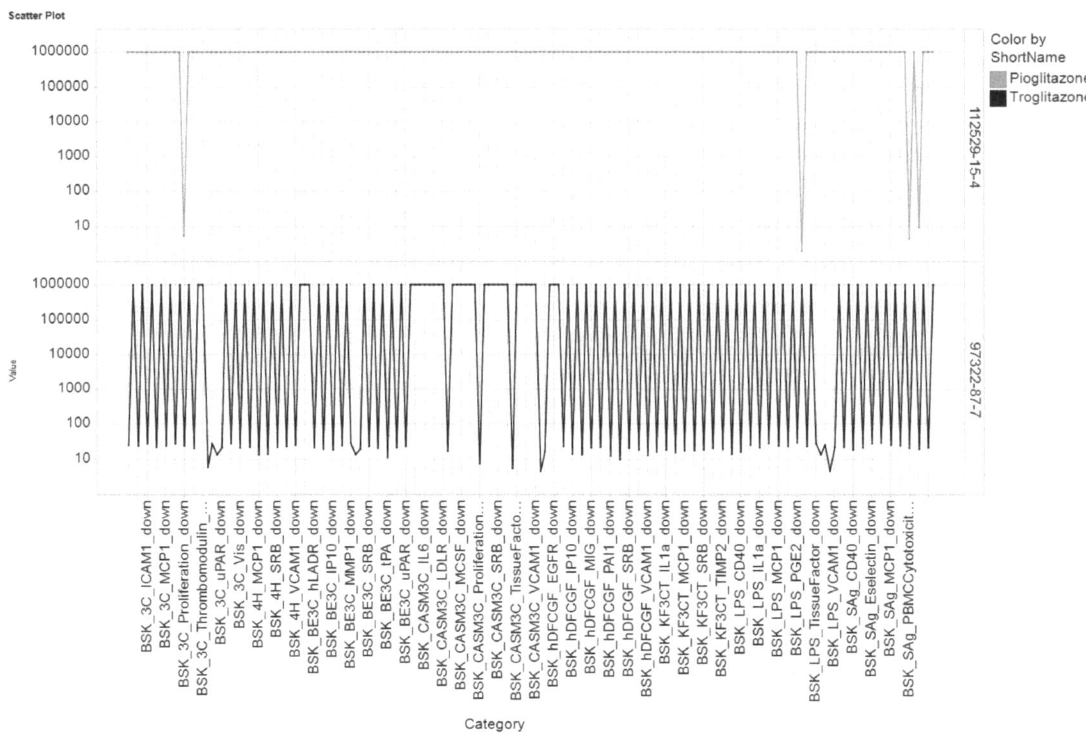

FIGURE 12-7 BioSeek panel assay profiles for Glitazone-T (Trogliatazone) and P-ThZD (Pioglitazone). NOTE: Data in figure are from EPA ToxCast Initiative; figure generated using Spotfire. Conclusions from the In Vitro Safety Data.

litter size and neonatal viability were observed. Postnatal growth retardation that was reversible after puberty was also seen. The no-effect dose for effects on the placenta, embryo, and offspring was 0.2 mg/kg/day in rats and 15 mg/kg/day in rabbits. Fertility was decreased at a dose of 40 mg/kg per day, and estrous cyclicity was altered at 2 mg/kg per day, but these effects were not noted at doses less than 0.2 mg/kg per day. These effects were attributed to altered plasma levels of progesterone and estradiol. Based on this information, the reproductive and developmental hazard categorizations are High.

Neurotoxicity: In rats given amorphous Glitazone-T at 6, 25, 100, or 400 mg/kg by gavage for 13 weeks, there were no deaths or drug-related clinical signs. Based on this information, the neurotoxicity hazard categorization is considered Low. No functional or behavioral toxicity was observed in offspring of rats given oral doses up to 80 mg/kg of P-ThZD. Based on this information the neurotoxicity hazard categorization is considered Moderate.

In a 13-week dietary range-finding study, mice were given R-ThZD at doses of 0, 0.4, 2, 10, or 20 mg/kg/day by dietary admixture. There was no mortality. No remarkable clinical signs were noted except firm, but palpable, swellings in the scapular areas noted in 14/16 animals in the high-dose group and 6/16 animals dosed at 10 mg/kg/day. Based on this information, the neurotoxicity hazard categorization is considered Moderate.

Repeated dose toxicity: In 13-week studies with Glitazone-T, dose-related increases in absolute and relative liver weight of 21%-75% in male rats at 400 mg/kg and 14%-48% in female rats at 50 mg/kg were observed. Heart weight and its body weight ratios in female rats increased 28%-53% at 200 and 400 mg/kg at week 13, respectively. No effects in dogs or monkeys given up to 400 mg/kg/day for 28 days. Based on this information, the repeat dose hazard categorization is Moderate.

Anemia with reduced erythrocytes, hematocrit and hemoglobin concentration, and splenic extramedullary hematopoiesis were present in rats after 13 weeks of oral administration of P-ThZD at doses of 100 or 300 mg/kg. The toxicological no effect dose might be near 30 mg/kg. Based on this information, the repeat dose hazard categorization is Moderate.

In a 13-week study, there was a dose-related increase in scapular adipose tissue weight in female mice at 2 mg/kg/day. Because this brown adipose tissue is not found in people, this response is not considered relevant. In males, there was a slight increase of 10% in kidney weight at 2 mg/kg/day and above. An increase of up to 16% in heart weight at 10 and 20 mg/kg/day was noted. This end point was chosen as the point of departure because heart effects were noted in longer-term studies in multiple species. Based on this information, the repeat dose hazard categorization is Moderate.

Respiratory and skin sensitization: No information is available to assess the respiratory and skin sensitization hazards associated with Glitazone-T, P-ThZD, or R-ThZD. Therefore, the respiratory and skin sensitization hazard categorization is Unknown.

Eye and Skin irritation/corrosivity: No information is available to assess the eye and skin irritation and the corrosive hazards associated with Glitazone-T, P-ThZD, or R-ThZD. Therefore the eye and skin irritation and corrosivity hazard categorization is Unknown.

Mammalian toxicity summary: Table 12-17 summarizes the mammalian toxicity assessment based on the GreenScreen® classification system.

Step 6.2: Assess Ecotoxicity (Chemical Hazards)

This section compares the environmental toxicity of three compounds: P-ThZD, R-ThZD, and Glitazone-T. There is sufficient experimental data for P-ThZD to characterize the aquatic toxicity by comparing the measured toxic end points to the thresholds described in several chemical alternatives assessments. There is, however, a lack of directly measured empirical data to characterize the aquatic toxicity of R-ThZD or Glitazone-T. The toxicity of the latter two chemicals, compared to P-ThZD, was estimated based on the chemical properties and reactivity of these chemicals. There is, however, uncertainty in any conclusions when comparing a chemical with an experimentally well-defined toxicity (P-ThZD) relative to the other two alternatives, which have no direct measurements of aquatic toxicity. The latter is a significant data gap in making any comparison.

There is no terrestrial toxicity data for any of these three compounds. However, the mammalian toxicity data generated to estimate human toxicity (see Mammalian Toxicity Summary)can be used to compare the toxicity of these three compounds to small mammals.

TABLE 12-17 Summary of Mammalian Toxicity Assessment

Alternative	Acute mammalian				Carcinogenicity				Mutagenicity/ Genotoxicity				Reproductive Toxicity				Developmental Toxicity				Neurotoxicity			Repeated Dose			Respiratory and Skin Sensitization
	VH	H	M	L	VH	H	M	L	VH	H	M	L	H	M	L	VL	H	M	L	VL	H	M	L	H	M	L	
Glitazone-T				[green]				[yellow]			[orange]				[yellow]					[yellow]			[orange]		[green]		DG
P-ThZD	[yellow]						[yellow]					[green]		[green]				[green]			[orange]				[yellow]		DG
R-ThZD		[yellow]					[yellow]				[green]		[green]				[green]				[orange]				[yellow]		DG

Note: Toxicity data has been benchmarked using the GreenScreen® system. The uncertainty associated with each toxicological finding is depicted by colors (green = minimal uncertainty, yellow= moderate uncertainty, orange = highly uncertain, gray = data gap)

Aquatic Toxicity

Weltman et al. (2011) provide an assessment of the environmental fate and effects of P-ThZD conducted as a higher-tier assessment triggered by exceeding screening criteria under a preliminary evaluation based on "Guideline on the Environmental Risk Assessment of Medicinal Products from Human Use"(EMA 2006). The data generated included various physical-chemical parameters (e.g., biodegradation, K_{ow}, aerobic transformation in sediments, K_{oc}) and toxicity to sewage microorganisms. The aquatic toxicity was characterized based on toxicological testing with a freshwater algae (species not provided), a freshwater invertebrate (*Daphnia magna*), and an early life stage fish (species not provided).

The algal test was a 72-hour exposure that measured the algal response as average specific growth rate and yield (as cell number) over a range of concentrations. The testing provided a no observed effect concentration (NOEC) and percentage effect (relative to controls) of EC_{10}, EC_{20}, and EC_{50}. The EC_{10} was the lowest effect level measured. The invertebrate test was a 21-day test that measured the parental mortality and reproduction (as neonates per female) over a range of concentrations. The testing provided a NOEC for reproduction, an overall NOEC (reproduction and mortality), and a percentage mortality (relative to controls and measured as immobile adults) of EC_{20}, EC_{40}, and EC_{50}. The EC_{20} was the lowest effect measured.

The fish early life stage test derived a NOEC in a 21-day test (range of concentrations) based on larval survival (post-hatch) and growth of larvae over the course of the test.

Several of the existing chemical assessment alternatives reviewed in this report (Chapter 7) use the type of ecological toxicity test data measured in this study of P-ThZD to characterize the acute and chronic toxicity of a chemical based on a range of thresholds. Table 12-18 summarizes the thresholds and categories provided by the four chemical alternatives assessments that provide quantitative characterizations of toxicity.

Characterization of Aquatic Toxicity

The various categories in Table 12-18 were applied to the toxicity data from Weltman et al. (2011) to characterize the aquatic toxicity of P-ThZD. The toxicity data for algae included the following:

- Algal toxicity (growth rate) had a measured EC_{10} at 0.702 mg/L, but there was no further response at higher concentrations. The authors report the EC_{50} at some concentration above 0.851 mg/L. Therefore the characterization of the EC_{50} for growth rate as very high toxicity is a conservative (i.e., environmentally protective) characterization. The actual EC_{50} may be higher.
- Algal toxicity (yield) had a measured EC_{10} at 0.189 mg/L and a measured EC_{20} at 340 mg/L, but there was no further response at higher concentrations. The authors report the EC_{50} at some concentration above 0.851 mg/L. Therefore, the characterization of the EC_{50} for yield as very high toxicity is a conservative (i.e., environmentally protective) characterization. The actual EC_{50} may be higher.
- Weltman et al. (2011) estimate the overall NOEC for algae at 0.189 mg/L, which was the EC_{10} for the yield end point. They did not estimate a LOEC. However, we used the EC_{20} for yield (the first measured response above the NOEC), 0.340 mg/L as the LOEC.

The toxicity data for invertebrates were chronic end points (21- day test) and included the following:

- Invertebrate mortality (measured as invertebrate mobility) had a chronic LOEC of 0.0387 mg/l based on a LC_{20} for adult mobility.
- A NOEC of 0.296 mg/l and a LOEC of 0.530 for reproduction measured as the number of offspring produced per adult *D. magna*.
- An estimated overall NOEC of 0.7530 (Weltman et al. 2011).

The fish early life stage toxicity tests indicated no response in survival of fry over the course of the test (32 days). The estimated NOEC and LOEC for body weight were 0.0584 mg/L and 0.1296 mg/L, respectively.

Table 12-19 provides this comparison. The aquatic toxicity for P-ThZD is generally characterized as high toxicity, with the exception of the characterization of NOEL under the P20ASys.

In terms of structure, the difference in the compounds is only in the substitution of pyridine rings, which would have a minor effect on the reactivity. This analysis indicates that the toxicity of the two compounds is likely to be similar.

TABLE 12-18 Aquatic Toxicological End Points and Assigned Category from Chemical Alternatives Assessments

	Acute Toxicity End point	End point Thresholds (mg/L)	Assigned Category	**Chronic Toxicity** Endpoint	End point Thresholds (mg/L)	Assigned Category
DfE	EC_{50} or LC_{50}	<1	Very High	LOEC	<0.1	Very High
		1 - 10	High		0.1 - 1	High
		10 - 100	Moderate		>1 - 10	Moderate
		>100	Low		>10	Low
IC2	96 hr LC_{50} (fish); 48 hr EC_{50} (crustacean); 72 hr or 96 hr ER_{50} (algae or aquatic plants)	<1	Very High			
		1 - 10	High			
		10 - 100	Moderate			
		>100	Low			
TURI P2OASys	LC_{50} (aquatic)	<0.1	10	NOAEC (fish)	<0.00002	10
		0.1 - 1	8		0.0002	8
		1 - 50	6		0.002	6
		50 - 1000	4		0.02	4
		> 1000	2		<0.2	2
	LC_{50} (plant)	<0.1	10			
		0.1 - 1	8			
		1 - 10	6			
		10 - 100	4			
		> 100	2			
Guide on Sustainable Chemicals				NOEC	<0.01	Not Toxic

Step 6.3: Conduct Comparative Exposure Assessment

Measurement of plasma protein binding in human serum showed that Glitazone-T was greater than 99.9% bound to protein, whereas P-ThZD and R-ThZD were 99.2% and 99.7% bound, respectively. Therefore, the free concentration available for the intended pharmacological action will be approximately seven times greater in the case of P-ThZD and two times greater for R-ThZD. These differences in free concentrations and absorption result in lower concentrations being required of both P-ThZD and R-ThZD to achieve the same biological effect compared to Glitazone-T, assuming equivalent potency against the PPARγ receptor across all three chemicals.

In vitro experiments have shown a 50% increase in PPARγ activity following exposure of transfected HepG2 cells with 0.72 μM Glitazone-T. In 3T3-L1, adipocytes Glitazone-T was shown to reduce the uptake of 2-deoxyglucose by 50% at a concentration of 2 μM. The bioavailability (the amount entering the bloodstream) of Glitazone-T is approximately 58%; for product effectiveness, it is necessary to have relatively high concentrations. As a result, it is estimated that the maximum adult human daily exposure to the active ingredient is in the region of 400 mg through the normal use of products containing Glitazone-T.

During in vitro experiments, R-ThZD and P-ThZD were shown to increase the activity of PPARγ by 50% at concentrations of 0.082 μM and 0.81 μM, respectively, when tested in transfected HepG2 cells. In 3T3-L1, adipocytes R-ThZD and P-ThZD were shown to reduce the uptake of 2-deoxyglucose by 50% at concentrations of 50 nM and 3 μM, respectively. The bioavailability of P-ThZD and R-ThZD is 81% and 60%, respectively, and the free concentrations in plasma are seven times greater for P-ThZD and two times greater for R-ThZD when

TABLE 12-19 Summary of Toxicity Data for P-ThZD

End Point	Measured Value	Toxic Category By Chemical Alternatives Assessment Method			
		DfE	IC2	P2OAsys	Guide on Sustainable Chemicals
Algae End Points					
Algae 72- hour EC_{50} (growth rate)	>0.851 mg/L	Very High	Very High	8	NA
Algae 72- hour EC_{50} (yield)	>0.851 mg/L	Very High	Very High	8	NA
Algae 72- hour NOEC	0.189 mg/L	NA	NA	NA	Toxic
Algae 72- hour LOEC	0.340 mg/L	High	NA	NA	NA
Invertebrate End Points					
D. magna LOEC (LC_{20} for adult mobility)	0.0387 mg/L	Very High	NA	NA	NA
D. magna NOEC (for reproduction)	0.296 mg/L	NA	NA	NA	Toxic
D. magna LOEC (for reproduction)	0.530 mg/L	Very High	NA	NA	NA
D. magna overall NOEC	0.0753 mg/l	NA	NA	NA	Toxic
Fish End Points					
Fish 32- day NOEC (early life stage body weight)	0.0584 mg/L			2	Toxic
Fish 32- day LOEC (early life stage body weight)	0.1296 mg/L	Very High			Toxic

compared to Glitazone-T. Based on these data, it is anticipated that concentrations of the biological ingredient in products will be substantially reduced; the anticipated maximum daily exposure to the active ingredient will be in the region of 45 mg in the case of P-ThZD and 4 mg in the case of R-ThZD.

Step 7: Are Alternatives Considered Safer?

Based on the available data, there are numerous ways to visualize and compare the profiles of the chemicals. No one way is considered as the preferred method. In all cases, one effect has not been deliberately ranked over another. Table 12-18 shows one approach incorporating the data into a single rank ordering of alternatives.

Another way to visualize and rank order these compounds would be to use the ToxPi, software, as explained in Appendix C and Reif et al. 2013. This software allows the categories of data to be grouped and weighted, if desired, to give a graphic comparison of chemicals. In addition to the graphic comparison, ToxPi software can be used to calculate an overall score for each chemical, using all the domains of data. In addition, the impact of giving more weight to some evidence categories on the overall ranking of compounds can easily be explored.

For the purpose of illustrating the effect that relative weightings can have on an overall assessment and ranking, data were grouped into seven logical categories or slices as outlined in Table 12-21. For the purpose of the illustration, the individual data points were rescaled to fall between 0 and 1, where "1" represents the most favorable value of the three for the data point in question and the rest are converted to a fraction of this data point. It should be noted that for some properties, lower numbers are considered more favorable than higher ones. For this reason, the calculations were adjusted to compensate for this directionality. Finally, no absolute thresholds were defined for an assay or property values because this was beyond the scope of the committee.

In Figure 12-8, the different slices of the pie charts represent the different components of the physicochemical properties, in vitro data, and in vivo and in silico predictions. In this example of data integration, the in vivo safety and exposure assessments carry the highest weighting, as illustrated by the lengths of the arcs for each slice. Preclinical ADME and in vitro data were the next highest weightings, with off-target activity, in silico predictions and physicochemical properties given the lowest weightings.

The relative ranking of each chemical can be seen in the three data points that the arrows point to. The higher ToxPi score represents a more favorable compound. In this example, P-ThZD had the best score, with R-ThZD in second place, and Glitazone-T the least favorable. As shown by the relative size of each slice, Glitazone-T was ranked last because of lower (unfavorable) scores in exposure, in vitro safety, off-target activities, and physicochemical properties.

In Figure 12-9, greater emphasis was placed on the in vivo (e.g., animal) safety assessments, increasing this to contribute 50% of the overall score for each compound. This was done to illustrate the effect of putting greater weight on the safety of a product over the functional use of the alternatives. In this case, Glitazone-T was the most favorable option, with P-ThZD second and R-ThZD the least favored.

In these analyses, the committee recognizes that there are varying levels of confidence in the different end point categorizations. In the illustration with mammalian toxicity data, uncertainty was considered and handled using a Missing-Data-Neutral approach (see Chapter 9 for more details). In this approach, the presence of uncertainty and missing data are noted, but would not exclude, or otherwise demote, the alternative at this point in the selection process.

Step 8: Life Cycle Thinking

In Step 8 (Life Cycle Thinking), it is first important to map the product system. For an agent like Glitazone-T, the key elements of the product system include: (a) transportation and storage of raw materials; (b) initial production of the active ingredient; (c) secondary processing resulting in the production of the product formulation; (d) product storage and distribution; (e) auxiliary operations, including disposal of production waste products; (f) therapeutic usage; and (g) post-consumer disposal and environmental fate of the drug and its metabolites (Mata et al. 2012). Life Cycle Thinking did not identify a significant difference in these areas, when the life cycle of the original chemical was compared to that of the alternative. Thus, additional screening life cycle analyses or more quantitative analyses were not required.

Step 10: Identify Acceptable Alternatives

In Figure 12-10, ToxPi was used to integrate different types of information (as discussed in Chapter 9). Specifically, ToxPi is used to combine the data from the human exposure assessments with the functional efficacy of each compound at the PPAR receptor, to incorporate a measure of functional performance into the weighting and ranking process. In addition, the relative contribution from the exposure and performance slices were increased to give exposure and performance the greatest emphasis, followed by in vivo safety, in vitro safety, preclinical ADME with in silico predictions, and physicochemical properties, off-target activities having the lowest weight. This illustrates the impact that weighting of functional performance as the highest criteria for selection can have at the integration step and how it may influence the outcome of an alternatives assessment. In this case, R-ThZD was the most favorable option, with P-ThZD in second place, and Glitazone-T least favored.

CONCLUSION

From these examples, it becomes clear that each of the three chemicals can be ranked as the most favorable, depending on the relative emphasis placed on the data points available. Depending on the entity performing the alternatives assessment, subtle differences in a chemical's attributes and rankings may lead to selection (or deselection) of an alternative. In this case, each of the alternatives has one or more human health or ecological hazards that may be desirable to avoid. Therefore, some framework users may initiate additional research and development efforts (Step 13).

The real-life outcome was that Glitazone-T

(Troglitazone) was withdrawn from the market in 2000, less than three years after regulatory approval, as a result of cases of severe liver injury in patients taking the drug. R-ThZD (Rosiglitazone) was approved in 1999 and reached peak sales of $2.5 billion in 2007, but was finally withdrawn in 2012, after reports linked the drug to cardiac toxicity. P-ThZD (pioglitazone) was approved by the FDA in 1999 and achieved sales worth $2.4 billion in 2008. It is still prescribed today, but has been withdrawn in some markets because of concerns with its association to bladder cancer after extended periods of treatment. Additional research and development efforts have led to the development of novel pharmaceutical treatment options for type 2 diabetes mellitus.

TABLE 12-20 Incorporation of Data into a Single Rank Ordering of Alternatives.
Note: A relative level of preference is assigned where "1" is the most preferable and "3" is the least preferable.

Category of Data	End Point	Glitazone-T	P-ThZD	R-ThZD
Exposure assessment	Estimated daily exposure	3	2	1
Physicochemical data	Log*P*/Log*D*	3	2	1
	Polar surface area	1	1	1
	Aqueous solubility	3	2	1
Predicted properties	Cytotoxicity LC_{50}	3	1	1
	hERG IC_{50}	1	1	1
	Volume of distribution	1	1	1
	Free fraction in human plasma	3	1	2
	Passive permeability	2	1	1
	MDR efflux	1	1	1
	Structural alerts	1	1	1
	Mitochondrial dysfunction	3	2	2
	BSEP inhibition @ 100μM	3	3	3
In vitro safety assays	Cytotoxicity in THLE & HepG2 cells	3	1	1
	XBP1 reporter assay (ER Stress)	2	1	3
	BSEP inhibition	1	3	2
	Mitochondrial uncoupling	3	1	2
	Mitochondrial inhibition	3	1	1
	Off-target pharmacology	3	1	1
Mammalian exposure	Bioavailability	3	1	2
	Protein binding	3	1	2
Mammalian toxicity	Acute toxicity	1	2	2
	Carcinogenicity	2	2	2
	Mutagenicity/Genotoxicity	2	1	2
	Reproductive toxicity	1	2	3
	Developmental toxicity	1	2	3
	Neurotoxicity	1	2	2
	Repeated dose toxicity	2	2	2
Ecological toxicity	Aquatic Toxicity	3	3	3
TOTAL SCORES		62	45	50

TABLE 12-21 Components of ToxPi Slices in Case Study Illustration

Category or Slice	Properties, assays or data
Functional efficacy	Human: $T_{1/2}$, T_{max}, AUC, PPB, C_{max}, projected human exposure or dose
Preclinical ADME	Bioavailability (rat, monkey and dog), VD_{ss}, Rat C_{max}, Rat Tmax, Rat AUC
In silico predictions	BSEP inhib, hERG inhib, MDR and RRCK, calculated rat PPB, THLE cytotoxicity, calculated human PPB
Off-target activity	% inhib @ 10μM values for the following Cerep targets: COX2, Dopamine Transporter, 5-HT transporter, PPAR gamma, PDE3, Na channel, Ca Channel, CB1, M1, Glucocorticoid, GABAA, Mu, Beta2, D1, H1, Alpha1, NE Transporter, 5HT2b.
In vitro safety	Cytotoxicity LC_{50} in HepG2 cells at 24 hrs in glucose and galactose, XBP1 activation assay, Caspase 3/7 activation, Mitochondrial inhibition and uncoupling, BSEP inhibition, cytotoxicity LC_{50} in HepG2 and THLE cells at 72 hrs in glucose containing media
In vivo safety	Assessments of mutagenicity, carcinogenicity, reproductive toxicity, neurotoxicity, repeat dose toxicity, acute toxicity, developmental toxicity
Physicochemical properties	Log*P*, Log*D*, PSA, PSA/MW, cSolubility, Acidic pK_a, Basic pK_a

NOTE: AUC = area under receiver operating characteristic curve; PPB = parts per billion; C_{max} = maximum concentration; VD_{ss} = volume of distribution at steady-state; Tmax = time of maximum plasma concentration; BSEP = bile salt export pump; hERG = human Ether-à-go-go-Related Gene; MDR = multi-drug resistant; THLE = T-antigen-immortalized human liver epithelial; COX2 = cyclooxygenase-2; 5-HT = serotonin transporter; PPAR = peroxisome proliferator-activated receptor; PDE3 = phosphodiesterase 3; Na = sodium; Ca = calcium; CB1 = cannabinoid receptor 1; GABAA = γ-Aminobutyric acid a; NE = norepinephrine; 5HT2b = 5-Hydroxytryptamine receptor 2B; LC_{50} = lethal concentration 50; XBP1 = X-box binding protein 1; LogP = partition coefficient; LogD = distribution coefficient; PSA = prostate-specific antigen

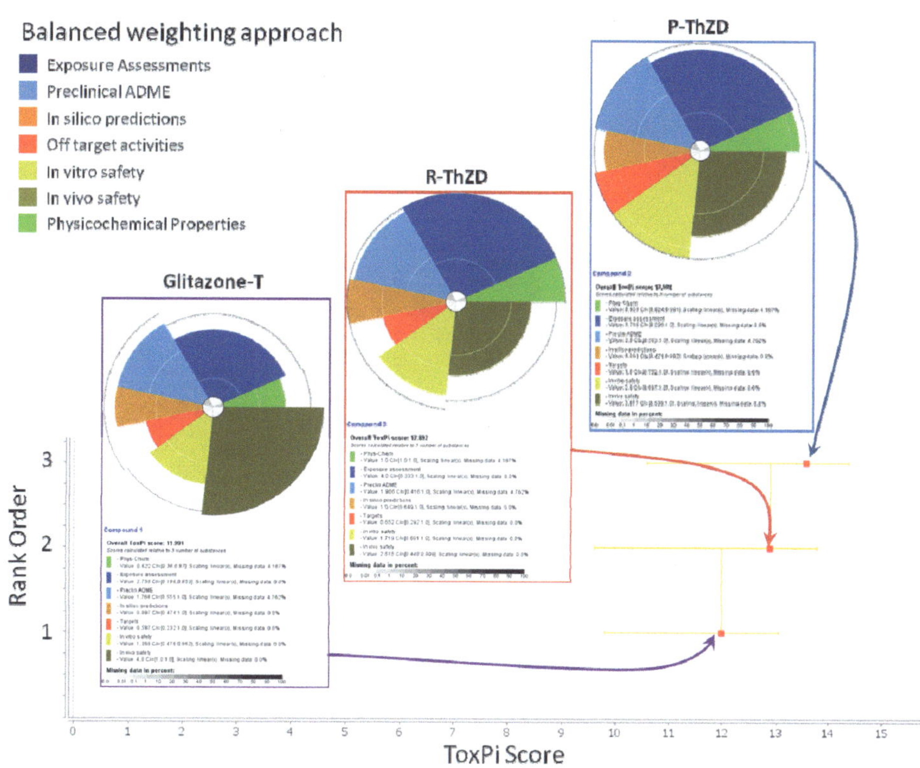

FIGURE 12-8 ToxPi visualization of data by data type and resultant rank ordering of chemicals.

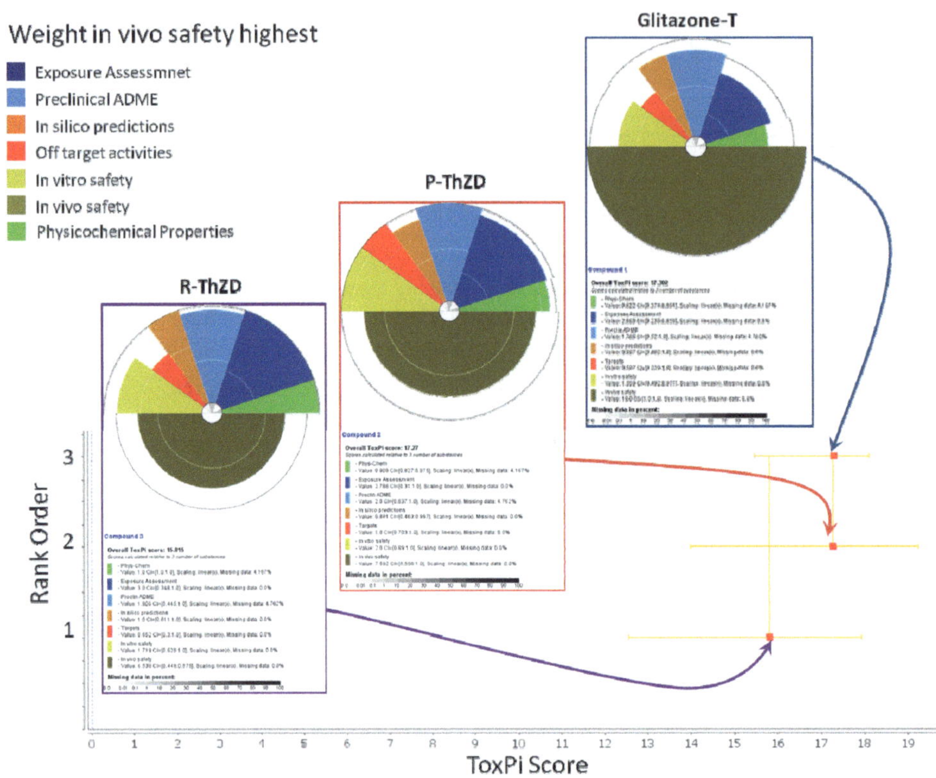

FIGURE 12-9: ToxPi visualization of data with in vivo safety heavily weighted.

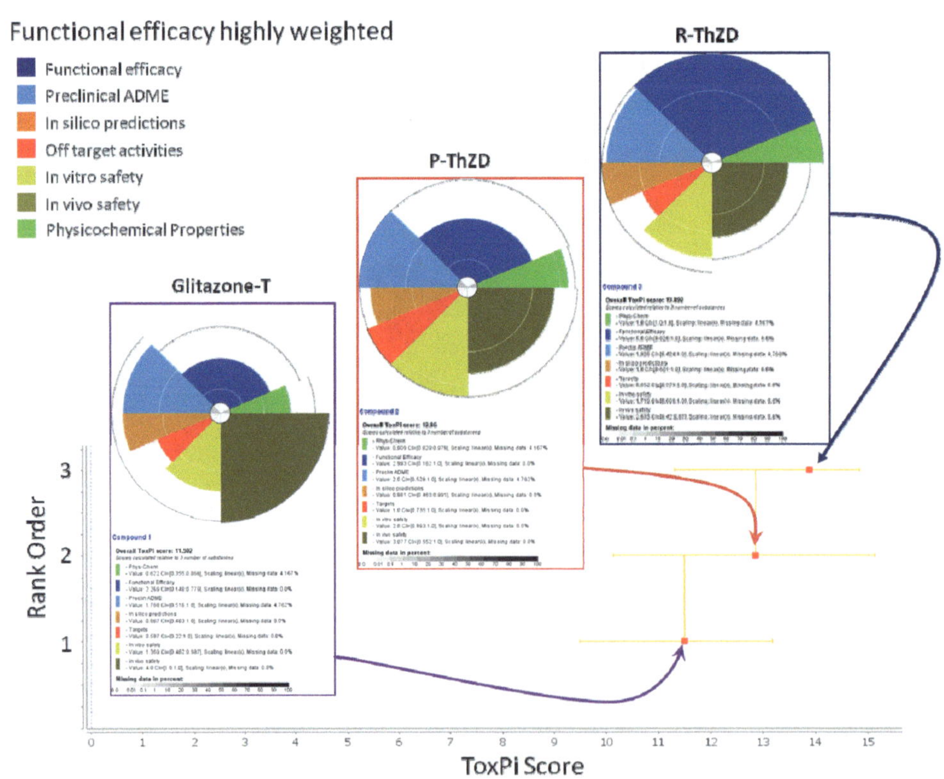

FIGURE 12-10 ToxPi visualization of data with functional efficacy heavily weighted.

13

Chemical Design: An Opportunity for Innovation

Alternatives assessment as described in this report typically begins with the recognition that a particular chemical is problematic from a health, safety, or environmental standpoint (Step 1 in the committee's framework), followed by a comparative assessment of potential alternatives. In many cases, alternatives assessment only considers chemical substitutes that have been commercialized, can be readily obtained and, typically, have known physicochemical properties or information about their effects that can be compared. De novo design of new chemicals is a less common, but important approach to finding safer alternatives to existing chemicals.

This chapter illustrates how the scientific concepts applied to alternatives assessment and described in earlier chapters can also be applied to the process of designing new alternatives de novo—Step 13, see Figure 13-1 and Box 13-1. Multidisciplinary teams are commonly tasked with this effort. While the term *de novo design* is used here, the concept of designing chemicals to be inherently safer is often referred to as "green chemistry." Green chemistry is a proactive approach to reducing the potential for unwanted health and environmental impacts early in chemical design or discovery.

De novo chemical designs begin as drawings of chemical structures on paper or on the computer. At this point, chemical designs are only conceptual; therefore, the properties or effects of the different chemicals cannot be compared through empirical measurement and testing. Actually synthesizing the designed chemical can take many resources and an extended period of time (months to years). Thus, compared to evaluation of existing chemicals that can actually be tested, a different assessment strategy is needed for these conceptual chemicals. The goal is to get rapid, if imperfect, feedback that guides innovators away from candidates that are likely to have undesirable properties or impacts. Such feedback enables innovators to focus on alternatives that are more likely to be successful

BOX 13-1

DESIGN AND INNOVATION AT A GLANCE

1. Chemicals of concern can be addressed by developing a new chemical to meet the functional needs or by developing an innovative concept that addresses the problem in a different way.
2. The design of new chemicals is an opportunity to address the lack of satisfactory alternatives.
3. During the design process, it is important to consider the environmental and health impacts in parallel with performance criteria.
4. During the consideration of novel alternative structures, before they have been synthesized, rules of thumb, or general principles; computational methods; and expert systems can be used to predict both physicochemical properties and biological impacts so that the structures selected for further development are the least likely to fail later on because of poor environmental or toxicity performance.
5. For newly synthesized candidates, physicochemical properties should be determined to identify which candidates are predicted by these properties to have poor environmental or health performance. Avoid these candidates and use this information as feedback to design.
6. In the future, newly synthesized candidates could be screened through a battery of in vitro tests, like those in ToxCast or Tox21, to provide a baseline of information about initial compounds' potential hazards and effective concentration at a relatively low cost. This would allow triaging and focus on the most promising candidates.
7. Potential impacts, health or environmental, should continue to be considered as chemical designs are changed to address performance weaknesses identified later in product development.

and to reallocate resources and effort away from those associated with negative environmental or safety concerns. This chapter describes the design of new chemicals as an opportunity to develop safer chemicals and outlines considerations for scientists who design new chemicals.

INNOVATION WITHIN THE COMMITTEE'S ALTERNATIVES ASSESSMENT FRAMEWORK

In the search for alternatives, there will be cases where alternatives assessment is not, by itself, sufficient to identify a viable option. Considered alternatives may fail on performance, economic, safety, or other grounds. Or, entrepreneurs (and innovators inside a company) may see the alternatives assessment process as an opportunity to create a new compound or an entirely new product concept to satisfy the desired needs of the customer base. In either of these cases, the framework should include information that aids such innovators in their quest to find compounds that offer both better performance and improved environmental and human health attributes compared to the initial chemical of concern. The committee acknowledges that scientists within select companies may practice some, or all, of the suggested approaches described; however, teams tasked with alternatives assessment often have not incorporated these approaches.

Within the Committee's framework, there are several steps where consideration of de novo designs (Step 13) is important:

1. At the decision point in Step 4, if no alternatives are available, or if there is a business opportunity to consider novel alternatives, de novo design should be considered.
2. Innovators may also enter Step 13 based on a business opportunity to develop a safer alternative that is not necessarily driven by the identification of a chemical of concern. (This is indicated by the direct point of entry into Step 13 of the committee's framework diagram.)
3. Finally, de novo design may be required (or motivated) by the results of testing at decision points that occur in Steps 7 or 10. Two types of outcomes are likely:
   a. The determination that alternatives have undesirable properties or impacts, leading to additional efforts toward de novo design.
   b. Information from testing provides feedback to inform further optimization of innovative alternatives.

BOX 13-2

LESSONS FROM THE PHARMACEUTICAL INDUSTRY

Consideration of environmental and health consequences of chemical structures and physicochemical properties of new chemicals does not usually take place until the very late in the process, if at all. One example of this reality can be found in the pharmaceutical industry, where in the early 1990s, when the primary focus of development work was developing a potent inhibitor or activator of an intended protein target. Little regard was given to the physicochemical properties that would allow the new drug to be readily absorbed into the bloodstream. In the late 1990s (highlighted by the publication of Lipinski's "Rule of 5" in 1997 (Lipinski et al. 1997)), awareness of the properties that differentiated compounds with good oral bioavailability from those that were poorly absorbed became a central part of medicinal chemistry thinking. (See Chapter 5 for more details on physicochemical properties and their relationship to bioavailability.)

After solving the problem of bioavailability, the pharmaceutical industry began to realize that safety-related issues were now a significant cause of failure for new drug candidates. As result, much effort has been put into trying to understand the relationships between chemical structure and the toxicity observed for a given compound. By considering what is known about chemical structures and physicochemical properties early in the design process, these problems can be avoided. However, mechanisms of toxicity are often complex and poorly understood, so success in avoiding these problems altogether has been limited. Progress has been made, however, through the use of in silico models and in vitro assays, which can help identify the best compounds to put forward for further development. The thinking is that using these methods can at least improve the odds of success if not guarantee it (see Chapter 8 for more detail).

Despite this increased understanding of the importance of the safety profile that constitutes a successful drug candidate; medicinal chemists will often focus *first* on optimizing the potency and bioavailability of the molecule, rapidly narrowing down the search to within a single chemical series. Only then will they search for the one with the fewest safety liabilities within a narrow range of available substrates. Perhaps if safety were considered when there were still choices about which option was the optimal chemical series, then it might be possible to select molecules that had the ideal balance of target impact, bioavailability, and toxicity avoidance, leading to higher success rates and increased productivity.

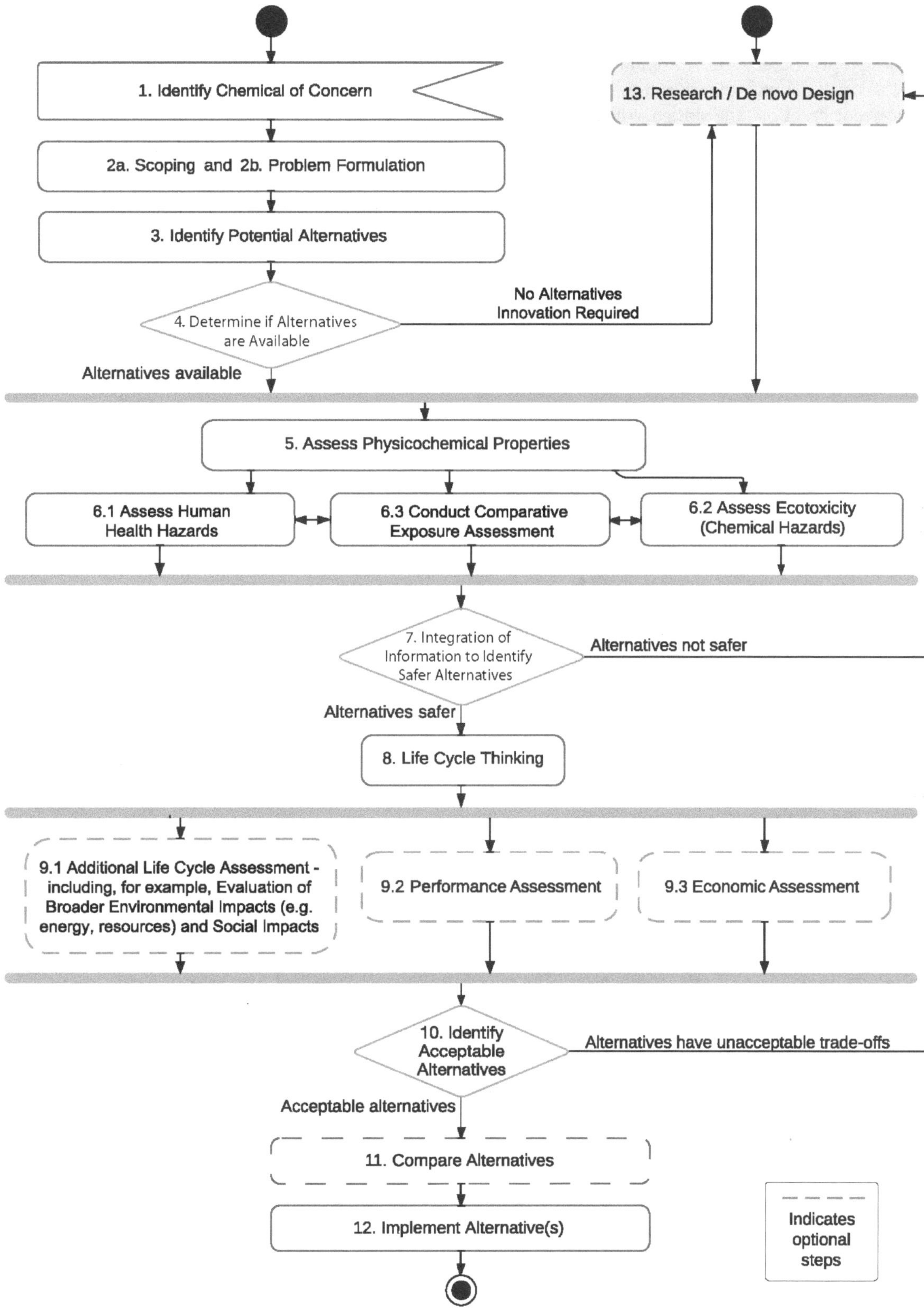

FIGURE 13-1 Committee's framework highlighting where design and innovation occur.

Figure 13-2 shows a typical "front end" of the innovation process that has been modified (shaded boxes) to incorporate early inclusion of safer chemical design principles. The system includes three main activities: opportunity identification and analysis, concept creation, and design. Although innovators typically do not include aspects of safer chemical design in these early stages, the committee believes that this approach can help reduce problems during later stages. A key aspect to finding opportunity (in the business sense) is to truly understand customers' desired outcomes; as shown in Figure 13-2, such understanding also helps to identify concepts that result in safer products.

As noted, one of the crucial early steps in the innovation process is to develop a deep understanding of "who your customers are" and "what their desired outcomes are" in the context of one's product or service. For example, when regulatory bodies in Europe raised concerns over the use of phthalate plasticizers in polyvinylchloride-containing toys, one approach to the problem was to create more benign plasticizers (such as the cyclohexyl analog to a phthalate synthesized by BASF). However, customers are not interested in plasticizer design per se, but rather, in a safe, flexible material for use in children's toys. Focus on this desired outcome can lead one to many possible solutions, such as Dow's Insite® polymers (thermoplastic elastomers made from ethylene and propylene that are inherently soft and pliable without any need for plasticizer). Successful product design firms typically use a combination of ethnography and voice of the customer analyses to uncover desired customer outcomes, which prove critical to prototype fabrication.

Once a business understands its customer base, structured brainstorming can be used to generate novel solutions. In the case of safer chemical product design, one of Goldberg's rules of thumb (Goldberg et al., 2003) can be borrowed: innovation by elimination to help create safer products. For example, in the plasticized polyvinylchloride case or the case of brominated flame retardants, removing the need for the problematic chemical while satisfying desired outcomes (an inherently soft material vs. softness through plasticizer or an inherently flame-retardant material vs. addition of flame retardant compounds) can lead to safer products.

Finally, once a promising concept has been generated, it is useful to examine the expected life cycle of a chemical as a way to check for red flags that might appear in the early stages of a product's lifetime (see Life Cycle Thinking, Chapter 10).

In each of these instances, a consideration of the human health hazards and ecotoxicity is needed, alongside consideration of other environmental impacts and product performance attributes, as early as possible in the design process (ideally, when concepts are being penned to paper). In the traditional approach to innovation, health and environmental concerns are considered, if at all, near the end of the innovation process—only after significant time and resources has been committed to product development and the satisfaction of customer-centric performance criteria. If the goal is to reduce undesirable health and environmental impacts, these issues must be considered early in the design process. Ramani et al. (2010) and others have proposed that many health and safety impacts are "locked in" at the concept stage (before any significant bench work has begun). Consequently, considering these impacts early in the process is necessary to create true eco-innovations, products and services that promise enhanced performance with a reduced footprint.

Although the strategies and tools for safer chemical design provide primarily qualitative guidance, these approaches, when used early and often, can steer innovators away from products unlikely to meet safety criteria. A recent example of this comes from the use of heavy metal-containing nanoparticles (Bystrzejewska-Pitrowska et al. 2009). Despite the exceptional fluorescence properties of CdSe and PbS nanoparticles, each contains heavy metal cations. The presence of those cations might not pose environmental or health and safety concerns for macroscopic thin films embedded in electronic devices, but it is a different story if they are used to cover extremely high surface areas. Then cations from these nanoparticles are more readily released, potentially posing a hazard in many applications. It is now clear that these types of nanoparticles have limited potential due to the toxicity of their constituent elements (Schrand et al. 2010). By considering the safety concerns earlier in the innovation process, development time and resources might have been applied to solutions with environmental and health safety performance on par with their other performance attributes.

OPPORTUNITIES FOR INNOVATION

Figure 13-2 shows a typical flow diagram for the early stages of the innovation process and how innovators can eliminate potentially problematic

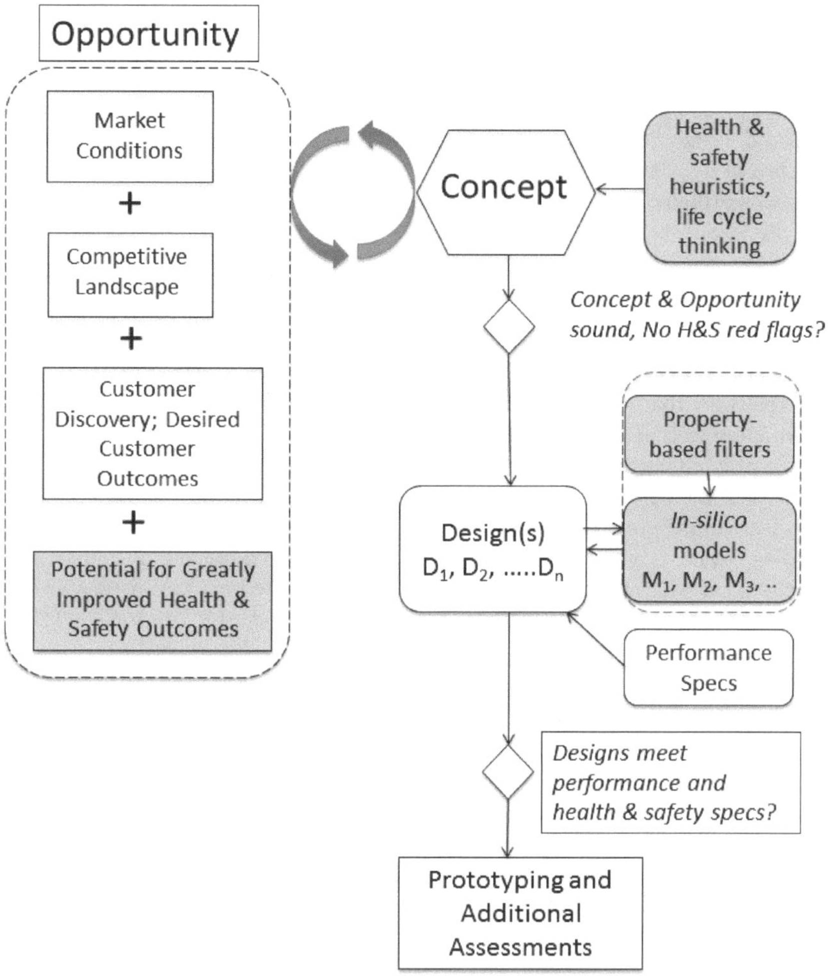

FIGURE 13-2. Flow diagram for the innovation process with the addition of alternatives assessment at the early stages. The traditional front end of the process is indicated by the white boxes, with proposed addition of tools (in shaded boxes) that can aid in design of safer products. Strong opportunity includes identification of a group of customers whose desired outcomes are not being met, a significant market, and typically failings among the competition. The potential for improvements to product safety can be included as a contributor to opportunity as well. Concepts are created to service opportunities. This is known as creating good product-market fit. Designs are then the physical manifestation of a concept (see Box 13-3 for detail). Further, as shown in this chapter, both creation of entirely new concepts or the de novo design of chemicals can benefit from inclusion of safer chemical principles.

design choices. From the perspective of the entrepreneur, the front end of the innovation process usually begins either with the realization that an unfilled, yet lucrative, opportunity exists and/or the identification of a novel concept or solution. Often, there is a gap between the desired outcomes of a significant customer segment (or segments) and current offerings.

Another type of "gap" that could lead to new opportunities could thus be the failings of current offerings due to environmental, health, and human health deficits. Indeed, identification of such opportunities for "green chemistry" or "eco-innovation" is a potential outcome of Steps 3 and 4 of the committee's framework. A manufacturer with a product containing a compound of concern may see a lack of satisfactory chemical offerings as a problem that needs to be dealt with, but an innovator will view this same "problem" as the rationale for new concept or business creation. It all depends on how a chemical of concern is perceived. One example of an opportunity created by a substance of concern is California's effort to phase out perchloroethylene in dry cleaning because of toxicity issues, resulting in the development of a spate of new dry cleaning technologies in the 1990s (Sabanadesan and Vanderlinden 2007). Likewise, emerging problems related to Bisphenol A use in polycarbonates created the opportunity for a non-Bisphenol A transparent thermoplastic with high-use temperature. In response to this opportunity, Eastman Chemical created Tritan copolyesters, while other companies invested in new, transparent, high-use temperature polyolefins (Nelson and Long 2012).

New Concepts and Chemical Designs

There are two primary approaches to developing innovative solutions that go beyond the consideration of known chemicals. The first involves design and synthesis of a new chemical to directly replace a known chemical of concern, or starting with the "design" step shown in Figure 13-2. This approach typically involves evaluating the function and structure of that chemical and modifying its structure to meet the functional need while reducing the impacts of concern (as illustrated in Case Study 1 on DecaDBE, Chapter 12). The second approach starts in either the "opportunity" or "concept" box in Figure 13-2. It involves identifying or developing novel approaches that seek to duplicate the *function* of the chemical of concern, not just the chemical itself. One might expect established companies that currently manufacture chemicals or chemical formulations to focus on the first approach (de novo *design* of a replacement chemical), given the constraints imposed by a mature business model that itself depends upon certain feedstocks or plant configurations. Similarly, one might expect start-up companies or downstream users of chemicals to instead focus on *new concepts*—providing the desired function without necessarily duplicating the original chemical. For an illustration of the difference between concept and design, see Box 13-3.

In either of these approaches—new design or new concept—innovators should proactively check to see whether there are any environmental, health, or other red flags related to chemical hazard in the design. They should use rules of thumb, structure/function relationships, computational tools, safer chemical lists and guides, and other early indicators to guide design at each stage of innovation. By identifying the functional use clearly early in the process, it may be possible to identify particular areas of concern (e.g., inhalational toxicity for a chemical that will be used as a fragrance or flammability for a product often used near open flames or heat sources) that can be considered during the design process. As noted in Figure 13-2, these early checks can be conducted at each stage of the innovation process, regardless of which approach is used.

BOX 13-3
CONCEPT VS. DESIGN

A concept is a top-level response that fulfills the desired outcomes of customers, while a design is a more specific manifestation of the concept. It is possible to use health and safety screening tools at both the concept and the design stage. Below are two examples.

Example 1: If the desired customer outcome is "a surface free of bacteria," one might have:

Concept 1: An antibacterial spray
Design 1A: A spray of triclosan and ethanol
Design 1B: A spray of lactic acid in water

Concept 2: A surface that prevents bacterial colonization.
Design 2A: A silver-functional acrylic coating that kills bacteria on contact.
Design 2B: A shark-scale biomimetic coating that prevents bacteria from sticking.

Example 2: If the desired customer outcome is a "fabric with bright color," one might have:

Concept 1: Use a dye to color the fabric.
Design 1A: Use a metal-based dye.
Design 1B: Use a dye extracted from a plant or animal.

Concept 2: Use reflection from surfaces to create the illusion of color.
Design 2A: Layers of polymer to mimic the Morpho Butterfly (Teijin Fibers, MorphoTex)
Design 2B: Rolled layers to mimic the plant *Margaritaria Nobilis* (Kolle et al. 2013)

Guidance for New Concept Creation

Generally, in its early stages the innovation process is strongly influenced by the needs of the market, and concept creation is guided by an understanding of these market needs (and the competitive landscape). Whereas early inclusion of health, ecotoxicity, and physicochemical principles, as well as Life Cycle Thinking, would be valuable in this process, this is not common. The committee recommends that such inclusions occur early in the process. For example, at the concept stage, use of Life Cycle Thinking can be useful in avoiding undesirable building blocks and stimulating thinking about a novel way to reduce the environmental footprint (for example, creating inherently flame-retardant materials vs. the use of chemical flame retardants).

Guidance for De Novo Design of Alternative Compounds

In the early design stage, there are a number of approaches (including the descriptions in Chapter 5 about physicochemical properties), which innovators should consider to guide chemical designers and help them select from a number of potential chemical structures. When de novo chemical design is required, consideration of both 1) physicochemical properties and 2) potential biological activities will reduce the likelihood of new chemicals encountering issues as development and further testing proceeds.

The following stages can be used to guide the design of new chemicals. They are tiered and based on the speed with which they can be applied and increasing sensitivity.

Stage 1: Apply qualitative structure-based[55] *design filters.* At this stage, it is useful to screen for chemical functional groups or other structural features that are highly likely to be associated with particular hazards. This can be done before a chemical is synthesized, while it is still in the conceptual phase. A common example of an undesirable feature is the presence of an unhindered aromatic amine, which is strongly associated with carcinogenicity (Benigni and Passerini 2002). Box 13-4 lists various overlapping approaches for qualitative structure-based screening.

Stage 2: Apply qualitative property-based design filters (see Box 13-5) to eliminate chemicals highly likely to exhibit hazards associated with particular undesirable physicochemical properties. As soon as samples of chemicals are synthesized, these physicochemical properties can be measured, or these properties can be predicted based on computational models when chemicals are still in the concept phase.

Stage 3: Apply a more refined set of in silico tools and quantitative models to further assess toxicity hazards. Such models can be either based on structure (Quantitative Structure - Activity Relationships, QSARs) or spectra (Quantitative Spectroscopic Data Activity Relationships, QSDARs). These models will allow screening for additional human and ecotoxicity end points, such as carcinogenicity, mutagenicity, endocrine disruption, etc. For a more information, see QSAR discussions in Chapters 7 and 8. Discretion must be applied to use these models in a way that provides meaningful results. If a candidate chemical is predicted to have high toxicity for one or more end points, it should either be screened out, given a low priority, or redesigned and fed back through the workflow.

Stage 4: Apply mechanistic prediction tools for end points that are available. For the remaining candidates, use of more complex novel high throughput testing and computational models, such as those described in Chapter 8, may further decrease the probability that the candidates proposed will cause unintended consequences. While such models are routinely used in the pharmaceutical industry in drug design to avoid unintended consequences (see Box 13-2), they are underutilized in the rational design of commercial chemicals. The mechanistic underpinning of these models allows a more refined prediction for some end points, such as skin sensitization and carcinogenicity.

Stages 1-4 provide guidance for improving environmental and health attributes, by using available tools before the chemical synthesis stage. In addition to being used to screen out less desirable chemicals in the design stage, the information gathered can inform future designs of analogous alternatives. Although these steps are described in a linear fashion for the sake of simplicity, a strong and continuous flow of information, from the analysis of chemical structure to a description of physicochemical properties, is needed as feedback to guide design of safer alternatives. This type of feedback is key to developing more robust structure/activity relationships for chemical classes.

[55] Note: While structure-based filters and physicochemical property-based filters are described here separately, physicochemical properties obviously stem from structure.

BOX 13-4

QUALITATIVE STRUCTURE-BASED DESIGN FILTERS: FEATURES ASSOCIATED WITH UNWANTED BIOLOGICAL ACTIVITY

Qualitative screening for chemical functional groups or other features that are highly likely to be associated with particular hazards can be done before a chemical is synthesized, while it is still in the conceptual phase. Design filters are listed here with common names. In reality, the approaches listed here overlap in the concepts they cover.

"Rules of Thumb": Principles developed from experience that have broad application but are not intended to be strictly accurate or reliable for every situation. They should be used to qualitatively screen for structural features associated with high probability of hazard. Two examples widely used in pharmaceutical chemistry, but not widely applied by those engaged in alternatives assessment, include:

- Avoid unhindered aromatic amines, which are strongly associated with carcinogenicity (Benigni and Passerini 2002).
- Lipinski's rule of five for drug design[56] (Lipinski et al. 1997):
  - Number of hydrogen bond donors (nitrogen or oxygen atoms with one or more hydrogen atoms < 5).
  - Number of hydrogen bond acceptors (all nitrogen or oxygen atoms < 10).
  - Molecular mass < 500 daltons.
  - Lipophilicity ($\log P \leq 5$).

Computational predictive approaches: This refers to computational approaches that strive to predict activity from structural information. These approaches would typically involve the use of various computational methods to calculate structures, properties, or impacts.

Expert rules: Structure- or mechanism-based decision-making approaches that are typically computerized and aim to mimic the integrative analysis that an "expert" would provide. Expert rules may incorporate both rules of thumb and computational learning about toxicity prediction. Expert rules should be used to qualitatively screen for structural features associated with high probability of hazard. One example is DEREK:

- "DEREK is a knowledge-based expert system comprising a number of structural rules that aim to encode structure-toxicity information with an emphasis on mechanisms. The toxicity predictions made by DEREK are the result of two processes. The program checks whether any alerts in the knowledge base match toxicophores in the query structure. The reasoning engine then assesses the likelihood of a structure being toxic. There are 9 levels of confidence: certain, probable, plausible, equivocal, doubted, improbable, impossible, open, contradicted. The reasoning model considers the following information:
  - The toxicological end point.
  - The alerts that match toxicophores in the query structure.
  - The physicochemical property values calculated for the query structure.
  - The presence of an exact match between the query structure and a supporting example within the knowledge base" (Saliner et al. 2005).

Structure activity relationships: These are relationships that intend to link specific structural features with biological activity.

[56] These rules of thumb are associated with increased likelihood of oral activity in humans. Avoiding them in chemical design would reduce the likelihood of unwanted oral activity.

BOX 13-5

QUALITATIVE PROPERTY-BASED DESIGN FILTERS: DESIRABLE/UNDESIRABLE PROPERTIES

Structure-property relationships. These are relationships that intend to link specific structural features with particular chemical properties (physicochemical properties).

Physicochemical property-based design guidelines (see also Chapter 5)

Examples of established property-based design guidelines are listed below, but it is clear that there is a need to develop additional guidelines that address materials safety and additional biological end points.

- Rules of thumb for increasing biodegradation according to Williams and Williams (Williams and Williams 2012) are to avoid:
    - "Halogens, especially chlorine and fluorine and especially if there are more than three in a small molecule (iodine and (probably) bromine contribute to a lesser extent);
    - Chain branching if extensive (quaternary C is especially problematic);
    - Tertiary amine, nitro, nitroso, azo, and arylamino groups;
    - Polycyclic residues (such as in polycyclic aromatic hydrocarbons), especially with more than three fused rings; heterocyclic residues, for example, imidazole); and
    - Aliphatic ether bonds (except in ethoxylates)."
- Criteria for human bioavailability by different exposure routes: If a chemical meets all of the property limits associated with skin, oral, respiratory, or ocular bioavailability, it is likely to pose higher risk of exhibiting human toxicity. While this may not be detrimental, it is reasonable that chemicals with low bioavailability are given higher preference.
- Criteria for aquatic toxicity: If an organic chemical meets the criteria for high risk of acute and/or chronic aquatic toxicity, it should be redesigned, screened out, or given low priority.
- Criteria for physical hazard: These include flammability, flash point, corrosivity, etc.

Redesign of an Existing Chemical

The considerations required for redesigning an *existing* chemical to minimize hazard while retaining function overlap partly with those outlined in the previous section. Structural optimization to tune biological activity is not uncommon in the pharmaceutical industry, but it is not typically utilized in the rational design of commercial chemicals. This process starts with the identification of the structural core of a chemical that is associated with function. In cases where this is not obvious, the functional core can be identified by understanding how the chemical exerts the desired function. Identifying this motif will allow for the identification of the non-essential structural features of the molecule that could be modified. The possible analogs can then be generated to obtain a set of candidates. These candidates are fed through the above process starting at Stage 2, and proceeding to the end. The result of these workflows will be a number of candidate chemicals that can be carried through the alternatives assessment workflow described earlier in this report.

Looking Forward: New Tools for Early Insights into Toxicity

The stages describe how to use what is known about chemical structures and physicochemical properties to design chemicals that avoid unfavorable characteristics. The structure-based prediction can be conducted before a chemical is even synthesized. Physicochemical properties may be predicted and/or measured. The advent of high throughput testing of chemicals through a large battery of tests designed to identify a number of common toxicity end points is likely to yield yet another opportunity for early insight into toxicity. As described in Chapter 8, computational toxicologists, who evaluate the results of such high throughput robotic testing through hundreds of assays for various end points, are working to discern what type of information they can glean from these approaches, such as the Tox21 or ToxCast batteries of assays developed by EPA, NIH, and FDA collaborations. While the assays have shortcomings, there are indications that batteries of assays may be useful for predicting particular end points.

Furthermore, there are hints that the assays may be even more valuable in predicting the chemical concentration at which biological activity occurs. As the toxicology community moves toward a common understanding about the value that can be gleaned from these assays, it is likely that chemical designers who can synthesize their compounds in a pure enough form to avoid artifacts from the assays could benefit from the ability to quickly screen compounds they are developing.

SUMMARY

Where no alternatives exist and a new chemical must be rationally designed, a series of qualitative structure-based or physicochemical property-based design filters can be used to assess chemical designs while they are still conceptual or have only small amounts synthesized, to minimize health and ecotoxicity issues. Then, more refined tools, such as in silico modeling of mechanisms and QSAR and QSDAR, should be used to guide designs that meet environmental and health requirements as well as functional performance. The most important aspect is to consider attributes that increase ecological or health risks, in tandem with other performance attributes, as early as possible in the design process.

The staged evaluation of these novel alternatives is tiered and based on the speed with which they can be applied and increasing sensitivity. The advantage of this approach is that fatally flawed alternatives may be eliminated from consideration earlier in the process. Innovation time and resources can then be focused on viable alternatives, and when more of the actual compound is available for testing, additional information can be obtained.

References

Adler, S., D. Basketter, S. Creton, O. Pelkonen, J. van Benthem, V. Zuang, K.E. Andersen, A. Angers-Loustau, A. Aptula, A. Bal-Price, E. Benfenati, U. Bernauer, J. Bessems, F.Y. Bois, A. Boobis, E. Brandon, S. Bremer, T. Broschard, S. Casati, S. Coecke, R. Corvi, M. Cronin, G. Daston, W. Dekant, S. Felter, E. Grignard, U. Gundert-Remy, T. Heinonen, I. Kimber, J. Kleinjans, H. Komulainen, R. Kreiling, J. Kreysa, S.B. Leite, G. Loizou, G. Maxwell, P. Mazzatorta, S. Munn, S. Pfuhler, P. Phrakonkham, A. Piersma, A. Poth, P. Prieto, G. Repetto, V. Rogiers, G. Schoeters, M. Schwarz, R. Serafimova, H. Tahti, E. Testai, J. van Delft, H. van Loveren, M. Vinken, A. Worth, and J.M. Zaldivar. 2011. Alternative (non-animal) methods for cosmetics testing: Current status and future prospects-2010. Arch. Toxicol. 85(5):367-485.

Adriano, D.C. 2001. Trace Elements in Terrestrial Environments: Biogeochemistry, Bioavailability and Risks of Metals, 2nd Ed. New York: Springer.

Agius, R.M., J. Nee, B. McGovern, and A. Robertson. 1991. Structure activity hypotheses in occupational asthma caused by low molecular weight substances. Ann. Occup. Hyg. 35(2):129-137.

Agius, R.M., R.A. Elton, L. Sawyer, and P. Taylor. 1994. Occupational asthma and the chemical properties of low molecular weight organic substances. Occup. Med. 44(1):34-36.

An, N., F. Van Der Mei, and A.M. Voutchkova-Kostal. 2014. Global model for octanol-water partition coefficients from Proton Nuclear Magnetic Resonance Spectra. Mol. Inform. 33(4):286-292.

Arnot, J., T. Gouin, and D. Mackay. 2005. Development and Application of Models of Chemical Fate in Canada: Practical Methods for Estimating Environmental Biodegradation Rates. CEMN Report No. 200503. Trent University, Peterborough, Ontario. March 31, 2005 [online]. Available: http://www.trentu.ca/academic/aminss/envmodel/CEMNReport200503.pdf [accessed March 29, 2014].

Ashby, J., and R.W. Tennant. 1991. Definitive relationships among chemical structure, carcinogenicity and mutagenicity for 301 chemicals tested by the U.S. NTP. Mutat. Res. 257(3):229-306.

Atlee, J. 2012. NSF Offers GreenScreen to Promote Safer Chemicals. Environmental Building News [online]. Available: http://www2.buildinggreen.com/article/nsf-offers-greenscreen-promote-safer-chemicals [accessed July 7, 2014].

ATSDR (Agency for Toxic Substances and Disease Registry). 2004. Toxicological Profile for Polybrominated Biphenyls and and Polybrominated Diphenyl Ethers. U.S. Department of Health and Human Services, Public Health Service, Agency for Toxic Substances and Disease Registry [online]. Available: http://www.atsdr.cdc.gov/toxprofiles/tp68.pdf [accessed July 21, 2014].

ATSDR (Agency for Toxic Substances and Disease Registry). 2013. Minimal Risk Levels (MRLs) [online]. Available: http://www.atsdr.cdc.gov/mrls/index.asp [accessed April 28, 2014].

Bartlett, I.W., A.J.P. Dalton, A. McGuinness, and H. Palmer. 1999. Substitution of organic solvent cleaning agents in the lithographic printing industry. Ann. Occup Hyg. 43(6):83-90.

Basketter, D.A., H. Clewell, I. Kimber, A. Rossi, B. Blaauboer, R. Burrier, M. Daneshian, C. Eskes, A. Goldberg, N. Hasiwa, S. Hoffmann, J. Jaworska, T.B. Knudsen, R. Landsiedel, M. Leist, P. Locke, G. Maxwell, J. McKim, E.A. McVey, G. Ouedraogo, G. Patlewicz, O. Pelkonen, E. Roggen, C. Rovida, I. Ruhdel, M. Schwarz, A. Schepky, G. Schoeters, N. Skinner, K. Trentz, M. Turner, P. Vanparys, J. Yager, J. Zurlo, and

T. Hartung. 2012. A roadmap for the development of alternative (non-animal) methods for systemic toxicity testing - t4 report*. ALTEX 29(1):3-91.

Belton, V., and T. Stewart. 2002. Multiple Criteria Decision Analysis: An integrated Approach. New York: Springer.

Benbow, J.W., J. Aubrecht, M.J. Banker, D. Nettleton, and M.D. Aleo. 2010. Predicting safety toleration of pharmaceutical chemical leads: Cytotoxicity correlations to exploratory toxicity studies. Toxicol. Lett. 197(3):175-182.

Benigni, R. 2013. Evaluation of the toxicity forecasting capability of EPA's ToxCast Phase I data: Can ToxCast in vitro assays predict carcinogenicity? J. Environ. Sci. Health C Environ. Carcinog. Ecotoxicol. Rev. 31(3):201-212.

Benigni, R., and A. Giuliani. 2003. Putting the predictive toxicology challenge into perspective: Reflections on the results. Bioinformatics 19(10):1194-1200.

Benigni, R., and L. Passerini. 2002. Carcinogenicity of the aromatic amines: From structure-activity relationships to mechanisms of action and risk assessment. Mutat. Res. 511(3):191-206.

Benigni, R., C. Bossa, O. Tcheremenskaia, and A. Giuliani. 2010. Alternatives to the carcinogenicity bioassay: In silico methods, and the in vitro and in vivo mutagenicity assays. Expert Opin. Drug Metab. Toxicol. 6(7):809-819.

Besser, J.M., W.G. Brumbaugh, N.E. Kemble, C.D. Ivey, J.L. Kunz, C.G. Ingersoll, and D. Rudel. 2011. Toxicity of Nickel-Spiked Freshwater Sediments to Benthic Invertebrates-Spiking Methodology, Species Sensitivity, and Nickel Bioavailability. Scientific Investigations Report No. 2011-5225. U.S. Geological Survey [online]. Available: http://pubs.usgs.gov/sir/2011/5225/pdf/sir2011-5225.pdf [accessed June 23, 2014].

Birnbaum, L.S. 2013. State of the science of endocrine disruptors. Environ. Health Perspect. 121(4):A107.

BizNGO. 2013. The Commons Principles for Alternatives Assessment: Addressing Chemicals of Concern to Human Health or the Environment. October 14, 2013 [online]. Available: http://www.bizngo.org/resources/entry/commons-principles-for-alternatives-assessment [accessed June 25, 2014].

Blaauboer, B.J. 2010. Biokinetic modeling and in vitro-in vivo extrapolations. J. Toxicol. Environ. Health B Crit. Rev. 13(2-4):242-252.

Blaauboer, B.J., K. Boekelheide, H.J. Clewell, M. Daneshian, M.M. Dingemans, A.M. Goldberg, M. Heneweer, J. Jaworska, N.I. Kramer, M. Leist, H. Seibert, E. Testai, R.J. Vandebriel, J.D. Yager, and J. Zurlo. 2012. The use of biomarkers of toxicity for integrating in vitro hazard estimates into risk assessment for humans. ALTEX 29(4):411-425.

Boethling, R.S., and D. Mackay. 2000. Handbook of Property Estimation Methods for Chemicals. Boca Raton, FL: CRC Press.

Boethling, R.S., and J.V. Nabholz. 1997. Environmental assessment of polymers under the U.S. Toxic Substances Control Act. Pp. 187-234 in Ecological Assessment of Polymers Strategies for Product Stewardship and Regulatory Programs, J.D. Hamilton, and R. Sutcliffe, eds. New York: Van Nostrand Reinhold.

Boethling, R.S., D.G. Lynch, J.S. Jaworska, J.L. Tunkel, G.C. Thom, and S. Webb. 2004. Using Biowin, Bayes, and batteries to predict ready biodegradability. Environ. Toxicol. Chem. 23(4):911-920.

Burkhard, L.P., J.A. Arnot, M.R. Embry, K.J. Farley, R.A. Hoke, M. Kitano, H.A. Leslie, G.R. Lotufo, T.F. Parkerton, K.G. Sappington, G.T. Tomy, and K.B. Woodburn. 2012. Comparing laboratory and field measured bioaccumulation endpoints. Integr. Environ. Assess. Manag, 8(1):17-31.

Bus, J.S., and R.A. Becker. 2009. Toxicity testing in the 21st century: A view from the chemical industry. Toxicol. Sci. 112(2):297-302.

Bystrzejewska-Piotrowska, G., J. Golimowski, and P.L. Urban. 2009. Nanoparticles: Their potential toxicity, waste and environmental management. Waste Manage. 29(9):2587-2595.

CA DTSC (California Department of Toxic Substances Control). 2010. Chemical List: Candidate Chemical Database [online]. Available: http://www.dtsc.ca.gov/SCP/ChemList.cfm [accessed June 26, 2014].

CA DTSC (California Department of Toxic Substances Control). 2013a. Proposed Regulation. Safer Consumer Products Regulation [online]. Available: https://dtsc.ca.gov/LawsRegsPolicies/Regs/SCPA.cfm [accessed March 8, 2014].

CA DTSC (California Department of Toxics Substances Control). 2013b. Safer Consumer Products (SCP): Summary of Revised Proposed Regulations, R-2011-02. April 2013. [online]. Available: https://dtsc.ca.gov/SCPRegulations.cfm [accessed April 2, 2014].

CA DTSC (California Department of Toxic Substances Control). 2013c. Safer Consumer Products: Division 4.5, Title 22, California Code of Regulations, Chapter 55 [online]. Available: https://dtsc.ca.gov/LawsRegsPolicies/Regs/SCPA.cfm [accessed March 8, 2014].

CA DTSC (California Department of Toxic Substances Control). 2014. Safer Consumer Products [online]. Available: https://dtsc.ca.gov/scp/index.cfm [accessed March 29, 2014].

CDC (Centers for Disease Control and Prevention). 2009. Fourth Report on Human Exposure to Environmental Chemicals. U.S. Department of Health and Human Services, Centers for Disease Control and Prevention, Atlanta, GA [online]. Available: http://www.cdc.gov/exposurereport/pdf/fourthreport.pdf [accessed August 14, 2014].

ChemSpider. 2014a. Troglitazone [online]. Available: http://www.chemspider.com/Chemical-Structure.5389.html [accessed July 22, 2014].

ChemSpider. 2014b. 5-{4-[2-(5-Ethyl-2-pyridinyl)ethoxy]benzyl}-1,3-thiazolidine-2,4-dione [online]. Available: http://www.chemspider.com/Chemical-Structure.21230467.html [accessed July 22, 2014].

ChemSpider. 2014c. Rosiglitazone [online]. Available: http://www.chemspider.com/Chemical-Structure.70383.html [accessed July 22, 2014].

Chen, M., H. Hong, H. Fang, R. Kelly, G. Zhou, J. Borlak, and W. Tong. 2013. Quantitative structure-activity relationship models for predicting drug-induced liver injury based on FDA-approved drug labeling annotation and using a large collection of drugs. Toxicol. Sci. 136(1):242-249.

Chiu, W.A., K.Z. Guyton, K. Hogan, and J. Jinot. 2012. Approaches to human health risk assessment based on the signal-to-noise crossover dose. Environ. Health Perspect. 120(7):a264.

Christensen, J. R., M. MacDuffee, R.W. Macdonald, M. Whiticar, and P.S. Ross. 2005. Persistent organic pollutants in British Columbia grizzly bears: Consequence of divergent diets. Environ. Sci. Technol. 39(18):6952-6960.

Christiansen, K, ed. 1997. Simplifying LCA: Just a Cut? SETAC EUROPE- LCA Screening and Streamlining Working Group. Society of Environmental Toxicology and Chemistry, Brussels. March 1997.

Clean Production Action. 2011. GreenScreen for Safer Chemicals v 1.2 Benchmarks. October 2011 [online]. Available: http://www.cleanproduction.org/static/ee_images/uploads/resources/GreenScreen_v1-2_Benchmarks_REV.pdf [accessed July 21, 2014].

Clean Production Action. 2012. Guide to Safer Chemicals. Implementing the BizNGO Principles for Safer Chemicals. Version 1.0. http://www.glgc.org/wp-content/uploads/2014/03/GuideToSaferChemicals_BizNGO.pdf [accessed April 27, 2014].

Clean Production Action. 2013. GreenScreen® for Safer Chemicals Chemical Hazard Assessment Procedure V 1.2 Final [online]. Available: http://tcodevelopment.com/files/2014/04/GreenScreenv1-2_Guidance_Assessment_Procedure_FINAL_2013_9_18.pdf [accessed April 27, 2014].

Clean Production Action. 2014. Green Screen or Safer Chemicals [online]. Available:

http://www.greenscreenchemicals.org/ [accessed June 26, 2014].

Cleek, R.L., and A.L. Bunge. 1993. A new method for estimating dermal absorption from chemical exposure. 1. General approach. Pharm. Res. 10(4):497-506.

Clode, S.A. 2006. Assessment of in vivo assays for endocrine disruption. Best Pract. Res. Clin. Endocrinol. Metab. 20(1):35-43.

Collins, F.S., G.M. Gray, and J.R. Bucher. 2008. Toxicology. Transforming environmental health protection. Science 319(5865):906-907.

Corton, J.C., M.L. Cunningham, B.T. Hummer, C. Lau, B. Meek, J.M. Peters, J.A. Popp, L. Rhomberg, J. Seed, and J.E. Klaunig. 2014. Mode of action framework analysis for receptor-mediated toxicity: The peroxisome proliferator-activated receptor alpha (PPARalpha) as a case study. Crit. Rev. Toxicol. 44(1):1-49.

Costanza, J., D.G. Lynch, R.S. Boethling, and J.A. Arnot. 2012. Use of the bioaccumulation factor to screen chemicals for bioaccumulation potential. Environ. Toxicol. Chem. 31(10):2261-2268.

Cronin, M.T. 2010. Quantitative structure–activity relationships (QSARs)–applications and methodology. Pp. 3-12 in Recent Advances in QSAR Studies: Methods and Applications, T. Puzyn, J. Leszczynski, and M.T.D. Cronin, eds. New York: Springer.

Crump, K.S., C. Chen, and T.A. Louis. 2010. The future use of in vitro data in risk assessment to set human exposure standards: Challenging problems and familiar solutions. Environ. Health Perspect. 118(10):1350-1354.

Cullen, A.C., and H.C. Frey. 1999. Probabilistic Techniques in Exposure Assessment: A Handbook for Dealing with Variability and Uncertainty in Models and Inputs. New York: Plenum Press.

Davies, C., M. Adams, E. Connor, E. Sommer, C. Baier-Andersen, E. Lavoie, L. Romano, and D. Difiore. 2013. U.S. Environmental Protection Agency's Design for the Environment (DfE) Alternatives Assessment Program. Pp. 198-229 in Chemical Alternatives Assessments, R.E. Hester, and R.M. Harrison, eds. London, UK: Royal Society of Chemistry.

Davies, J.C. 1996. Comparing Environmental Risks: Tools for Setting Government Priorities. New York: Resources for the Future.

Davis, A.P., C.G. Murphy, R. Johnson, J.M. Lay, K. Lennon-Hopkins, C. Saraceni-Richards, D. Sciaky, B.L. King, M.C. Rosenstein, T.C. Wiegers, and C.J. Mattingly. 2013. The Comparative Toxicogenomics Database: Update 2013. Nucleic Acids Res. 41(Database issue):D1104-1114.

Davis, G.A., M. Swanson, and S. Jones. 1994. Comparative Evaluation of Chemical Ranking and Scoring Methodologies. University of Tennessee, Center for Clean Products and Clean Technologies [online]. Available: http://web.utk.edu/~isse2006/ccp/pubs/pdfs/CECRSM.pdf [accessed July 18, 2014].

de Wit, C.A. 2002. An overview of brominated flame retardants in the environment. Chemosphere 46(5):583-624.

Dearden, J.C. 2002. Prediction of environmental toxicity and fate using quantitative structure-activity relationships (QSARs). J. Braz. Chem. Soc. 13(6):754-762.

Delmaar, J.E., M.V.D.Z. Park, J.G.M. van Engelen. 2005. ConsExpo 4.0 Consumer Exposure and Uptake Models: Program Manual. RIVM report 320104004/2005 [online]. Available: http://rivm.openrepository.com/rivm/bitstream/10029/7307/1/320104004.pdf [accessed July 18, 2014].

DeVito, S.C., and R.L. Garrett. 1996. Designing Safer Chemicals: Green Chemistry for Pollution Prevention. ACS Symposium Vol. 640. Washington, DC: American Chemical Society.

Di Toro, D.M., H.E. Allen, H.L. Bergman, J.S. Meyer, P.R. Paquin, and R.S. Santore. 2001. Biotic ligand model of the acute toxicity of metals. 1. Technical basis. Environ. Toxicol. Chem. 20(10):2383-2396.

Diamond, M., D. Mackay, R.J. Cornett, and L.A. Chant. 1990. A model of the exchange of inorganic chemicals between water and sediments. Environ. Sci. Technol. 24(5):713-722.

Dimitrov, S., T. Pavlov, D. Nedelcheva, P. Reuschenbach, M. Silvani, R. Bias, M. Comber, L. Low, C. Lee, T. Parkerton, and O. Mekenyan. 2007. A kinetic model for predicting biodegradation. SAR QSAR Environ Res. 18(5-6):443-457.

Dix, D.J., K.A. Houck, R.S. Judson, N.C. Kleinstreuer, T.B. Knudsen, M.T. Martin, D.M. Reif, A.M. Richard, I. Shah, N.S. Sipes, and R.J. Kavlock. 2012. Incorporating biological, chemical, and toxicological knowledge into predictive models of toxicity. Toxicol. Sci. 130(2):440-441.

Doak, S.H., B. Manshian, G.J. Jenkins, and N. Singh. 2012. In vitro genotoxicity testing strategy for nanomaterials and the adaptation of current OECD guidelines. Mutat. Res. 745(1-2):104-111.

Dodder, N. G., B. Strandberg, and R.A. Hites. 2002. Concentrations and spatial variations of polybrominated diphenyl ethers and several organochlorine compounds in fishes from the Northeastern United States. Environ. Sci. Technol. 36(2):146-151.

Dragos, H., M. Gilles, and V. Alexandre. 2009. Predicting the predictability: A unified approach to the applicability domain problem of QSAR models. J. Chem. Inf. Model. 49(7):1762-1776.

Drug Bank. 2013a. Pioglitazone [online]. Available: http://www.drugbank.ca/drugs/DB01132 [accessed July 21, 2014].

Drug Bank. 2013b. Rosiglitazone [online]. Available: http://www.drugbank.ca/drugs/DB00412 [accessed July 21, 2014].

Durbin, R.M., G.R. Abecasis, D.L. Altshuler, A. Auton, L.D. Brooks, R.A. Gibbs, M.E. Hurles, and G.A. McVean. 1000 Genome Project Consortium. 2010. A map of human genome variation from population-scale sequencing. Nature 467(7319):1061-1073.

EC (European Commission). 2003. Directive 2002/95/EC of the European Parliament and of the Council of 27 January 2003 on the Restriction of the Use of Certain Hazardous Substances in Electrical and Electronic Equipment. OJ EU L37:19-L283/51 [online]. Available: http://ec.europa.eu/environment/waste/rohs_eee/legis_en.htm [accessed July 22, 2014].

EC (European Commission). 2007. Corrigendum to Regulation (EC) No 1907/2006 of the European Parliament and of the Council of 18 December 2006 Concerning the Registration, Evaluation, Authorisation and Restriction of Chemicals (REACH), Establishing a European Chemicals Agency. OJ EU I.136 [online]. Available: http://www.reach-compliance.eu/english/legislation/docs/launchers/launch-2006-1907-EC.html [accessed March 6, 2014].

EC (European Commission). 2011. Commission Regulation (EU) No 286/2011 of 10 March 2011 Amending, for the Purposes of its Adaptation to Technical and Scientific Progress, Regulation (EC) No 1272/2008 of the European Parliament and of the Council on Classification, Labelling and Packaging of Substances and Mixtures. OJ EU L83 [online]. Available: http://ehs.selerant.com/ehs/LinkClick.aspx?fileticket=8obGbumkuno%3D&tabid=1506 [accessed July 14, 2014].

ECHA (European Chemical Agency). 2011. Guidance on the Preparation of an Application for Authorization. European Chemical Agency, Helsinki, Finland. January 2011 [online]. Available: http://echa.europa.eu/documents/10162/13643/authorisation_application_en.pdf [accessed June 26, 2014].

ECHA (European Chemical Agency). 2012. Guidance on Information Requirements and Chemical Safety Assessment: Chapter R.11: PBT Assessment, Version 1.1. November 2012 [online]. Available: http://echa.europa.eu/documents/10162/13632/information_requirements_r11_en.pdf [accessed March 29, 2014].

ECHA (European Chemical Agency). 2014a. Candidate List of Substances of Very High Concern for Authorisation [online]. Available: http://echa.europa.eu/candidate-list-table [accessed June 26, 2014].

ECHA (European Chemical Agency). 2014b. REACH [online]. Available: http://echa.europa.eu/regulations/reach [accessed June 26, 2014].

ECHA (European Chemical Agency). 2014c. Guidance on Information Requirements and Chemical Safety Assessment: Chapter R.7a: Endpoint Specific Guidance, Version 2.4.

February 2014 [online]. Available: http://echa.europa.eu/documents/10162/13632/information_requirements_r7a_en.pdf [accessed March 29, 2014].

ECHA (European Chemical Agency). 2014d. Topical Scientific Workshop on Risk Assessment for the Sediment Compartment: May 7-8, 2013, ECHA Workshop, Helsinki, Finland [online]. Available: http://echa.europa.eu/view-article/-/journal_content/title/topical-scientific-workshop-on-risk-assessment-for-the-sediment-compartme-1 [accessed March 30, 2014].

ECHA (European Chemicals Agency). 2014e. C&L Inventory Database [online]. Available: http://echa.europa.eu/web/guest/information-on-chemicals/cl-inventory-database [accessed June 26, 2014].

Edwards, S. 2009. A New Way of Thinking: The Lowell Center Framework for Sustainable Products. University of Massachusetts Lowell. November 2009 [online]. Available: http://www.sustainableproduction.org/downloads/LowellCenterFrameworkforSustainableProducts11-09.09.pdf [accessed March 28, 2014].

Edwards, S., J. Tickner, Y. Torrie, M. Coffin, and L. Kernan. 2011. A Compendium of Methods and Tools for Chemical Hazard Assessment. University of Massachusetts Lowell. May 2011 [online]. Available: http://www.sustainableproduction.org/downloads/Methods-ToolsforChemHazardAss5-2011.pdf [accessed March 28, 2014].

Edwards, W., R.F. Miles, and D. von Winterfeldt. 2007. Advances in Decision Analysis: From Foundations to Applications. Cambridge: Cambridge University Press.

Ellison, C.M., R. Sherhod, M.T. Cronin, S.J. Enoch, J.C. Madden, and P.N. Judson. 2011. Assessment of methods to define the applicability domain of structural alert models. J. Chem. Inf. Model. 51(5):975-985.

EMA (European Medicines Agency) 2006. Guideline on the Environmental Risk Assessment of Medicinal Products from Human Use. EMEA/CHMP/SWP/4447/00. Committee for Medicinal Products for Human Use (CHMP) [online]. Available: http://www.ema.europa.eu/docs/en_GB/document_library/Scientific_guideline/2009/10/WC500003978.pdf [accessed July 21, 2014].

EMA (European Medicines Agency). 2010. Question and Answers on the "Guideline of the Limits of Genotoxic Impurities". EMA/CHMP/SWP/431994/2007 Rev.3/ Committee for Medicinal Products for Human Use [online]. Available: http://www.ema.europa.eu/docs/en_GB/document_library/Scientific_guideline/2009/09/WC500002907.pdf [accessed September 14, 2014].

Enoch, S.J., M.T. Cronin, T.W. Schultz, and J.C. Madden. 2008. Quantitative and mechanistic read across for predicting the skin sensitization potential of alkenes acting via Michael addition. Chem. Res. Toxicol. 21(2):513-520.

Enoch, S.J., M.T.D. Cronin, and C.M. Ellison. 2011. The use of a chemistry-based profiler for covalent DNA binding in the development of chemical categories for read-across for genotoxicity. ATLA 39(2):131-145.

Enoch, S.J., M.J. Seed, D.W. Roberts, M.T. Cronin, S.J. Stocks, and R.M. Agius. 2012. Development of mechanism-based structural alerts for respiratory sensitization hazard identification. Chem. Res. Toxicol. 25(11):2490-2498.

Environment Canada. 2006. Ecological Screening Assessment Report on Polybrominated Diphenyl Ethers (PBDEs). June 2006 [online]. Available: http://www.ec.gc.ca/lcpe-cepa/09F567A7-B1EE-1FEE-73DB-8AE6C1EB7658/sar_pbde-eng.pdf [accessed July 18, 2014].

Environment Canada. 2010. State of the Science Report on Decabromodiphenyl Ether (decaBDE): Bioaccumulation and Transformation [online]. Available: http://www.ec.gc.ca/lcpe-cepa/documents/substances/decabde/ess_report_decabde-eng.pdf [accessed July 18, 2014].

EPA (U.S. Environmental Protection Agency). 1991. Guidelines for Developmental Toxicity Risk Assessment. EPA/600/FR-91/001. Risk Assessment Forum, U.S. Environmental Protection Agency [online]. Available: http://cfpub.epa.gov/ncea/cfm/recordisplay.cfm?deid=23162 [accessed July 14, 2014].

EPA (U.S. Environmental Protection Agency). 1994. Lead bioavailability. Chapter 4 in

Guidance Manual for the Integrated Exposure Uptake Biokinetic Model for Lead in Children. EPA 540-R-93-081. Office of Solid Waste and Emergency Response, U.S. Environmental Protection Agency [online]. Available: http://www.epa.gov/superfund/health/contaminants/lead/products/ch04.pdf [accessed June 30, 2014].

EPA (U.S. Environmental Protection Agency). 1996. Guidelines for Reproductive Toxicity Risk Assessment. EPA/630/R-96/009. Risk Assessment Forum, U.S. Environmental Protection Agency [online]. Available: http://www.epa.gov/raf/publications/pdfs/REPRO51.PDF [accessed July 14, 2014].

EPA (U.S. Environmental Protection Agency). 1998. Guidelines for Neurotoxicity Risk Assessment. EPA/630-95/001F. Risk Assessment Forum, U.S. Environmental Protection Agency [online]. Available: http://www.epa.gov/raf/publications/pdfs/NEUROTOX.PDF [accessed July 14, 2014].

EPA (U.S. Environmental Protection Agency). 2000. Bioaccumulation Testing and Interpretation for the Purpose of Sediment Quality Assessment: Status and Needs. EPA-823-R-00-001. Office of Water, Office of Solid Waste, U.S. Environmental Protection Agency, Washington, DC [online]. Available; http://water.epa.gov/polwaste/sediments/cs/upload/bioaccum.pdf [accessed June 27, 2014].

EPA (U.S. Environmental Protection Agency). 2006. Life Cycle Assessment: Principles and Practice. EPA/600/R-06/060. National Risk Management Research Laboratory, Office of Research and Development, U.S. Environmental Protection Agency, Cincinnati, OH [online]. Available: http://www.epa.gov/nrmrl/std/lca/pdfs/chapter4lca101.pdf [accessed June 26, 2014].

EPA (U.S. Environmental Protection Agency). 2007. Public Process to Screening Information Data Sets (SIDS) [online]. Available: http://www.epa.gov/oppt/sids/pubs/overview.htm [accessed September 9, 2014].

EPA (U.S. Environmental Protection Agency). 2009. Polybrominated Diphenyl Ethers (PBDEs) Action Plan [online]. Available: http://www.epa.gov/opptintr/existingchemicals/pubs/actionplans/pbdes_ap_2009_1230_final.pdf [accessed November 18, 2013].

EPA (U.S. Environmental Protection Agency). 2010a. TSCA New Chemicals Program (NCP) Chemical Categories. Office of Pollution Prevention and Toxics, U.S. Environmental Protection Agency, Washington, DC [online]. Available: http://www.epa.gov/oppt/newchems/pubs/npcchemicalcategories.pdf [accessed April 3, 2014].

EPA (U.S. Environmental Protection Agency). 2010b. An Exposure Assessment of Polybrominated Diphenyl Ethers. EPA/600/R-08/)86F. National Center for Environmental Assessment, Office of Research and Development, U.S. Environmental Protection Agency, Washington, DC. May 2010 [online]. Available: http://cfpub.epa.gov/ncea/cfm/recordisplay.cfm?deid=210404 [accessed July 18, 2014].

EPA (U.S. Environmental Protection Agency). 2011a. Design for the Environment Program Alternatives Assessment Criteria for Hazard Evaluation, Version 2.0. Office of Pollution Prevention and Toxics, U.S. Environmental Protection Agency, August 2011 [online]. Available: http://www.epa.gov/dfe/alternatives_assessment_criteria_for_hazard_eval.pdf [accessed June 26, 2014].

EPA (U.S. Environmental Protection Agency). 2011b. Interpretive Assistance Document for Assessment of Discrete Organic Chemicals. Sustainable Futures Summary Assessment. Office of Pollution, Prevention and Toxics, U.S. Environmental Protection Agency, Washington, DC.

EPA (U.S. Environmental Protection Agency). 2012a. Priority Chemicals. The National Waste Minimization Program. U.S. Environmental Protection Agency [online]. Available: http://www.epa.gov/osw/hazard/wastemin/priority.htm [accessed June 26, 2014].

EPA (U.S. Environmental Protection Agency). 2012b. Existing Chemicals Action Plans [online]. Available: http://www.epa.gov/oppt/existingchemicals/pubs/ecactionpln.html [accessed March 28, 2014].

EPA (U.S. Environmental Protection Agency). 2012c. Human Health Risk- Step 4-Risk Characterization [online]. Available: http://www.epa.gov/risk_assessment/risk-characterization.htm [accessed September 9, 2014].

EPA (U.S. Environmental Protection Agency). 2012d. EPA's DfE Standard for Safer Peoducts (DfE Standard). U.S. Environmental Protection Agency, September 2012 [online]. Available: http://www.epa.gov/dfe/pubs/projects/gfcp/standard-for-safer-products.pdf [accessed July 7, 2014].

EPA (U.S. Environmental Protection Agency). 2012e. Essential Principles for Reform of Chemicals Management Legislation. Existing Chemicals, U.S. Environmental Protection Agency [online]. Available: http://www.epa.gov/oppt/existingchemicals/pubs/principles.html [accessed March 29, 2014].

EPA (U.S. Environmental Protection Agency). 2012f. Glossary of Technical Terms [online]. Available: http://www.epa.gov/oust/cat/tumgloss.htm [accessed June 27, 2014].

EPA (U.S. Environmental Protection Agency). 2012g. Deca BDE Phase-Out Initiative [online]. Available online: http://www.epa.gov/oppt/existingchemicals/pubs/actionplans/deccadbe.html [accessed September 11, 2014].

EPA (U.S. Environmental Protection Agency). 2013a. Enhancing EPA's Chemical Management Program [online]. Available: http://www.epa.gov/opptintr/existingchemicals/pubs/enhanchems.html [accessed June 26, 2014].

EPA (U.S. Environmental Protection Agency). 2013b. SPARC Performs Automated Reasoning in Chemistry [online]. Available: http://www.epa.gov/athens/research/sparc.html [accessed June 30, 2014].

EPA (U.S. Environmental Protection Agency). 2013c. Draft OCSPP Test Guidelines [online]. Available: http://www.epa.gov/ocspp/pubs/frs/home/draftguidelines.htm [accessed July 7, 2014].

EPA (U.S. Environmental Protection Agency). 2013d. OCSPP Harmonized Test Guidelines: Series 850 - Ecological Effects Test Guidelines [online]. Available: http://www.epa.gov/ocspp/pubs/frs/publications/Test_Guidelines/series850.htm [accessed July 7, 2014].

EPA (U.S. Environmental Protection Agency). 2013e. Next Generation Risk Assessment: Incorporation of Recent Advances in Molecular, Computational, and Systems Biology (External Review Draft). EPA/600/R-13/14A. National Center for Environmental Assessment, Office of Research and Development, U.S. Environmental Protection Agency Washington, DC [online]. Available: http://cfpub.epa.gov/ncea/risk/recordisplay.cfm?deid=259936 [accessed January 14, 2014].

EPA (U.S. Environmental Protection Agency). 2013f. SAB Advice on Advancing the Application of Computational Toxicology Research for Human Health Risk Assessment. Science Advisory Board, U.S. Environmental Protection Agency Washington, DC [online]. Available: http://yosemite.epa.gov/sab/sabproduct.nsf/0/F3315D0EE2EDC11285257B0A005F5B7E/$File/CompTox-edited+1-29-13.pdf [accessed January 14, 2014].

EPA (U.S. Environmental Protection Agency). 2014a. Significant New Alternatives Policy (SNAP) Program [online]. Available: http://www.epa.gov/ozone/snap/ [accessed June 28, 2014].

EPA (U.S. Environmental Protection Agency). 2014b. Basics of Green Chemistry [online]. Available: http://www2.epa.gov/green-chemistry/basics-green-chemistry [accessed March 28, 2014].

EPA (U.S. Environmental Protection Agency). 2014c. Alternatives Assessments. Design for the Environment [online]. Available: http://www.epa.gov/dfe/alternative_assessments.html [accessed March 28, 2014].

EPA (U.S. Environmental Protection Agency). 2014d. EPA-Expo-Box (A Toolbox for Exposure Assessors) [online]. Available: http://www.epa.gov/risk/expobox/index.htm [accessed June 12, 2014].

EPA (U.S. Environmental Protection Agency). 2014e. EPA-Expo-Box: Indirect Estimation (Scenario Evaluation) [online]. Available: http://www.epa.gov/risk_assessment/expob

ox/approaches/ie-ef.htm [accessed September 22, 2014].

EPA (U.S. Environmental Protection Agency). 2014f. Technical Overview of Ecological Risk Assessment. Analysis Phase: Ecological Effects Characterization [online]. Available: http://www.epa.gov/oppefed1/ecorisk_ders/toera_analysis_eco.htm [accessed July 7, 2014].

EPA (U.S. Environmental Protection Agency). 2014g. Toxicity Estimation Software Tool (TEST). Risk Management Sustainable Technology [online]. Available: http://www.epa.gov/nrmrl/std/qsar/qsar.html#TEST [accessed July 9, 2014].

EPA (U.S. Environmental Protection Agency). 2014h. Ecological Structure Activity Relationships (ECOSAR) [online]. Available: http://www.epa.gov/oppt/newchems/tools/21ecosar.htm [accessed July 9, 2014].

EPA (U.S. Environmental Protection Agency). 2014i. An Alternatives Assessment for the Flame Retardant Decabromodiphenyl Ether (DecaBDE). Final Report. U.S. Environmental Protection Agency, January 2014 [online]. Available: http://www.epa.gov/dfe/pubs/projects/decaBDE/deca-report-complete.pdf [accessed July 18, 2014].

European Chemicals Bureau. 2002. European Union Risk Assessment Report. Bis(Pentabromophenyl) Ether: Risk Assessment. CAS No. 1163-19-5. EINECS No. 214-604-9. Luxemborg: Office for Official Publications of the European Communities [online]. Available: http://echa.europa.eu/documents/10162/da9bc4c4-8e5b-4562-964c-5b4cf59d2432 [accessed July 18, 2014].

Fairbrother, A. 2000. Comparative aspects of estrogen functions and measurements in oviparous and viviparous vertebrates. Hum. Ecol. Risk Assess. 6(1):73-102.

Farley, K.J., R.F. Carbonaro, K.J. Rader, C.J. Fanelli, R. Costanzo, and D.M. Di Toro. 2011. TICKET-UWM: A coupled kinetic, equilibrium and transport screening model for metals in lakes. Environ. Toxicol. Chem. 30(6):1278-1287.

Figueira, J., S. Greco, and M. Ehrgott, eds. 2005. Multiple Criteria Decision Analysis: State of the Art Surveys. Boston: Springer.

Finkel, A.M. 1995. Toward less misleading comparisons of uncertain risks: The example of aflatoxin and alar. Environ. Health Perspect. 103(4):376-385.

Finkel, A.M., and D. Golding. 1995. Worst Things First? The Debate Over Risk-based National Environmental Priorities. Washington, DC: Resources for the Future.

Fjodorova, N., M. Vracko, M. Tusar, A. Jezierska, M., Novic, and G. Schurmann. 2010. Quantitative and qualitative models for carcinogenicity prediction for non-congeneric chemicals using CP ANN method for regulatory uses. Mol. Divers. 14(3):581-594.

Florig, H.K., M.G. Morgan, K.M. Morgan, K.E. Jenni, B. Fishhoff, P.S. Fishbeck, and M.L. DeKay. 2001. A deliberative method for ranking risks (I): Overview and test bed development. Risk Anal. 21(5):913-921.

Fostel, J.M. 2008. Towards standards for data exchange and integration and their impact on a public database such as CEBS (Chemical Effects in Biological Systems). Toxicol. Appl. Pharmacol. 233(1):54-62.

Fox, J.T., S. Sakamuru, R. Huang, N. Teneva, S.O. Simmons, M. Xia, R.R. Tice, C.P. Austin, and K. Myung. 2012. High-throughput genotoxicity assay identifies antioxidants as inducers of DNA damage response and cell death. Proc. Natl. Acad. Sci. U.S.A. 109(14):5423-5428.

Fraunhofer Institute. 2007. Literature Study: Effects of Molecular Size and Lipid Solubility on Bioaccumulation Potential. Report No. FKZ 360 01 043. Fraunhofer Institute for Molecular Biology and Applied Ecology, Schmallenberg, Germany. February 15, 2007 [online]. Available: http://www.reach-info.de/dokumente/bcf.pdf [accessed March 30, 2014].

Fraysse, B., R. Mons, and J. Garric. 2006. Development of a zebrafish 4-day embryo-larval bioassay to assess toxicity of chemicals. Ecotoxicol. Environ. Saf. 63(2):253-267.

Fukui, K., T. Yonezawa, and H. Shingu. 1952. A molecular orbital theory of reactivity in aromatic hydrocarbons. J. Chem. Phys. 20(4):722-725.

Ganesan, A. 2008. The impact of natural products upon modern drug discovery. Curr. Opin. Chem. Biol. 12(3):306-317.

GAO (U.S. Government Accountability Office). 2005. Chemical Regulation: Approaches in the United States, Canada, and the European Union. GAO-06-217R [online]. Available: http://www.chemicalspolicy.org/downloads/GAO11-4-05.pdf [accessed June 25, 2014].

Gerberick, F., M. Aleksic, D. Basketter, S. Casati, A.T. Karlberg, P. Kern, I. Kimber, J.P. Lepoittevin, A. Natsch, J.M. Ovigne, C. Rovida, H. Sakaguchi, and T. Schultz. 2008. Chemical reactivity measurement and the predictive identification of skin sensitisers. The report and recommendations of ECVAM Workshop 64. Altern. Lab Anim. 36(2):215-242.

Goldenberg, J., R. Horowitz, A. Levav, and D. Mazursky. 2003. Finding your innovation sweet spot. Harvard Bus. Rev.81(3):120-129.

Gombar, V.K., and S.D. Hall. 2013. Quantitative structure-activity relationship models of clinical pharmacokinetics: Clearance and volume of distribution. J. Chem. Inf. Model. 53(4):948-957.

Goracci, L., M. Ceccarelli, D. Bonelli, and G. Cruciani. 2013. Modeling phospholipidosis induction: Reliability and warnings. J. Chem. Inf. Model. 53(6):1436-1446.

Government of Canada. 2014. Chemical Management Plan [online]. Available: http://www.chemicalsubstanceschimiques.gc.ca/plan/index-eng.php [accessed June 26, 2014].

Gramatica, P. 2007. Principles of QSAR models validation: Internal and external. QSAR Comb. Sci. 26(5):694-701.

Green Seal. 2009. Green Seal Pilot Sustainability Standard for Companies, Part 1. Product Manufacturers. Green Seas, Washington, DC. August 28, 2009 [online]. Available: http://www.labormarketinfo.edd.ca.gov/contentpub/GreenDigest/Green-Seal-Pilot-Company-Certification.pdf [accessed July 7, 2014].

Greene, N., M.D. Aleo, S. Louise-May, D.A. Price, and Y. Will. 2010a. Using an in vitro cytotoxicity assay to aid in compound selection for in vivo safety studies. Bioorg. Med. Chem. Lett. 20(17):5308-5312.

Greene, N., L. Fisk, R.T. Naven, R.R. Note, M.L. Patel, and D.J. Pelletier. 2010b. Developing structure-activity relationships for the prediction of hepatotoxicity. Chem. Res. Toxicol. 23(7):1215-1222.

Groothuis, F.A., M.B. Heringa, B. Nicol, J.L. Hermens, B.J. Blaauboer, and N.I. Kramer. in press. Dose metric considerations in in vitro assays to improve quantitative in vitro-in vivo dose extrapolations. Toxicology.

GSK (GlaxoSmithKline Inc.). 2012. Avandia®: Rosiglitazone (as Rosiglitazone Maleate). Product Monograph. GlaxoSmithKline Inc., Mississauga, Ontario [online]. Available: http://www.gsk.ca/english/html/our-products/avandia.html [accessed July 22, 2014].

Guha, R., and P. Jurs. 2004. The development of QSAR models to predict and interpret the biological activity of artemisinin analogues. J. Chem. Inf. Comput. Sci. 44(4):1440-1449.

Hansson, S.O., and C. Ruden. 2007. The Substitution Principle. Royal Institute of Technology. March 29, 2007 [online]. Available: http://archives.eppo.int/MEETINGS/2009_conferences/comparative_assessment/background.htm [accessed June 25, 2014].

Harris, D, 2012. "Going Green". Emerging Science for Environmental Health Decisions Newsletter October 2012 [online]. Available: http://nas-sites.org/emergingscience/files/2011/05/green-chem-nl-25.pdf [accessed June 25, 2014].

Harvey, C., D. Mackay, and E. Webster. 2007. Can the unit world model concept be applied to hazard assessment of both organic chemicals and metal ions? Environ. Toxicol. Chem. 26(10):2129-2142.

Hatamura, Y. 2006. Decision-Making in Engineering Design. London: Springer.

Hehre, W.J., L. Radom, P. von R. Schleyer, and J. Pople. 1986. AB INITIO Molecular Orbital Theory. Hoboken, NJ: Wiley.

Heine, L.G., and S.A. Franjevic. 2013. Chemical Hazard Assessment and the GreenScreen for Safer Chemicals. Pp. 129-156 in Chemical Alternatives Assessments, R.E.

Hester, and R.M. Harrison, eds. London, UK: Royal Society of Chemistry.

Hewitt, M., and C.M. Ellison. 2010. Developing the applicability domain of in silico models: Relevance, importance and methodology. Pp. 301-333 in In Silico Toxicology: Principles and Applications, M.T. Cronin, and J.C. Madden, eds. Cambridge, UK: Royal Society of Chemistry.

Hewitt, M., C.M. Ellison, C.J. Enoch, J.C. Madden, and M.T. Cronin. 2010. Integrating (Q)SAR models, expert systems and read-across approaches for the prediction of developmental toxicity. Reprod. Toxicol. 30(1):147-160.

Hewitt, M., S.J. Enoch, J.C. Madden, K.R. Przybylak, and M.T. Cronin. 2013. Hepatotoxicity: A scheme for generating chemical categories for read-across, structural alerts and insights into mechanism(s) of action. Crit. Rev. Toxicol. 43(7):537-558.

Hillebrecht, A., W. Muster, A. Brigo, M. Kansy, T. Weiser, and T. Singer. 2011. Comparative evaluation of in silico systems for Ames test mutagenicity prediction: Scope and limitations. Chem. Res. Toxicol. 24(6):843-854.

Hogue, C. 2013. Assessing alternatives to toxic chemicals. C&EN 91(50):19-20.

Holder, J.W. 2012. Physical and physicochemical factors effecting transport of chlorohydrocarbon gases from lung alveolar air to blood as measured by the causation of narcosis. J. Environ. Sci. Health C Environ. Carcinog. Ecotoxicol. Rev. 30(1):42-80.

Hollósy, F., K. Valkó, A. Hersey, S. Nunhuck, G. Kéri, and C. Bevan. 2006. Estimation of volume of distribution in humans from high throughput HPLC-based measurements of human serum albumin binding and immobilized artificial membrane partitioning. J. Med. Chem. 49(24):6958-6971.

Holt, M.P., and C. Ju. 2006. Mechanisms of drug-induced liver injury. AAPS J. 8(1):E48-E54.

Howard, P.H., and D.C. Muir. 2010. Identifying new persistent and bioaccumulative organics among chemicals in commerce. Environ. Sci. Technol. 44 (7):2277-2285.

Howard, P.H., and D.C. Muir. 2011. Identifying new persistent and bioaccumulative organics among chemicals in commerce. II. Pharmaceuticals. Environ. Sci. Technol. 45(16):6938-6946.

Howard, P.H., and D.C. Muir. 2013. Identifying new persistent and bioaccumulative organics among chemicals in commerce. III: Byproducts, impurities, and transformation products. Environ. Sci. Technol. 47(10):5259-5266.

Hsiao, Y.W., U. Fagerholm, and U. Norinder. 2013. In silico categorization of in vivo intrinsic clearance using machine learning. Mol. Pharm. 10(4):1318-1321.

Huang, R., N. Southall, M.H. Cho, M. Xia, J. Inglese, and C.P. Austin. 2008. Characterization of diversity in toxicity mechanism using in vitro cytotoxicity assays in quantitative high throughput screening. Chem. Res. Toxicol. 21(3):659-667.

Huang, R., M. Xia, M.H. Cho, S. Sakamuru, P. Shinn, K.A. Houck, D.J. Dix, R.S. Judson, K.L. Witt, R.J. Kavlock, R.R. Tice, and C.P. Austin. 2011. Chemical genomics profiling of environmental chemical modulation of human nuclear receptors. Environ. Health Perspect. 119(8):1142-1148.

Huggett, D.B., C.M. Foran, B.W. Brooks, J. Weston, B. Peterson, K.E. Marsh, T.W. La Point, and D. Schlenk. 2003. Comparison of in vitro and in vivo bioassays for estrogenicity in effluent from North American municipal wastewater facilities. Toxicol. Sci. 72(1):77-83.

Hughes, J.D., J. Blagg, D.A. Price, S. Bailey, G.A. DeCrescenzo, R.V. Devraj, E. Ellsworth, Y.M. Fobian, M.E. Gibbs, R.W. Gilles, N. Greene, E. Huang, T. Krieger-Burke, J. Loesel, T. Wager, L. Whiteley, and Y. Zhang. 2008. Physiochemical drug properties associated with in vivo toxicological outcomes. Bioorg. Med. Chem. Lett. 18(17):4872-4875.

Ibrahim, M., and L. Garcia-Contreras. 2013. Mechanisms of absorption and elimination of drugs administered by inhalation. Ther. Deliv. 4(8):1027-1045.

Ibrahim, R., J.M. Nitsche, and G.B. Kasting. 2012. Dermal clearance model for epidermal bioavailability calculations. J. Pharm. Sci. 101(6):2094-2108.

IC2 (Interstate Chemicals Clearinghouse). 2013. Interstate Chemicals Clearinghouse Alternatives Assessment Guide, Version 1.0. November 2013 [online]. Available: http://www.newmoa.org/prevention/ic2/IC2_AA_Guide-Version_1.pdf [accessed June 26, 2014].

IC2 (Interstate Chemicals Clearinghouse). 2014. IC2 Safer Alternatives Assessments [online]. Available: http://www.ic2saferalternatives.org/ [accessed June 26, 2014].

Illinois Environmental Protection Agency. 2006. DecaBDE Study: A Review of Available Scientific Research. A Report to the General Assembly and the Governor In Response to Public Act 94-100 [online]. Available: http://www.epa.state.il.us/reports/decabde-study/available-research-review.pdf [accessed July 18, 2014].

IOM (Institute of Medicine). 2006. Valuing Health for Regulatory Cost-Effectiveness Analysis, W. Miller, L.A. Robinson, and R.S. Lawrence, eds. Washington, DC: The National Academies Press.

ISO (International Organization for Standardization). 2006a. 14040. Environmental Management-Life Cycle Assessment- Principles and Framework. International Organization for Standardization (ISO), Geneva, Switzerland.

ISO (International Organization for Standardization). 2006b. 14044. Environmental Management-Life Cycle Assessment-Requirements and Guidelines. International Organization for Standardization (ISO), Geneva, Switzerland.

IUPAC (International Union of Pure and Applied Chemistry). 1993. Glossary for chemists of terms used in toxicology. Pure Appl. Chem. 65(9):2003-2122.

Janssen, C.R., D.G. Heijerick, K.A. De Schamphelaere, and H.E. Allen. 2003. Environmental risk assessment of metals: Tools for incorporating bioavailability. Environ. Int. 28(8):793-800.

Jarvis, J., M.J. Seed, R. Elton, L. Sawyer, and R. Agius. 2005. Relationship between chemical structure and the occupational asthma hazard of low molecular weight organic compounds. Occup. Environ. Med. 62(4):243-250.

Jaspers, V. L., A. Covaci, S. Voorspoels, T. Dauwe, M. Eens, and P. Schepens. 2006. Brominated flame retardants and organochlorine pollutants in aquatic and terrestrial predatory birds of Belgium: Levels, patterns, tissue distribution and condition factors. Environ. Pollut. 139(2):340-352.

Jaworska, J., M. Comber, C. Auer, and C. J. van Leeuwen. 2003. Summary of workshop on regulatory acceptance of (Q)SARs for human health and environmental endpoints. Environ. Health Perspect. 111(10):1358-1360.

Johnson-Restrepo, B., and K. Kannan. 2009. An assessment of sources and pathways of human exposure to polybrominated diphenyl ethers in the United States. Chemosphere 76(4):542-548.

Johnson-Restrepo, B., K. Kannan, R. Addink, and D.H. Adams. 2005. Polybrominated diphenyl ethers and polychlorinated biphenyls in a marine foodweb of coastal Florida. Environ. Sci. Technol. 39(21):8243-8250.

Jonnalagadda, S.B., and P.V. Rao. 1993. Toxicity, bioavailability and metal speciation. Comp. Biochem. Physiol. C 106(3):585-595.

Jørgensen, A., A. Le Bocq, L. Nazarkina, and M. Hauschild. 2008. Methodologies for social life cycle assessment. Int. J. LCA 13(2):96-103.

Jorgensen, W.L., and E.M. Duffy. 2002. Prediction of drug solubility from structure. Adv. Drug Deliver. Rev. 54(3):355-366.

Judson, P.N. 2009. Towards establishing good practice in the use of computer prediction. Qual. Assur. J. 12(3-4):120-125.

Judson, R.S., R.J. Kavlock, R.W. Setzer, E.A. Cohen Hubal, M.T. Martin, T.B. Knudsen, K.A. Houck, R.S. Thomas, B.A. Welmore, and D.J. Dix. 2011. Estimating toxicity-related biological pathway altering doses for high-throughput chemical risk assessment. Chem. Res. Toxicol. 24(4):451-462.

Judson, R.S., M.T. Martin, P. Egeghy, S. Gangwal, D.M. Reif, P. Kothiya, M. Wolf, T. Cathey, T. Transue, D. Smith, J. Vall, A. Frame, S. Mosher, E.A. Cohen Hubal, and A.M.

Richard. 2012. Aggregating data for computational toxicology applications: The U.S. Environmental Protection Agency (EPA) Aggregated Computational Toxicology Resource (ACToR) System. Int. J. Mol. Sci. 13(2):1805-1831.

Kamber, M., S. Fluckiger-Isler, G. Engelhardt, R. Jaeckh, and E. Zeiger. 2009. Comparison of the Ames II and traditional Ames test responses with respect to mutagenicity, strain specificities, need for metabolism and correlation with rodent carcinogenicity. Mutagenesis 24(4):359-366.

Kang, G.S., P.A. Gillespie, A. Gunnison, H. Rengifo, J. Koberstein, and L.C. Chen. 2011. Comparative pulmonary toxicity of inhaled nickel nanoparticles; role of deposited dose and solubility. Inhal. Toxicol. 23(2):95-103.

Kaplowitz, N. 2005. Idiosyncratic drug hepatotoxicity. Nat. Rev. Drug Discov. 4(6):489-499.

Kar, S., O. Deeb, and K. Roy. 2012. Development of classification and regression based QSAR models to predict rodent carcinogenic potency using oral slope factor. Ecotoxicol. Environ. Saf. 82:85-95.

Kavlock, R., K. Chandler, K. Houck, S. Hunter, R. Judson, N. Kleinstreuer, T. Knudsen, M. Martin, S. Padilla, D. Reif, A. Richard, D. Rotroff, N. Sipes, and D. Dix. 2012. Update on EPA's ToxCast program: Providing high throughput decision support tools for chemical risk management. Chem. Res. Toxicol. 25(7):1287-1302.

Keeney, R.L., and H. Raiffa. 1976. Decisions with Multiple Objectives: Preferences and Value Tradeoffs. New York: John Wiley & Sons.

Kelsey, J.W., and M. Alexander. 1997. Declining bioavailability and inappropriate estimation of risk of persistent compounds. Environ. Toxicol. Chem. 16(3):582-585.

KEMI (Swedish Chemicals Agency). 2014. A Non-Toxic Environment [online]. Available: http://www.kemi.se/en/Start/A-Non-Toxic-Environment/ [accessed June 28, 2014].

Khetan, S.K., and T.J. Collins. 2007. Human pharmaceuticals in the aquatic environment: A challenge to green chemistry. Chem. Rev. 107(6):2319-2364.

Kirkland, D., L. Reeve, D. Gatehouse, and P. Vanparys. 2011. A core in vitro genotoxicity battery comprising the Ames test plus the in vitro micronucleus test is sufficient to detect rodent carcinogens and in vivo genotoxins. Mutat. Res. 721(1):27-73.

Kleinstreuer, N.C., D.J. Dix, K.A. Houck, R.J. Kavlock, T.B. Knudsen, M.T. Martin, K.B. Paul, D.M. Reif, K.M. Crofton, K. Hamilton, R. Hunter, I. Shah, and R.S. Judson. 2013. In vitro perturbations of targets in cancer hallmark processes predict rodent chemical carcinogenesis. Toxicol. Sci. 131(1):40-55.

Knudsen, T., M. Martin, K. Chandler, N. Kleinstreuer, R. Judson, and N. Sipes. 2013. Predictive models and computational toxicology. Methods Mol. Biol. 947:343-374.

Kola, I., and J. Landis. 2004. Can the pharmaceutical industry reduce attrition rates? Nat. Rev. Drug Discov. 3(8):711-715.

Kolle, M., A. Lethbridge, M. Kreysing, J.J. Baumberg, J. Aizenberg and P. Vukusic. 2013. Bio-inspired band-gap tunable elastic optical multilayer fibers. Adv. Mater. 25(15):2239-2245.

Koslov-Davino, E., X. Wang, and T. Schroeter. 2013. Target promiscuity and physicochemical properties contribute to pharmacologically induced ER-stress. Toxicol. In Vitro 27(1):204-210.

Kostal, J., A. Voutchkova-Kostal, P.T. Anastas, and J.B. Zimmerman. In press. Identifying and designing chemicals with minimal acute aquatic toxicity. Proc. Natl. Acad. Sci. U.S.A.

Krewski, D., M. Westphal, M. Al Zoughool, M.C. Croteau, and M.E. Andersen. 2011. New directions in toxicity testing. Annu. Rev. Public Health 32:161-178.

Kreyling, W.G., M. Semmler-Behnke, S. Takenaka, and W. Möller. 2013. Differences in the biokinetics of inhaled nano-versus micrometer-sized particles. Acc. Chem. Res. 46(3):714-722.

Krupnick, A., R. Morgenstern, M. Batz, P. Nelson, D. Burtraw, J.S. Shih, and M. McWilliams. 2006. Not a Sure Thing: Making Regulatory Choices under Uncertainty. Washington, DC: Resources for the Future. February 2006 [online].

Available: http://www.rff.org/rff/Documents/RFF-Rpt-RegulatoryChoices.pdf [accessed July 17, 2014].

Kuczenski, B., R. Geyer, and B. Hall. 2010. Chemical Alternatives Analysis: Methods, Models, and Tools: Revised Final Report to the Department of Toxic Substance Control. California Department of Toxic Substance Control. August 2010 [online]. Available: https://migreenchemistry.org/wp-content/uploads/2011/09/CA-DTSC_Chemical-Assessments-Report_August-2010.pdf [accessed June 27, 2014].

Kuruto-Niwa, R., R. Nozawa, T. Miyakoshi, T. Shiozawa, and Y. Terao. 2005. Estrogenic activity of alkylphenols, bisphenol S, and their chlorinated derivatives using a GFP expression system. Environ. Toxicol. Pharmacol. 19(1):121-130.

Lagalante, A.F., T.D. Oswald, and F.C. Calvosa. 2009. Polybrominated diphenyl ether (PBDE) levels in dust from previously owned automobiles at United States dealerships. Environ. Int. 35(3):539-544.

Lange, I.G., A. Hartel, and H.H. Meyer. 2002. Evolution of oestrogen functions in vertebrates. J. Steroid Biochem. Mol. Biol. 83(1-5):219-226.

Lassen, C., S. Havelund, A. Leisewitz, and P. Maxson. 2006. Deca-BDE and Alternatives in Electrical and Electronic Equipment (Environmental Project No. 1141 2006). Danish Environmental Protection Agency [online]. Available: http://infohouse.p2ric.org/ref/42/41725.pdf [accessed July 18, 2014].

Lavoie, E.T., L.G. Heine, H. Holder, M.S. Rossi, R.E. Lee, E.A. Connor, M.A. Vrabel, D.M. Difiore, and C.L. Davies. 2010. Chemical alternatives assessment: Enabling substitution to safer chemicals. Environ. Sci. Technol. 44(24):9244-9249.

Leo, A.J. 1995. Critique of recent comparison of log P calculation methods. Chem. Pharm. Bull. 43(3):512-513.

Leo, A.J. 2000. Octanol/water partition coefficient. Pp. 89-114 in Handbook of Property Estimation Methods for Chemical, D. Mackay, and R.S. Boething, eds. Boca Raton: CRC Press.

Leung, M.C., P.L. Williams, A. Benedetto, C. Au, K.J. Helmcke, M. Aschner, and J.N. Meyer. 2008. *Caenorhabditis elegans*: An emerging model in biomedical and environmental toxicology. Toxicol. Sci. 106(1):5-28.

Lindberg, P., U. Sellström, L. Häggberg, and C.A. de Wit. 2004. Higher brominated diphenyl ethers and hexabromocyclododecane found in eggs of peregrine falcons (*Falco peregrinus*) breeding in Sweden. Environ. Sci. Technol. 38(1):93-96.

Linkov, I., F.K. Satterstrom, G. Kiker, C. Batchelor, T. Bridges, and E. Ferguson. 2006. From comparative risk assessment to multi-criteria decision analysis and adaptive management: Recent developments and applications. Environ. Int. 32(8):1072-1093.

Lipinski, C.A., F. Lombardo, B.W. Dominy, and P.J. Feeney. 1997. Experimental and computational approaches to estimate solubility and permeability in drug discovery and development settings. Adv. Drug Deliv. Rev. 46(1-3):3-26.

Lock, E.F., N. Abdo, R. Huang, M. Xia, O. Kosyk, S.H. O'Shea, Y.H. Zhou, A. Sedykh, A. Tropsha, C.P. Austin, R.R. Tice, F.A. Wright, and I. Rusyn. 2012. Quantitative high-throughput screening for chemical toxicity in a population-based in vitro model. Toxicol. Sci. 126(2):578-588.

Lofstedt, R. 2014. The substitution principle in chemical regulation: A constructive critique. J. Risk Res. 17(5):543-564.

Lombardo, F., R.S. Obach, M.Y. Shalaeva, and F. Gao. 2002. Prediction of volume of distribution values in humans for neutral and basic drugs using physicochemical measurements and plasma protein binding data. J. Med. Chem. 45(13):2867-2876.

Lorber, M. 2008. Exposure of Americans to polybrominated diphenyl ethers. J. Expo. Sci. Environ. Epidemiol. 18(1):2-19.

Lounkine, E., M.J. Keiser, S. Whitebread, D. Mikhailov, J. Hamon, J.L. Jenkins, P. Lavan, E. Weber, A.K. Doak, S. Cote, B.K. Shoichet, and L. Urban. 2012. Large-scale prediction and testing of drug activity on side-effect targets. Nature 486(7403):361-367.

Lowell. 2005. Designing Safer Alternatives: Chemical, Materials and Products: Workshop Summary, Lowell Center for

Sustainable Production, University of Massachusetts [online]. Available: http://www.chemicalspolicy.org/downloads/alternativesassessmentfinalreport_000.pdf [accessed June 27, 2014].

Luoma, S.N., and P.S. Rainbow. 2005. Why is metal bioaccumulation so variable? Biodynamics as a unifying concept. Environ. Sci. Technol. 39(7):1921-1931.

Lynch, A.M., J.C. Sasaki, R. Elespuru, D. Jacobson-Kram, V. Thybaud, M. De Boeck, M.J. Aardema, J. Aubrecht, R.D. Benz, S.D. Dertinger, G.R. Douglas, P.A. White, P.A. Escobar, A. Fornace Jr., M. Honma, R.T. Naven, J.F. Rusling, R.H. Schiestl, R.M. Walmsley, E. Yamamura, J. van Benthem, and J.H. Kim. 2011. New and emerging technologies for genetic toxicity testing. Environ. Mol. Mutagen. 52(3):205-223.

MacDonald, J.S., and R.T. Robertson. 2009. Toxicity testing in the 21st century: A view from the pharmaceutical industry. Toxicol. Sci. 110(1):40-46.

MacKay, C., M. Davies, V. Summerfield, and G. Maxwell. 2013. From pathways to people: Applying the adverse outcome pathway (AOP) for skin sensitization to risk assessment. ALTEX 30(4):473-486.

Mackay, D., and S. Paterson. 1991. Evaluating the regional multimedia fate of organic chemicals: A level III fugacity model. Environ. Sci. Technol. 25(3):427-436.

Mackay, D., J.A. Arnot, F.A. Gobas, and D.E. Powell. 2013. Mathematical relationships between metrics of chemical bioaccumulation in fish. Environ. Toxicol. Chem. 32(7):1459-1466.

Mahadevan, B., R.D. Snyder, M.D. Waters, R.D. Benz, R.A. Kemper, R.R. Tice, and A.M. Richard. 2011. Genetic toxicology in the 21st century: Reflections and future directions. Environ. Mol. Mutagen 52(5):339-354.

Malloy, T.F., P.J. Sinsheimer, A. Blake, and I. Linkov. 2011. Developing Regulatory Alternatives Analysis Methodologies for the California Green Chemistry Initiative. Sustainable Technology and Policy Program, University of California, Los Angeles, CA. October 20, 2011[online]. Available: http://www.stpp.ucla.edu/sites/default/files/Final%20AA%20Report.final%20rev.pdf [accessed June 26, 2014].

Malloy, T., P. Sinsheimer, A. Blake, and I. Linkov. 2013. Use of multi-criteria decision analysis in regulatory alternatives analysis: A case study of lead free solder. Integr. Environ. Assess. Manag. 9(4):652-664.

Mandalakis, M., E.G. Stephanou, Y. Horii, and K. Kannan. 2008. Emerging contaminant in car interiors: Evaluating the impact of airborne PBDEs and PBDD/Fs. Environ. Sci. Technol. 42(17):6431-6436.

Marchant, C.A., K.A. Briggs, and A. Long. 2008. In silico tools for sharing data and knowledge on toxicity and metabolism: Derek for Windows, Meteor, and Vitic. Toxicol. Mech. Methods 18(2-3):177-187.

Martin, M.T., R.S. Judson, D.M. Reif, R.J. Kavlock, and D.J. Dix. 2009. Profiling chemicals based on chronic toxicity results from the U.S. EPA ToxRef Database. Environ. Health Perspect. 117(3):392-399.

Martin, M.T., D.J. Dix, R.S. Judson, R.J. Kavlock, D.M. Reif, A.M. Richard, D.M. Rotroff, S. Romanov, A. Medvedev, N. Poltoratskaya, and M. Gambarian. 2010. Impact of environmental chemicals on key transcription regulators and correlation to toxicity end points within EPA's ToxCast program. Chem. Res. Toxicol. 23(3):578-590.

Martin, M.T., T.B. Knudsen, D.M. Reif, K.A. Houck, R.S. Judson, R.J. Kavlock, and D.J. Dix. 2011. Predictive model of rat reproductive toxicity from ToxCast high throughput screening. Biol. Reprod. 85(2):327-339.

Martin, M.T., T.B. Knudsen, R.S. Judson, R.J. Kavlock, and D.J. Dix. 2012. Economic benefits of using adaptive predictive models of reproductive toxicity in the context of a tiered testing program. Syst. Biol. Reprod. Med. 58(1):3-9.

Mata, T.M., A.A. Martins, N.B. Belmira, M.L. Martins, R.L.R. Salcedoa, and C.A.V. Costa. 2012. LCA tool for sustainability evaluations in the pharmaceutical industry. Chem. Eng. Trans. 26:261-266.

Matthews, E.J., N.L. Kruhlak, R.D. Benz, J. Ivanov, G. Klopman, and J.F. Contrera. 2007. A comprehensive model for reproductive and developmental toxicity hazard identification: II. Construction of QSAR models to predict activities of untested chemicals. Regul. Toxicol. Pharmacol. 47(2):136-155.

Matthews, J., T. Celius, R. Halgren, and T. Zacharewski. 2000. Differential estrogen receptor binding of estrogenic substances: A species comparison. J. Steroid Biochem. Mol. Biol. 74(4):223-234.

Maxwell, G., C. MacKay, R. Cubberley, M. Davies, N. Gellatly, S. Glavin, T. Gouin, S. Jacquoilleot, C. Moore, R. Pendlington, O. Saib, D. Sheffield, R. Stark, and V. Summerfield. 2014. Applying the skin sensitisation adverse outcome pathway (AOP) to quantitative risk assessment. Toxicol. In Vitro 28(1):8-12.

McDougal, J.N., and M.F. Boeniger. 2002. Methods for assessing risks of dermal exposures in the workplace. Crit. Rev. Toxicol. 32(4):291-327.

McGeer, J., G. Henningsen, R. Lanno, N. Fisher, K. Sappington, and J. Drexler. 2004. Issue Paper on the Bioavailability and Bioaccumulation of Metals. Submitted to Risk Assessment Forum, U.S. Environmental Protection Agency, Washington, DC, by Eastern Research Group, Inc., Lexington, MA. August 19, 2004 [online]. Available: http://www.epa.gov/raf/publications/pdfs/BIOFINAL81904.PDF[accessed June 23, 2014].

Meek, B., and J. Doull. 2009. Pragmatic challenges for the vision of toxicity testing in the 21st century in a regulatory context: Another Ames test? ...or a new edition of "the Red Book"? Toxicol. Sci. 108(1):19-21.

Mekenyan, O.G., G.Y. Patlewicz, C. Kuseva, I. Popova, A. Mehmed, S. Kotov, T. Zhechev, T. Pavlov, S. Temelkov, and D.W. Roberts. 2014. A mechanistic approach to modelling respiratory sensitization. Chem. Res. Toxicol. 27(2):219-239.

Meylan, W.M., and P.H. Howard. 1995. Atom/fragment contribution method for estimating octanol-water partition coefficients. J. Pharm. Sci. 84(1):83-92.

Meylan, W.M., R. Boethling, D. Aronson, P. Howard, and J. Tunkel. 2007. Chemical structure-based predictive model for anaerobic biodegradation potential. Environ. Toxicol. Chem. 26(9):1785-1792.

Minnesota Department of Health. 2010. Take-home Lead: A Preventable Risk for Your Family [online]. Available: http://www.health.state.mn.us/divs/eh/lead/fs/takehome.pdf [accessed July 18, 2014].

Modi, S., M. Hughes, A. Garrow, and A. White. 2012. The value of in silico chemistry in the safety assessment of chemicals in the consumer goods and pharmaceutical Industries. Drug Discov. Today 17(3/4):135-142.

Mora, C., D.P. Tittensor, S. Adl, A.G.B. Simpson, and B. Worm. 2011. How many species are there on earth and in the ocean? PLoS Biol. 9(8):e1001127.

Morgan, K.T., and T.M. Monticello. 1990. Airflow, gas deposition, and lesion distribution in the nasal passages. Environ. Health Perspect. 85:209-218.

Morgan, K.M., M.L. DeKay, P.S. Fischbeck, M.G. Morgan, B. Fischhoff, and H.K. Florig. 2001. A deliberative method for ranking risks. 2. Evaluation of validity and agreement among risk managers. Risk Anal. 21(5):923-938.

Morgan, M.G., and M. Henrion. 1990. Uncertainty: A Guide to Dealing with Uncertainty in Quantitative Risk and Policy Analysis. Cambridge: Cambridge University Press.

Muehlbacher, M., A. El Kerdawy, C. Kramer, B. Hudson, and T. Clark. 2011. Conformation-dependent QSPR models: logPOW. J. Chem. Inf. Model. 51(9):2408-2416.

NASA. 2009. Whisker Failures [online]. Available: http://nepp.nasa.gov/Whisker/failures/index.htm [accessed September 22, 2014].

Naven, R.T., S. Louise-May, and N. Greene. 2010. The computational prediction of genotoxicity. Expert Opin. Drug Metab. Toxicol. 6(7):797-807.

Naven, R.T., N. Greene, and R.V. Williams. 2012. Latest advances in computational genotoxicity prediction. Expert Opin. Drug Metab. Toxicol. 8(12):1579-1587.

Naven, R.T., R. Swiss, J. K. McLeod, Y. Will, and N. Greene. 2013. The development of structure-activity relationships for mitochondrial dysfunction: Uncoupling of oxidative phosphorylation. Toxicol. Sci. 131(1):271-278.

Nelson, A.M., and T.E. Long. 2012. A perspective on emerging polymer technologies for bisphenol-A replacement. Polym. Int. 61(10):1485-1491.

Neubig, R.R., M. Spedding, T. Kenakin, and A. Christopoulos. 2003. International Union of Pharmacology Committee on Receptor Nomenclature and Drug Classification. XXXVIII. Update on terms and symbols in quantitative pharmacology. Pharmacol. Rev. 55(4):597-606.

NIOSH (National Institute for Occupational Safety and Health). 2005. Pocket Guide to Chemical Hazards. DHHS (NIOSH) Publication No. 92-100 [online]. Available: http://www.cdc.gov/niosh/npg/default.html [accessed April 27, 2014].

NIST (National Institute of Standards and Technology). 2011. NIST Chemistry WebBook [online]. Available: http://webbook.nist.gov/chemistry/form-ser.html [accessed June 17, 2014].

NRC (National Research Council). 1994. Science and Judgment in Risk Assessment. Washington, DC: National Academy Press.

NRC (National Research Council). 1996. Understanding Risk: Informing Decisions in a Democratic Society. Washington, DC: National Academy Press.

NRC (National Research Council). 2007. Toxicity Testing in the 21st Century: A Vision and a Strategy. Washington, DC: The National Academies Press.

NRC (National Research Council). 2009. Science and Decisions: Advancing Risk Assessment. Washington, DC: The National Academies Press.

NRC (National Research Council). 2012. Science for Environmental Protection: The Road Ahead. Washington, DC: The National Academies Press.

NTP (National Toxicology Program). 2011. 12th Report on Carcinogens. U.S. Department of Health and Human Services, Public Health Service National Toxicology Program [online]. Available: http://ntp.niehs.nih.gov/ntp/roc/twelfth/roc12.pdf [accessed January 7, 2014].

Oberdorster, G. 1996. Significance of particle parameters in the evaluation of exposure-dose-response relationships of inhaled particles. Inhal. Toxicol. 8(suppl.):73-89.

OECD (Organisation for Economic Co-operation and Development). 2005. Manual for Investigation of HPV Chemicals, Chapter 2. Data Gathering and Testing: SIDS, The SIDS Plan and the SIDS Dossier [online]. Available: http://www.oecd.org/env/ehs/risk-assessment/chapter2datagatheringandtestingsidsthesidsplanandthesidsdossier.htm [accessed March 12, 2014].

OECD (Organisation for Economic Co-operation and Development). 2007. Guidance on Grouping of Chemicals. Series on Testing and Assessment No. 80. ENV/JM/MONO(2007)28 [online]. Available: http://search.oecd.org/officialdocuments/displaydocumentpdf/?doclanguage=en&cote=env/jm/mono(2007)28 [accessed March 30, 2014].

OECD (Organization for Economic Cooperation and Development). 2012a. Descriptions of Existing Models and Tools Used for Exposure Assessment: Results of OECD Survey. ENV/JM/MONO/(2012)37. Series on Testing and Assessment No. 182, December 19, 2012 [online]. Available: http://search.oecd.org/officialdocuments/displaydocumentpdf/?cote=env/jm/mono(2012)37&doclanguage=en [accessed June 12, 2014].

OECD (Organisation for Economic Co-operation and Development). 2012b. QSAR Toolbox [online]. Available: http://www.qsartoolbox.org/ [accessed July 9, 2014].

OECD (Organisation for Economic Co-operation and Development). 2013a. Current Landscape of Alternatives Assessment Practice: A Meta Review. ENV/JM/MONO(2013)24. Series on Risk Management No. 26 [online]. Available: http://search.oecd.org/officialdocuments/publicdisplaydocumentpdf/?cote=ENV/JM/MONO(2013)24&docLanguage=En [accessed March 28, 2014].

OECD (Organisation for Economic Co-operation and Development). 2013b. Guidance Document on Developing and

Assessing Adverse Outcome Pathways. Series on Testing and Assessment No. 184. ENV/JM/MONO(2013)6. Paris: OECD [online]. Available: http://www.oecd.org/officialdocuments/publicdisplaydocumentpdf/?cote=env/jm/mono(2013)6&doclanguage=en [accessed July 11, 2014].

OECD (Organisation for Economic Co-operation and Development). 2014a. OECD Guidelines for the Testing of Chemicals, Section 2. Effects on Biotic Systems [online]. Available: http://www.oecd-ilibrary.org/environment/oecd-guidelines-for-the-testing-of-chemicals-section-2-effects-on-biotic-systems_20745761 [accessed July 9, 2014].

OECD (Organisation for Economic Co-operation and Development). 2014b. OECD Guidelines for the Testing of Chemicals, Section 1: Physical-Chemical Properties [online]. Available: http://www.oecd-ilibrary.org/environment/oecd-guidelines-for-the-testing-of-chemicals-section-1-physical-chemical-properties_20745753 [accessed May 10, 2014].

OEHHA (Office of Environmental Health Hazard Assessment). 2014. OEHHA Acute, 8-hour and Chronic Reference Exposure Levels as of January 2014. Office of Environmental Health Hazard Assessment, State of California [online]. Available: http://www.oehha.ca.gov/air/allrels.html [accessed April 28, 2014].

OSHA (Occupational Safety and Health Administration). 2014. Transitioning to Safer Chemicals: Explore the Steps [online]. Available: https://www.osha.gov/dsg/safer_chemicals/step1_engage.html [accessed July 18, 2014].

O'Shea, S.H., J. Schwarz, O. Kosyk, P.K. Ross, M.J. Ha, F.A. Wright, and I. Rusyn. 2011. In vitro screening for population variability in chemical toxicity. Toxicol. Sci. 119(2):398-407.

Padilla, S., D. Corum, B. Padnos, D.L. Hunter, A. Beam, K.A. Houck, N. Sipes, N. Kleinstreuer, T. Knudsen, D.J. Dix, and D.M. Reif. 2012. Zebrafish developmental screening of the ToxCast Phase I chemical library. Reprod. Toxicol. 33(2):174-187.

Pal, R., and J.P. Rai. 2010. Phytochelatins: Peptides involved in heavy metal detoxification. Appl. Biochem. Biotechnol. 160(3):945-963.

Parker, M.S., R. Sah, A. Balasubramaniam, E.A. Park, F.R. Sallee, and S.L. Parker. 2014. Dimers of G-protein coupled receptors as versatile storage and response units. Int. J. Mol. Sci. 15(3):4856-4877.

Pärt, P. 1989. Bioavailability and uptake of xenobiotics in fish. Pp. 113-127 in Chemicals in the Aquatic Environment: Advanced Hazard Assessment, L. Landner, ed. Springer Series on Environmental Management. New York: Springer.

Patlewicz, G., N. Ball, E.D. Booth, E. Hulzebos, E. Zvinavashe, and C. Hennes. 2013. Use of category approaches, read-across and (Q)SAR: General considerations. Regul. Toxicol. Pharmacol. 67(1):1-12.

Pavan, M., and A.P. Worth. 2008. Review of estimation models for biodegradation. QSAR Comb. Sci. 27(1):32-40.

Pessala, G.P., J. Aho, R. Lehti, I. Vehviläinen, M. Hjelt, E. Priha, T. Santonen, M. Koponen, B. Bäck, E.R. Hyytinen, and A. Kangas. 2012. Minimising Chemical Risk to Workers' Health and Safety through Substitution. Luxembourg: Publications Office of the European Union.

Petito Boyce, C., S.N. Sax, D.G. Dodge, M.C. Pollock, and J.E. Goodman. 2009. Human exposure to decabromodiphenyl ether, tertabromobisphenol A, and decabromodiphenyl ethane in indoor dust. J. Environ. Protect. Sci. 3:75-96.

Peukert, S., J. Nunez, F. He, M. Dai, N. Yusuff, A. DiPesa, K. Miller-Moslin, R. Karki, B. Lagu, C. Harwell, Y. Zhang, D. Bauer, J.F. Kelleher, and W. Egan. 2011. A method for estimating the risk of drug-induced phototoxicity and its application to smoothened inhibitors. MedChemComm 2(10):973-976.

Pfuhler, S., S. Albertini, R. Fautz, B. Herbold, S. Madle, D. Utesch, and A. Poth. 2007. Genetic toxicity assessment: Employing the best science for human safety evaluation part IV: Recommendation of a working group of the Gesellschaft fuer Umwelt-Mutationsforschung (GUM) for a simple and straightforward approach to genotoxicity testing. Toxicol. Sci. 97(2):237-240.

Platt, D.K. 2006. Biodegradable Polymers. Cambridge, UK: Woodhead Publishing.

Playle, R.C. 2004. Using multiple metal-gill binding models and the toxic unit concept to help reconcile multiple-metal toxicity results. Aquat. Toxicol. 67(4):359-370.

Potts, R.O., and R.H. Guy. 1992. Predicting skin permeability. Pharm. Res. 9(5):663-669.

Quinn, M.M., T.P. Fuller, A. Bello, and C.J. Galligan. 2006. Pollution prevention-occupational safety and health in hospitals: Alternatives and interventions. J. Occup. Environ. Hyg. 3(4):182-193.

Ramani, K., D. Ramanujan, W.Z. Bernstein, F. Zhao, J. Sutherland, C. Handwerker, J.K. Choi, H. Kim, and D. Thurston. 2010. Integrated sustainable lifecycle design: A review. J. Mech. Design 132(9):1-15.

Ran, Y., and S.H. Yalkowsky. 2001. Prediction of drug solubility by the General Solubility Equation (GSE). J. Chem. Inf. Comput. Sci. 41(2):354-357.

Reif, D.M., M.T. Martin, S.W. Tan, K.A. Houck, R.S. Judson, A.M. Richard, T.B. Knudsen, D.J. Dix, and R.J. Kavlock. 2010. Endocrine profiling and prioritization of environmental chemicals using ToxCast data. Environ. Health Perspect. 118(12):1714-1720.

Reif, D.M., M. Sypa, E.F. Lock, F.A. Wright, A. Wilson, T. Cathey, R.R. Judson, and I. Rusyn. 2013. ToxPi GUI: An interactive visualization tool for transparent integration of data from diverse sources of evidence. Bioinformatics 29(3):402-403.

Reihlen, A., D. Bunke, R. Gross, and D. Jepsen, eds. 2011. Guide on Sustainable Chemicals: A Decision Tool for Substance Manufacturers, Formulators and End Users of Chemicals. Federal Environmental Agency, Dessau-Rosslau, Germany. April 2011 [online]. Available: http://www.umweltbundesamt.de/sites/default/files/medien/publikation/long/4169.pdf [accessed June 26, 2014].

Roberts, S.C., P.D. Noyest, E.P. Gallagher, and H.M. Stapleton. 2011. Species-specific differences and structure-activity relationships in the debromination of PBDE congeners in three fish species. Environ. Sci. Technol. 45(5):1999-2005.

Rossi, M., and L. Heine. 2007. The Green Screen for Safer Chemicals: Evaluating Flame Retardants for TV Enclosures. Clean Production Action, Spring Brook, NY. March 2007 [online]. Available: http://www.chemicalspolicy.org/downloads/Green_Screen_Report.pdf [accessed March 29, 2014].

Rossi, M., J. Tickner, and K. Geiser. 2006. Alternatives Assessment Framework of the Lowell Center for Sustainable Production. Version 1.0. July 2006 [online]. Available: http://www.chemicalspolicy.org/downloads/FinalAltsAssess06.pdf [accessed June 26, 2014].

Rossi, M., C. Peele, and B. Thorpe. 2012. BizNGO Chemical Alternatives Assessment Protocol: How to Select Safer Alternatives to Chemicals of Concern to Human Health or to the Environment. The Business-NGO Working Group, April 12, 2012 [online]. Available: http://www.bizngo.org/static/ee_images/uploads/resources/BizNGOChemicalAltsAssessmentProtocol_V1.1_04_12_12-1.pdf [accessed June 25, 2014].

Rotroff, D.M., B.A. Wetmore, D.J. Dix, S.S. Ferguson, H.J. Clewell, K.A. Houck, E.L. Lecluyse, M.E. Andersen, R.S. Judson, C.M. Smith, M.A. Sochaski, R.J. Kavlock, F. Boellmann, M.T. Martin, D.M. Reif, J.F. Wambaugh, and R.S. Thomas. 2010. Incorporating human dosimetry and exposure into high-throughput in vitro toxicity screening. Toxicol. Sci. 117(2):348-358.

Rotroff, D.M., D.J. Dix, K.A. Houck, R.J. Kavlock, T.B. Knudsen, M.T. Martin, D.M. Reif, A.M. Richard, N.S. Sipes, Y.A. Abassi, C. Jin, M. Stampfl, and R.S. Judson. 2013a. Real-time growth kinetics measuring hormone mimicry for ToxCast Chemicals in T-47D human ductal carcinoma cells. Chem. Res. Toxicol. 26(7):1097-1107.

Rotroff, D.M., D.J. Dix, K.A. Houck, T.B. Knudsen, M.T. Martin, K.W. McLaurin, D.M. Reif, K.M. Crofton, A.V. Singh, M. Xia, R. Huang, and R.S. Judson. 2013b. Using in vitro high throughput screening assays to identify potential endocrine-disrupting chemicals. Environ. Health Perspect. 121(1):7-14.

Rusyn, I., D.M. Gatti, T. Wiltshire, S.R. Kleeberger, and D.W. Threadgill. 2010. Toxicogenetics: Population-based testing of drug and chemical safety in mouse models. Pharmacogenomics 11(8):1127-1136.

Rusyn, I., A. Sedykh, Y. Low, K.Z. Guyton, and A. Tropsha. 2012. Predictive modeling of chemical hazard by integrating numerical descriptors of chemical structures and short-term toxicity assay data. Toxicol. Sci. 127(1):1-9.

Rusyn, I., W.A. Chiu, L.H. Lash, H. Kromhout, J. Hansen, and K.Z. Guyton. 2014. Trichloroethylene: Mechanistic, epidemiologic and other supporting evidence of carcinogenic hazard. Pharmacol. Ther. 141(1):55-68.

Sabanadesan, S. and L. Vanderlinden. 2007. Reducing Health Impacts of Perchloroethylene from Dry Cleaning in Toronto. Toronto Public Health [online]. Available: http://www.toronto.ca/legdocs/mmis/2007/hl/bgrd/backgroundfile-3453.pdf [accessed August 15, 2014].

Saiakhov, R.D., and G. Klopman. 2010. Benchmark performance of MultiCASE Inc. software in Ames mutagenicity set. J. Chem. Inf. Model. 50(9):1521.

Saliner, A.G., G. Patlewicz, and A.P. Worth. 2005. A Similarity Based Approach for Chemical Category Classification. European Commission Joint Research Centre [online]. Available: http://ihcp.jrc.ec.europa.eu/our_labs/eurl-ecvam/laboratories-research/predictive_toxicology/doc/Report_Chemical_Similarity_for_Category_Classification.pdf [accessed July 22, 2014].

Sand, S., C.J. Portier, and D. Krewski. 2011. A signal-to-noise crossover dose as the point of departure for health risk assessment. Environ. Health Perspect. 119(12):1766-1774.

Sand, S., J. Ringblom, H. Hakansson, and M. Oberg. 2012. The point of transition on the dose-effect curve as a reference point in the evaluation of in vitro toxicity data. J. Appl. Toxicol. 32(10):843-849.

Santore, R.C., R. Matthew, P.R. Paquin, and D. DiToro. 2002. Application of the biotic ligand model to predicting zinc toxicity to rainbow trout, fathead minnow, and *Daphnia magna*. Comp. Biochem. Physiol. C Toxicol. Pharmacol. 133(1-2):271-285.

SC Johnson. 2014. Greenlist [online]. Available: http://www.scjohnson.ca/en/scj_greenlist.aspx [accessed July 7, 2014].

Schmid, D., D.L. Staudacher, H.G. Loew, P.G. Spieckermann, G.F. Ecker, S. Kopp, and P. Chiba. 2003. A subset of highly effective propafenone-type multidrug resistance modulators lacks effects on cardiac action potential and mechanical twitch parameters of rat papillary muscles. J. Pharm. Exp. Ther. 307(2):589-596.

Schrand, A.M., M.F. Rahman, S.M. Hussain, J.J. Schlager, D.A. Smith, and A.F. Syed. 2010. Metal-based nanoparticles and their toxicity assessment. Wiley Interdiscip. Rev. Nanomed. Nanobiotechnol. 2(5):544-68.

Schulte, P.A., L.T. McKernan, D.S. Heidel, A.H. Okun, G.S. Dotson, T.J. Lentz, C.L. Geraci, P.E. Heckel, and C.M. Branche. 2013. Occupational safety and health, green chemistry, and sustainability: A review of areas of convergence. Environ Health 12:31.

Seidler, J., S.L. McGovern, T.N. Doman, and B.K. Shoichet. 2003. Identification and prediction of promiscuous aggregating inhibitors among known drugs. J. Med. Chem. 46(21):4477-4486.

Shah, I., K. Houck, R.S. Judson, R.J. Kavlock, M.T. Martin, D.M. Reif, J. Wambaugh, and D.J. Dix. 2011. Using nuclear receptor activity to stratify hepatocarcinogens. PLoS One 6(2):e14584.

Shields, G.C., and P.G. Seybold. 2013. Computational Approaches for the Prediction of pKa Values. QSAR in Environmental and Health Sciences. Boca Raton, FL: CRC Press.

Shukla, S.J., R. Huang, C.P. Austin, and M. Xia. 2010. The future of toxicity testing: A focus on in vitro methods using a quantitative high-throughput screening platform. Drug Discov.Today 15(23-24):997-1007.

Siddall, J.N. 1972. Analytical Decision-Making in Engineering Design. Upper Saddle River, NJ: Prentice Hall.

Sipes, N.S., M.T. Martin, D.M. Reif, N.C. Kleinstreuer, R.S. Judson, A.V. Singh, K.J. Chandler, D.J. Dix, R.J. Kavlock, and T.B. Knudsen. 2011. Predictive models of prenatal developmental toxicity from ToxCast high-throughput screening data. Toxicol. Sci. 124(1):109-127.

Sipes, N.S., M.T. Martin, P. Kothiya, D.M. Reif, R.S. Judson, A.M. Richard, K.A. Houck, D.J. Dix, R.J. Kavlock, and T.B. Knudsen. 2013. Profiling 976 ToxCast chemicals across 331 enzymatic and receptor signaling assays. Chem. Res. Toxicol. 26(6):878-895.

Sirenko, O., C. Crittenden, N. Callamaras, J. Hesley, Y.W. Chen, C. Funes, I. Rusyn, B. Anson, and E.F. Cromwell. 2013. Multiparameter in vitro assessment of compound effects on cardiomyocyte physiology using iPSC cells. J. Biomol. Screen. 18(1):39-53.

Sjodin, A., L. Hagmar, E. Klasson-Wehler, K. Kronholm-Diab, E. Jakobsson, and A. Bergman. 1999. Flame retardant exposure: Polybrominated diphenyl ethers in blood from Swedish workers. Environ. Health Perspect. 107(8):643-648.

Smith, D.A., L. Di, and E.H. Kerns. 2010. The effect of plasma protein binding on in vivo efficacy: Misconceptions in drug discovery. Nat. Rev. Drug Discov. 9(12):929-939.

Smolders, E., K. Oorts, P. Van Sprang, I. Schoeters, C.R. Janssen, S.P. McGrath, and M.J. Mclaughlin. 2009. Toxicity of trace metals in soil as affected by soil type and aging after contamination: Using calibrated bioavailability models to set ecological soil standards. Environ. Toxicol. Chem. 28(8):1633-1642.

Song, W., A. Li, J.C. Ford, N.C. Sturchio, K.J. Rockne, D.R. Buckley, and W.J. Mills. 2005. Polybrominated diphenyl ethers in the sediments of the Great Lakes. 2. Lakes Michigan and Huron. Environ. Sci. Technol. 39(10):3474-3479.

SRC. 2014. Scientific Databases [online]. Available: http://www.srcinc.com/what-we-do/environmental/scientific-databases.html [accessed June 27, 2014].

Stapleton, H.M., M. Alaee, R.J. Letcher, and J.E. Baker. 2004. Debromination of the flame retardant decabromodiphenyl ether by juvenile carp (*Cyprinus carpio*). Environ. Sci. Technol. 38(1):112-119.

Stepan, A.F., D.P. Walker, J. Bauman, D.A. Price, T.A. Baillie, A.S. Kalgutkar, and M.D. Aleo. 2011. Structural alert/reactive metabolite concept as applied in medicinal chemistry to mitigate the risk of idiosyncratic drug toxicity: A perspective based on the critical examination of trends in the top 200 drugs marketed in the United States. Chem. Res. Toxicol. 24(9):1345-1410.

Sutherland, J.J., L.A. O'Brien, and D.F. Weaver. 2004. A comparison of methods for modeling quantitative structure−activity relationships. J. Med. Chem. 47(22) :5541-5554.

Sutherland, J.J., J.W. Raymond, J.L. Stevens, T.K. Baker, and D.E. Watson. 2012. Relating molecular properties and in vitro assay results to in vivo drug disposition and toxicity outcomes. J. Med. Chem. 55(14):6455-6466.

Sutter, A., A. Amberg, S. Boyer, A. Brigo, J.F. Contrera, L.L. Custer, K.L. Dobo, V. Gervais, S. Glowienke, J. van Gompel, N. Greene, W. Muster, J. Nicolette, M.V. Reddy, V. Thybaud, E. Vock, A.T. White, and L. Müller. 2013. Use of in silico systems and expert knowledge for structure-based assessment of potentially mutagenic impurities. Regul. Toxicol. Pharmacol. 67(1):39-52.

Swanson, M.B., and A.C. Socha, eds. 1997. Chemical Ranking and Scoring: Guidelines for Relative Assessments of Chemical: Proceedings of the Pellston Workshop on Chemical Ranking and Scoring, February 11-16, 1995, Sandestin. Florida: SETAC Press.

Takeda Canada, Inc. 2012. Actos (Pioglitazone Hydrochloride) Product Monograph [online]. Available: http://www.takedacanada.com/fr-ca/~/media/Countries/ca/Files/Product%20PDFs/Actos%20PM%20ENG%2018-May-2012.pdf [accessed March 21, 2014].

Takigamie, H., G. Suzuki, Y. Hirai, and S. Sakai. 2008. Transfer of brominated flame retardants from components into dust inside television cabinets. Chemosphere 73(2):161-169.

TCO Development AB. 2012. Certified Displays 6.0. TCO Development [online]. Available:

http://tcodevelopment.com/files/2013/04/TCO-Certified-Displays-6.0.pdf [accessed March 21, 2014].

Tessier, A., and D.R. Turner, eds. 1996. Metal Speciation and Bioavailability in Aquatic Systems. New York: John Wiley and Sons.

Teubner, W., A. Mehling, P.X. Schuster, K. Guth, A. Worth, J. Burton, B. van Ravenzwaay, and R. Landsiedel. 2013. Computer models versus reality: How well do in silico models currently predict the sensitization potential of a substance. Regul. Toxicol. Pharm. 67(3):468-485.

Thomas, R. 2014. Presentation at the Sixth Meeting on Design and Evaluation of Safer Chemical Substitutions, February 6, 2014, Washington, DC.

Thomas, R.S., M.B. Black, L. Li, E. Healy, T.M. Chu, W. Bao, M.E. Andersen, and R.D. Wolfinger. 2012. A comprehensive statistical analysis of predicting in vivo hazard using high-throughput in vitro screening. Toxicol Sci 128(2):398-417.

Thomas, R.S., M.A. Philbert, S.S. Auerbach, B.A. Wetmore, M.J. Devito, I. Cote, J.C. Rowlands, M.P. Whelan, S.M. Hays, M.E. Andersen, M.E. Meek, L.W. Reiter, J.C. Lambert, H.J. Clewell III, M.L. Stephens, Q.J. Zhao, S.C. Wesselkamper, L. Flowers, E.W. Carney, T.P. Pastoor, D.D. Petersen, C.L. Yauk, and A. Nong. 2013. Incorporating new technologies into toxicity testing and risk assessment: Moving from 21st century vision to a data-driven framework. Toxicol. Sci. 136(1):4-18.

Thompson, B., G.D. Coronado, J.E. Grossman, K. Puschel, C.C. Solomon, I. Islas, C.L. Curl, J.H. Shirai, J.C. Kissel, and R.A. Fenske. 2003. Pesticide take-home pathway among children of agricultural workers: Study design, methods, and baseline findings. J. Occup. Environ. Med. 45(1):42-53.

Thomsen, C., E. Lundanes, and G. Becher. 2001. Brominated flame retardants in plasma samples from three different occupational groups in Norway. Environ. Monit. 3(4):366-370.

Thuresson, K., P. Hoglund, L. Hagmar, A. Sjödin, A. Bergman, and K. Jakobsson. 2006. Apparent half-lives of hepta-to decabrominated diphenyl ethers in human serum as determined in occupationally exposed workers. Environ. Health Perspect. 114(2):176-181.

Tice, R.R., C.P. Austin, R.J. Kavlock, and J.R. Bucher. 2013. Improving the human hazard characterization of chemicals: A Tox21 update. Environ. Health Perspect. 121(7):756-765.

Tickner, J.A., K. Geiser, C. Rudisill, and J.N. Schifano. 2013. Alternatives assessment in regulatory policy: History and future directions. Pp. 256-295 in Chemical Alternatives Assessment, R.E. Hester and R.M. Harrison, eds. London: Royal Society of Chemistry.

Triantaphyllou, E. 2000. Multi-criteria decision making methods: A comparative study. Appl. Optim. 44:5-21.

Tropsha, A., P. Gramatica, and V.K. Gombar. 2003. The importance of being earnest: Validation is the absolute essential for successful application and interpretation of QSPR Models. QSAR Comb. Sci. 22(1):69-77.

Truong, L., D.M. Reif, L. St Mary, M.C. Geier, H.D. Truong, and R.L. Tanguay. 2014. Multidimensional in vivo hazard assessment using zebrafish. Toxicol. Sci. 137(1):212-233.

TURI (Toxics Use Reduction Institute), 2006a. Five Chemicals Study: Alternatives Assessment Process Guidance. March 2006 [online]. Available: http://www.turi.org/Our_Work/Research/Alternatives_Assessment/TURI_Alternatives_Assessment_Method_-_5_Chemicals_Study [accessed June 25, 2014].

TURI (Toxics Use Reduction Institute), 2006b. Five Chemicals Alternatives Assessment Study. University Massachusetts Lowell. June 2006 [online]. Available: http://www.turi.org/TURI_Publications/TURI_Methods_Policy_Reports/Five_Chemicals_Alternatives_Assessment_Study._2006/Full_Report [accessed June 26, 2014].

Uehara, T., A. Ono, T. Maruyama, I. Kato, H. Yamada, Y. Ohno, and T. Urushidani. 2010. The Japanese toxicogenomics project: Application of toxicogenomics. Mol. Nutr. Food Res. 54(2):218-227.

UNECE (United Nations Economic Commission for Europe). 2011. Globally Harmonized System of Classification and Labelling of

Chemicals (GHS), 4th Rev. Ed. New York: United Nations [online]. Available: http://www.unece.org/fileadmin/DAM/trans/danger/publi/ghs/ghs_rev04/English/ST-SG-AC10-30-Rev4e.pdf [accessed March 29, 2014].

UNECE (United Nations Economic Commission for Europe). 2013a. Annex 8. An example of classification in the Globally Harmonized Systems. Pp. 427-434 in Globally Harmonized System of Classification and Labelling of Chemicals (GHS), 5th Rev. Ed. New York: United Nations [online]. Available: http://www.unece.org/fileadmin/DAM/trans/danger/publi/ghs/ghs_rev05/English/ST-SG-AC10-30-Rev5e.pdf [accessed March 29, 2014].

UNECE (United Nations Economic Commission for Europe). 2013b. Globally Harmonized System of Classification and Labelling of Chemicals (GHS), 5th Rev. Ed. New York: United Nations [online]. Available: http://www.unece.org/trans/danger/publi/ghs/ghs_rev05/05files_e.html [accessed March 29, 2014].

UNECE (United Nations Economic Commission for Europe). 2013c. Globally Harmonized System for the Classification and Labeling of Chemicals (GHS). Part 3. Health Hazards. Geneva: United Nations [online]. Available: http://www.unece.org/fileadmin/DAM/trans/danger/publi/ghs/ghs_rev05/English/03e_part3.pdf [accessed July 9, 2014].

UNEP (United Nations Environment Programme). 2009. General Guidance on Considerations Related to Alternatives and Substitutes for Listed Persistent Organic Pollutants and Candidate Chemicals. UNEP/POPS/POPRC.5/10/Add.1. Persistent Organic Pollutants Review Committee, Fifth Meeting, October 12-16, Geneva [online]. Available: http://www.subsport.eu/images/stories/pdf_archive/substitution_tools/7_guidance_alt_pop_en.pdf [accessed June 26, 2014].

UNEP/SETAC (United Nations Environment Programme, Society of Environmental Toxicology and Chemistry). 2009. Guidelines for Social Life Cycle Assessment of Products [online]. Available: http://www.unep.fr/shared/publications/pdf/DTIx1164xPA-guidelines_sLCA.pdf [accessed July 7, 2014].

USGS (U.S. Geological Survey). 2014. Water Science Glossary of Terms [online], Available: http://water.usgs.gov/edu/dictionary.html#L [accessed June 27, 2014].

Valerio, L.G., Jr, and K.P. Cross. 2012. Characterization and validation of an in silico toxicology model to predict the mutagenic potential of drug impurities. Toxicol. Appl. Pharmacol. 260(3):209-221.

Valerio, L.G., Jr., and S. Choudhuri. 2012. Chemoinformatics and chemical genomics: Potential utility of in silico methods. J. Appl. Toxicol. 32(11):880-889.

Vallianatou, T., G. Lambrinidis, and A. Tsantili-Kakoulidou. 2013. In silico prediction of human serum albumin binding for drug leads. Expert Opin. Drug Discov. 8(5):583-595.

Varma, M.V., R.S. Obach, C. Rotter, H.R. Miller, G. Chang, S.J. Steyn, A. El-Kattan, and M.D. Troutman. 2010. Physicochemical space for optimum oral bioavailability: Contribution of human intestinal absorption and first-pass elimination. J. Med. Chem. 53(3):1098-1108.

Viñas, R., and C.S. Watson. 2013. Bisphenol S disrupts estradiol-induced nongenomic signaling in a rat pituitary cell line: Effects on cell functions. Environ. Health Perspect. 121(3):352-358.

Vocanson, M., J.F. Nicolas, and D. Basketter. 2013. In vitro approaches to the identification and characterization of skin sensitizers. Expert. Rev. Dermatol. 8(4):395-405.

Voorspoels, S., A. Covaci, P. Lepom, S. Escutenaire, and P. Schepens. 2006. Remarkable findings concerning the terrestrial top-predator red fox (*Vulpes vulpes*). Environ. Sci. Technol. 40(9): 2937-2943.

Vorkamp, K., M. Thomsen, K. Falk, H. Leslie, S. Møller, and P.B. Sørensen. 2005. Temporal development of brominated flame retardants in peregrine falcon (*Falco peregrinus*) eggs from South Greenland (1986-2003). Environ. Sci. Technol. 39(21):8199-8206.

Voutchkova, A.M., T.G. Osimitz, and P.T. Anastas. 2010. Toward a comprehensive molecular design framework for reduced hazard. Chem. Rev. 110(10):5845-5882.

Voutchkova, A.M., J. Kostal, J.B. Steinfeld, J.W. Emerson, B.W. Brooks, P. Anastas, and J.B. Zimmerman. 2011. Towards rational molecular design: Derivation of property guidelines for reduced acute aquatic toxicity. Green Chem. 13(9):2373-2379.

Voutchkova, A., J. Kostal, and P. Anastas. 2012. Property-based approaches to design rules for reduced toxicity. Pp. 349-373 in Handbook of Green Chemistry, Part 9. Designing Safer Chemicals, P. Anastas, ed. New York: Wiley.

Voutchkova-Kostal, A.M., J. Kostal, K.A. Connors, B.W. Brooks, P.T. Anastas, and J.B. Zimmerman. 2012. Towards rational molecular design for reduced chronic aquatic toxicity. Green Chem. 14(4):1001-1008.

Voutchkova-Kostal, A.M., F. Van Der Mei, and N. An. 2013. Method for Using Nuclear Magnetic Resonance Data to Predict Octanol-Water Partition Coefficient and Biological Activity. U.S. Application No. 61/836,430. U.S. Patent Office.

Vuorinen, A., A. Odermatt, and D. Schuster. 2013. In silico methods in the discovery of endocrine disrupting chemicals. J. Steroid Biochem. Mol. Biol.137:18-26.

WA Department of Ecology. 2014. The Reporting List of Chemicals of High Concern to Children (CHCC) [online]. Available: http://www.ecy.wa.gov/programs/swfa/cspa/chcc.html [accessed June 26, 2014].

Wambaugh, J.F., R.W. Setzer, D.M. Reif, S. Gangwal, J. Mitchell-Blackwood, J.A. Arnot, O. Joliet, A. Frame, J. Rabinowitz, T.B. Knudsen, R.S. Judson, P. Egeghy, D. Vallero, and E.A. Cohen Hubal. 2013. High-throughput models for exposure-based chemical prioritization in the ExpoCast project. Environ. Sci. Technol. 47(15):8479-8488.

Wang, J.X. 2002. What Every Engineer Should Know About Decision Making Under Uncertainty. New York: Marcel Dekker.

Wang, X., and N. Greene. 2012. Comparing measures of promiscuity and exploring their relationship to toxicity. Mol. Inf. 31(2):145-159.

Wang, Y., T. Suzek, J. Zhang, J. Wang, S. He, T. Cheng, B.A. Shoemaker, A. Gindulyte, and S.H. Bryant. 2014. PubChem BioAssay: 2014 update. Nucleic Acids Res. 42(Database issue):D1075-D1082.

Warne, M.A., J.K. Nicholson, J.C. Lindon, P.D. Guiney, and K.P. Gartland. 2009. A QSAR investigation of dermal and respiratory chemical sensitizers based on computational chemistry properties. SAR QSAR Environ. Res. 20(5-6):429-451.

Waters, M., S. Stasiewicz, B.A. Merrick, K. Tomer, P. Bushel, R. Paules, N. Stegman, G. Nehls, K.J. Yost, C.H. Johnson, S.F. Gustafson, S. Xirasagar, N. Xiao, C.C. Huang, P. Boyer, D.D. Chan, Q. Pan, H. Gong, J. Taylor, D. Choi, A. Rashid, A. Ahmed, R. Howle, J. Selkirk, R. Tennant, and J. Fostel. 2008. CEBS--Chemical Effects in Biological Systems: A public data repository integrating study design and toxicity data with microarray and proteomics data. Nucleic Acids Res. 36(Database issue):D892-D900.

Watkins, P.B., and R.W. Whitcomb. 1998. Hepatic dysfunction associated with troglitazone. N. Engl. J. Med. 338(13):916-917.

Welsh, M., L. Mangravite, M.W. Medina, K. Tantisira, W. Zhang, R.S. Huang, H. McLeod, and M.E. Dolan. 2009. Pharmacogenomic discovery using cell-based models. Pharmacol. Rev. 61(4):413-429.

Weltman, R., T. Hargraves, B. Knight, and L. Wardope. 2011. Assessment of the environmental fate and effects of the PPARgamma receptor agonist pioglitazone. Chemosphere 83(4):391-399.

Wetmore, B.A., J.F. Wambaugh, S.S. Ferguson, M.A. Sochaski, D.M. Rotroff, K. Freeman, H.J. Clewell III, D.J. Dix, M.E. Andersen, K.A. Houck, B.Allen, R.S. Judson, R. Singh, R.J. Kavlock, A.M. Richard, and R.S. Thomas. 2012. Integration of dosimetry, exposure, and high-throughput screening data in chemical toxicity assessment. Toxicol. Sci. 125(1):157-174.

Wetmore, B.A., J.F. Wambaugh, S.S. Ferguson, L. Li, H.J. Clewell III, R.S. Judson, K. Freeman,

W. Bao, M.A. Sochaski, T.M. Chu, M.B. Black, E. Healy, B. Allen, M.E. Andersen, R.D. Wolfinger, and R.S. Thomas. 2013. Relative impact of incorporating pharmacokinetics on predicting in vivo hazard and mode of action from high-throughput in vitro toxicity assays. Toxicol. Sci. 132(2):327-346.

Wheeler, H.E., and M.E. Dolan. 2012. Lymphoblastoid cell lines in pharmacogenomic discovery and clinical translation. Pharmacogenomics 13(1):55-70.

Whittaker, M.H., and L.G. Heine. 2013. Chemicals Alternatives Assessment (CAA): Tools for selecting less hazardous chemicals. Pp. 1-43 in Chemical Alternatives Assessment, R.E. Hester, and R.M. Harrison, eds. London: Royal Society of Chemistry.

Williams, D.P. 2006. Toxicophores: Investigations in drug safety. Toxicology 226(1):1-11.

Williams, R.T., and T.R. Williams. 2012. Environmental science; Guiding green chemistry, manufacturing, and product innovations. Pp. 33-66 in Green Techniques for Organic Synthesis and Medicinal Chemistry, W. Zhang, and B.W. Cue, eds Chichester, UK: John Wiley & Sons.

Willis, H.H, M.L. DeKay, M.G. Morgan, H.K. Florig, and P.S. Fischbeck. 2004. Ecological risk ranking: Development and evaluation of a method for improving public participation in environmental decision making. Risk Anal. 24(2):363-378.

Wilson, M.P., S.K. Hammond, M. Nicas, and A.E. Hubbard. 2007. Worker exposure to volatile organic compounds in the vehicle repair industry. J. Occup. Environ. Hyg. 4(5):301-310.

Wilson, A., D.M. Reif, and B.J. Reich. 2014. Hierarchical dose-response modeling for high-throughput toxicity screening of environmental chemicals. Biometrics 70(1):237-246.

Worth, A. 2008. QSAR and Chemical Read-Across within the Context of Non-Testing Strategies. Presentation at the 1st SETAC Europe Special Science Symposium: Integrated Testing Strategies for REACH, October 23-24, 2008. Brussells, Belgium.

Wu, S., J. Fisher, J. Naciff, M. Laufersweiler, C. Lester, G. Daston, and K. Blackburn. 2013. Framework for identifying chemicals with structural features associated with the potential to act as developmental or reproductive toxicants. Chem. Res. Toxicol. 26(12):1840-1861.

Xia, M., R. Huang, K.L. Witt, N. Southall, J. Fostel, M.H. Cho, A. Jadhav, C.S. Smith, J. Inglese, C.J. Portier, R.R. Tice, and C.P. Austin. 2008. Compound cytotoxicity profiling using quantitative high-throughput screening. Environ. Health Perspect. 116(3):284-291.

Yamamoto, K.N., K. Hirota, K. Kono, S. Takeda, S. Sakamuru, M. Xia, R. Huang, C.P. Austin, K.L. Witt, and R.R. Tice. 2011. Characterization of environmental chemicals with potential for DNA damage using isogenic DNA repair-deficient chicken DT40 cell lines. Environ. Mol. Mutagen. 52(7):547-561.

Yoon, M., A. Efremenko, B.J. Blaauboer, and H.J. Clewell. 2014. Evaluation of simple in vitro to in vivo extrapolation approaches for environmental compounds. Toxicol In Vitro 28(2):164-170.

Zeise, L., F.Y. Bois, W.A. Chiu, D. Hattis, I. Rusyn, and K.Z. Guyton. 2013. Addressing human variability in next-generation human health risk assessments of environmental chemicals. Environ. Health Perspect. 121(1):23-31.

Zhivkova, Z., and I. Doytchinova. 2013. Quantitative structure-clearance relationships of acidic drugs. Mol. Pharm. 10(10):3758-3768.

Zimmerman, H.J. 1999. Hepatotoxicity: The Adverse Effects of Drugs and Other Chemicals on the Liver. Philadelphia: Lippincott Williams & Wilkins.

Appendix A

Biographic Information of Committee Members

David C. Dorman *(chair)* is a professor of toxicology in the Department of Molecular Biosciences of North Carolina State University. The primary objective of his research is to provide a refined understanding of chemically induced neurotoxicity in laboratory animals that will lead to improved assessment of potential neurotoxicity in humans. Dr. Dorman's research interests include neurotoxicology, nasal toxicology, pharmacokinetics, and cognition and olfactory in military working dogs. He served as a member of the National Research Council Committee on Animal Models for Testing Interventions Against Aerosolized Bioterrorism Agents, as member and chair of two Committees on Emergency and Continuous Exposure Guidance Levels for Selected Submarine Contaminants and the Committee to evaluate Potential Health Risks from Recurrent Lead Exposure to DOD Firing Range Personnel, and as a member of the Committee to Review EPA's Draft IRIS Assessment of Formaldehyde. He received his D.V.M. from Colorado State University. He completed a combined Ph.D. and residency program in toxicology at the University of Illinois at Urbana-Champaign and is a diplomate of the American Board of Veterinary Toxicology and the American Board of Toxicology.

Peter Beak has made fundamental contributions to organic chemistry that have provided unifying concepts and opened new areas of investigation. His work has clarified the effect of molecular environment on structure-stability relationships, provided new reactions that are widely used for chemical synthesis, and identified novel reactive intermediates. His current research involves the determination of reaction trajectories in atom-transfer reactions and investigations of asymmetric reactions. He has held editorships, lectureships, and leadership positions in professional organizations. He has received a number of awards, lectured around the world, and served as research advisor for more than 100 graduate and postdoctoral students who are making significant independent contributions to their fields. Dr. Beak is a member of the National Academy of Sciences and the American Academy of Arts and Sciences. He received his B.A. from Harvard University in 1957 and his Ph.D. from Iowa State University in 1961 and then joined the faculty at Illinois. Dr. Beak's research interests are in synthetic, structural, and mechanistic organic chemistry, new reaction processes, synthetic methodology, and reactive intermediates.

Eric J. Beckman is the George M. Bevier Professor of Engineering at the University of Pittsburgh Department of Chemical and Petroleum Engineering. He is also a co-director of the Mascaro Sustainability Initiative, a center of engineering that focuses on the design of sustainable communities. Dr. Beckman's two main research areas are the use of carbon dioxide as either a solvent or raw material and polymer chemistry and processing. Recent work has focused on emulsion and dispersion polymerization on CO_2, copolymerization of CO_2 and cyclic ethers to form polycarbonates, generation of hydrogen peroxide in CO_2, and the extraction of heavy metals using CO_2 technology among other related studies. Dr. Beckman received his B.S. in chemical engineering from the Massachusetts Institute of Technology in 1980. After a period of employment with both Monsanto Plastics and Resins and the Union Carbide Corporation, Dr. Beckman went on to earn a Ph.D. in polymer chemistry and processing from the University of Massachusetts in 1988. As a postdoctoral student, Dr. Beckman held a research appointment at Battelle's Pacific Northwest Laboratory.

Jerome J. Cura is an ecological risk assessor and senior scientist for Woods Hole Group, Inc. in Falmouth, Massachusetts. He is an adjunct professor at Cape Cod Community College, where he teaches Fundamentals of Oceanography. He is an expert in the area of ecological risk analysis. He has conducted ecological risk analyses in various freshwater systems, marine and estuarine habitats, and terrestrial environments. He has developed guidance for conducting risk assessments at dredging sites for the U. S. Army Corps of Engineers, and he chaired the International Navigation Association's (PIANC) workgroup that developed international guidance. Dr. Cura's experience includes conducting assessments at CERCLA and RCRA sites (industrial and government facilities), providing technical advice on the design and execution of human health and ecological risk assessments, and providing expert testimony for law firms. Industry and government organizations frequently invite him to lead or participate in environmental conferences or symposia. Dr. Cura is a member of the Science Collaborative, a network of senior level environmental scientists. He was a founding partner of Cura Environmental and Menzie-Cura & Associates, Inc. Dr. Cura has published more than 30 peer-reviewed book chapters, technical papers, journal articles, and conference proceedings in the areas of risk assessment, environmental decision making, marine ecology, and dredged material disposal evaluation methods. Dr. Cura received his B.A. in biology from the College of the Holy Cross in 1971, his M.S. in biology from Northeastern University in 1974, and his Ph.D. in biological oceanography from the University of Maine in 1981.

Anne Fairbrother has more than 30 years of experience in ecotoxicology, wildlife toxicology, contaminated site assessment, and regulatory science. She has conducted small- and large-area (>100 sq. mile) risk assessments at contaminated sites in tropical, desert, and mountain ecosystems, determining risk thresholds for plants and wildlife. She provided consultation on future development of mine pit lakes, assessed the risk to livestock from use of wastewater on irrigated pasture during mine closure operations, and conducted several assessments of risk to terrestrial and aquatic organisms from mercury. She also assessed risks to wildlife at sites contaminated with organic chemicals, including DDT, PCBs, dioxins, and petroleum hydrocarbons in Delaware, Texas, Oregon, Washington, and California. Dr. Fairbrother has supported industry groups and government agencies in compiling and reviewing literature and industry reports in support of U.S., Canadian, and European regulatory processes for pesticide and chemical risk management. She has testified in front of boards of review and science advisory boards, and prepared expert testimony on environmental risks of pollutants for legal cases within the U.S. Dr. Fairbrother has published more than 90 peer-reviewed articles and book chapters, reflecting her expertise in wildlife toxicology, immunotoxicology, endocrine-disrupting chemicals, and ecological risk assessment. She serves on numerous scientific boards, expert panels, and editorial boards in support of scientific and regulatory issues. A veterinarian and certified wildlife biologist, Dr. Fairbrother served as president of the Society of Environmental Toxicology and Chemistry, American Association of Wildlife Veterinarians, and Wildlife Disease Association (WDA). She is the recipient of the WDA Distinguished Service Award (2002), and a gold medal for commendable service from EPA. Dr. Fairbrother holds an adjunct professorship at Oregon State University, Department of Environmental, and Molecular Toxicology. She earned her D.V.M. from University of California-Davis and her Ph.D. from University of Wisconsin.

Nigel Greene is an associate research fellow with Pfizer Global Research Company, specializing in compound safety prediction. His specific duties include establishing and managing a group of Ph.D. level scientists using computational modeling and analysis to help predict the safety profile of early discovery programs and aid in chemical series and compound selection prior to first in vivo studies by using chemical properties and in vitro assay profiles. Dr. Greene's other activities include mining internal and public databases of gene expression data to explore biological mechanisms of toxicity and helping in the development of new in vitro assays for safety profiling by conducting in vitro experiments to try to confirm the computational hypotheses derived from these transcriptional databases. Dr. Greene holds a Ph.D. in organometallic chemistry from the University of Leeds (1994) and a B.S. in chemistry and computational science from the University of Leeds (1991).

Carol J. Henry is a professorial lecturer at the George Washington University School of Public Health and Health Services, Department of Environmental and Occupational Health, and an advisor and consultant to public and private organizations. Her focus is on issues surrounding toxicology, public and environmental health, risk assessment and risk management, research management strategies, green chemistry, engineering technology, and sustainable practices. She was previously vice president of industry performance programs at the American Chemistry Council; director of the Health and Environmental Sciences Department at the American Petroleum Institute; associate deputy assistant secretary for science and risk policy at the U.S. Department of Energy; and director of the Office of Environmental Health Hazard Assessment (OEHHA) at the California Environmental Protection Agency. She is a diplomat of the American Board of Toxicology, certified in general toxicology. She was chair of the Federal Advisory Committee for the National Children's Study from 2010-2012. She is a member of the Joint Committee on the ANSI NSF Green Chemistry Institute Greener Chemical Products and Processes Standard Initiative, the Environmental Health Perspectives Editorial Board, and the National Research Council's Board on Chemical Sciences and Technology. She is an elected councilor for the American Chemical Society (ACS) and serves on the ACS Committee for Environmental Improvement. Dr. Henry received her undergraduate degree in chemistry from the University of Minnesota and doctorate in microbiology from the University of Pittsburgh.

Helen Holder is a master engineer at Hewlett-Packard (HP), where she leads the Global Environmental Materials (GEM) team. In her current role, she evaluates and qualifies materials for use in HP products, including plastics and additives, solders, fluxes, printed circuit board surface finishes, and other electronic materials. In this role, she has introduced environmental and human health criteria into technical specifications to complement traditional performance, cost, safety, and reliability requirements in materials selection. Ms. Holder started her career at HP in 1993, and has worked in a variety of manufacturing, materials, and procurement roles within the company. She received her B.S. from the Massachusetts Institute of Technology and her master's degree from the University of California at Berkeley, where she was an HP resident fellow.

James E. Hutchison joined the faculty at the University of Oregon (UO) in the fall of 1994. He now holds the Lokey-Harrington Chair in Chemistry. His research interests are in green chemistry, materials chemistry, and nanoscience. He led the development of the UO's nation-leading curriculum in "green" (environmentally benign) organic chemistry, launched the university's pioneering center in greener nanoscience, and is a member of the Governing Board of the ACS Green Chemistry Institute. He is a member of the leadership team for the Oregon Nanoscience and Microtechnologies Institute (ONAMI) and founded, and now directs, the ONAMI's Safer Nanomaterials and Nanomanufacturing Initiative (SNNI). He is the author of more than 100 refereed publications, three book chapters, and a textbook ("Green Organic Chemistry: Strategies, Tools, and Laboratory Experiments"). Dr. Hutchison received a B.S. from the University of Oregon in 1986, a Ph.D. in organic chemistry from Stanford University in 1991, and completed postdoctoral studies at the University of North Carolina at Chapel Hill from 1992 to 1994.

Greg Paoli serves as principal risk scientist and chief operating officer at Risk Sciences International, a consulting firm specializing in risk assessment, management, and communication in the field of public health, safety, and risk-based decision support. Mr. Paoli has experience in diverse risk domains, including toxicological, microbiological, and nutritional hazards, air and water quality, adaptation to climate change, safety of engineering devices, as well as emergency planning and response for natural and man-made disasters. He specializes in risk assessment methods, the development of risk-based decision-support tools, and comparative risk assessment. Mr. Paoli has served on a number of expert committees devoted to the risk sciences. He was a member of the National Research Council's Committee on Improving Risk Analysis Approaches used by the U.S. Environmental Protection Agency (EPA), which produced the 2009 report, *Science and Decisions: Advancing Risk Assessment.* He serves on an NRC Standing Committee on the use of public health data at the U.S. Food Safety and Inspection Service, and has served on several expert committees convened by the World Health Organization. He serves on the Standards Council of Canada Technical

Committee on Risk Management and served on advisory committees of the National Roundtable on the Environment and the Economy. Mr. Paoli completed a term as councilor of the Society for Risk Analysis (SRA) and is a member of the Editorial Board of Risk Analysis. Recently, Mr. Paoli was awarded the Sigma Xi - SRA Distinguished Lecturer award. Greg holds a B.A.Sc. in electrical and computer engineering and a M.A.Sc. in systems design engineering from the University of Waterloo.

Julia B. Quint is retired from the California Department of Public Health (CDPH), where she was a research scientist and chief of the Hazard Evaluation System and Information Service (HESIS), an occupational health program. She has a Ph.D. in biochemistry from the University of Southern California. Throughout her career as a public health scientist, Dr. Quint has initiated, developed, and contributed to projects, programs, and policies focused on protecting workers, communities, and the environment from toxic chemicals and promoting the development and use of safer alternatives to toxic chemicals. She has served on a number of scientific advisory committees, including committees of the National Academy of Sciences (NAS) on tetrachloroethylene, health impact assessment, and review of the Department of Labor's Site Exposure Matrix Database, the Cal/OSHA Health Expert Advisory Committee, and the Cal/EPA's Green Ribbon Science Panel. She currently serves on the Scientific Guidance Panel of the California Biomonitoring Program, the CDC/NIOSH World Trade Center Scientific and Technical Advisory Committee, the UCSF Program on Reproductive Health, the Environment's From Advancing Science to Ensuring Protection Advisory Group, the National Healthy Nail Salon Alliance Research Advisory Committee, and the CDPH Environmental Health Tracking Advisory Group. Dr. Quint has authored many peer-reviewed scientific articles and reports, and is the recipient of several awards for her work in public health.

Ivan Rusyn is a professor in the Department of Environmental Sciences and Engineering in the Gillings School of Public Health at the University of North Carolina at Chapel Hill. He directs the Laboratory of Environmental Genomics and the Carolina Center for Computational Toxicology. He also is a member of the Lineberger Comprehensive Cancer Center, Center for Environmental Health and Susceptibility, Bowles Center for Alcohol Studies, and the Carolina Center for Genome Sciences. Dr. Rusyn served on several working groups convened by the National Research Council and the WHO/IARC. Dr. Rusyn's laboratory has an active research portfolio funded by the National Institutes of Health and the EPA, with a focus on the mechanisms of action of environmental toxicants and the genetic determinants of the susceptibility to toxicant-induced injury. The Rusyn laboratory applies molecular, biochemical, genetic, and genomic approaches to understanding the mechanisms of environmental agent-related disease. His studies on health effects of environmental agents resulted in more than 135 peer-reviewed publications. Dr. Rusyn received his M.D. (with honors) from Ukrainian State Medical University in Kiev and his Ph.D. in toxicology from the University of North Carolina at Chapel Hill. He also trained at the University of Dusseldorf in Germany and at the Massachusetts Institute of Technology.

Kathleen Shelton is director of Crop Protection Research and Development. She is responsible for the leadership of the business discovery and development efforts, globally, and for ensuring that the business has a full and valuable pipeline of new products. In June 2013, she was selected by the Health and Environmental Sciences Institute (HESI) Board of Directors as a member of the Emerging Issues Committee. Dr. Shelton is also director of Central Research and Development, Enabling Technologies, and is responsible for leading the organizations that provide analytical, computational, and pilot scale services across DuPont. Dr. Shelton has worked at DuPont in various capacities since 1993. Recently, she was detailed to Geneva, Switzerland, where she led European advocacy efforts related to REACH (registration, evaluation, authorization and restriction of chemical substances) implementation and chemicals management, including participation in the Product Stewardship Programme Council of the European Chemical Industry Association (CEFIC) and the Strategic Approach to International Chemicals Management (SAICM, part of the United Nations Environmental Programs). Dr. Shelton has a B.S. in biology from the University of Notre Dame and a Ph.D. in microbiology and immunology from Hahnemann University (now part of Drexel University).

Joel A. Tickner is an associate professor in the Department of Community Health and Sustainability of the University of Massachusetts Lowell and a program director in The Lowell Center for Sustainable Production. He is interested in the development of innovative scientific methods and policies to implement a precautionary and preventive approach to decision making under uncertainty while advancing assessment and adoption of safer substitutes to chemicals and products of concern. His teaching and research interests include regulatory science and policy, risk assessment, pollution prevention, cleaner production, and environmental health. Dr. Tickner has served on several advisory boards and as an expert reviewer, most recently for the California Green Chemistry Initiative, the EPA's National Pollution Prevention and Toxics Advisory Committee, the NAS Panel on the Science for EPA's Future, and the First National Precautionary Principle Conference Advisory Committee. He is the recipient of several honors and awards, including the University of Massachusetts President's Award for Public Service, the National Pollution Prevention Roundtable Champion Award, and the North American Hazardous Waste Managers Policy Leader Award. Dr. Tickner earned an Sc.D. in cleaner production and pollution prevention from the University of Massachusetts Lowell.

Adelina Voutchkova is an assistant professor of chemistry in the Department of Chemistry at George Washington University. Dr. Voutchkova has substantial experience in the areas of molecular design, toxicology, and green chemistry. Her work aims at developing rational guidelines for the design of industrial chemicals that are acutely and chronically safe for a variety of aquatic families of species, including fish, crustaceans, and algae. Specific projects include the development of product design guidelines for chronic mammalian toxicity and the utilization of spectroscopic data to inform safer molecular design, with specific interests in minimizing skin sensitization from personal care products. Dr. Voutchkova earned her B.A. from Middlebury College and Ph.D. from Yale University in 2008.

Martin Wolf is director, Product Sustainability & Authenticity, for Seventh Generation Inc. In this capacity, Mr. Wolf is responsible for ensuring the sustainable design of products at Seventh Generation Inc., a manufacturer and distributor of ecological household and personal care products. He has more than 40 years of experience in industrial and environmental chemistry, initially performing environmental fate and metabolism studies for agricultural chemicals and later studying the occurrence of hazardous chemicals in the environment, conducting life cycle studies of industrial processes, and designing more sustainable household cleaning products. In addition to his work for Seventh Generation, he serves as chair of the Sustainability Committee, and has served as chair of the Strategic Advisory Committee (2009-2011), vice chair of the Sustainability Committee (2010-2012), and vice chair of the Asthma Task Group and the American Cleaning Institute (formerly the Soap & Detergent Association). Mr. Wolf received a 2010 EPA Environmental Merit Award for his work. He holds a master's degree in chemistry from Yeshiva University and a bachelor's in chemistry from Worcester Polytechnic Institute.

Appendix B
Ecotoxicity in Frameworks

This appendix provides a brief overview of methods for addressing ecological and environmental evaluation used by chemical alternatives assessment frameworks.

These include:

- BizNGO Alternatives Assessment Protocol (Rossi et al. 2012)
- California Safer Consumer Products Regulation (CA DTSC 2013)
- EPA's Design for the Environment (DfE) Chemical Alternatives Assessments (EPA 2011)
- Lowell Center Alternatives Assessment Framework (Rossi et al. 2006)
- REACH Authorization Analysis of Alternatives (ECHA 2011)
- TURI Alternatives Assessment Process Guidance (TURI 2006a)
- UNEP Persistent Organic Pollutants Review Committee General Guidance on Alternatives (UNEP 2009)
- Interstate Chemicals Clearinghouse (IC2) Alternatives Assessment Guidance (IC2 2013)
- German Guide on Sustainable Chemicals (Reihlen et al. 2011)
- UCLA Multi-Criteria Decision Analysis (Malloy et al. 2011)

Table B-1 summarizes the ecological assessment approach for the 10 chemical alternatives assessment frameworks.

TABLE B-1 Summary of Ecological Assessment Approach for Ten Chemical Assessment Alternative Methods.

	Level of Specificity	Risk Assessment Elements								Environmental Assessment Elements
		Hazard Assessment	*Hazard End Points*			*Exposure Assessment*	*Exposure Criteria*			
			Aquatic End Points	Terrestrial End Points	Other End Points		Bioaccumulation	Persistence	Mobility	
BizNGO	Non-specific, defers to other methods for specifics of analysis.	Precedes technical and economic feasibility.	Defers details of analysis to other methods.			None recommended	Assumes that where use patterns are similar, exposure will be equal.	None recommended	None recommended	Defers to vaguely defined "Life Cycle Thinking" to address assessment of potential for global warming, end-of-life management, and worker exposure.
EPA DfE	Provides explicit end points for assessing hazard.	This is largely a Hazard Evaluation Method.	See Table 7-1.	End points based on EPA Office of Pesticide Programs Ecotoxicity Categories for Terrestrial Organisms	Other toxicological end points to consider if data are available: epigenetic toxicity, lactational or transplacental transfer. Specific target organ toxicity—single exposure, wildlife developmental impairment, wildlife growth impairment, wildlife survival impairment, wildlife reproductive impairment, immunotoxicity.	Not robustly considered other than to provide criteria for assessing general aspects of environmental fate: persistence in water, sediment, and soil, and bioaccumulation.	See Table 5-2.	Based on data on ultimate degradation and persistence of degradation products. In absence of measured data, the requirement is to use information on analogs or estimated valued from models (EPA Suite or SPAEC). Considers persistence in soil, sediment, and water. Categorizes persistence into four categories based on half-life ranging from "readily biodegradable to 180 days half-life.	Not separately considered	Lists various end points to consider if data are available, including: domestic animal toxicity, mobility in the environmental media, ozone formation, eutrophication, global warming potential, loss of genetic diversity/biodiversity, non-target phytotoxicity.

German Guide on Sustainable Chemicals	A guide for selecting sustainable chemicals based on exposure as much as hazard.	The Guide uses the term "problematic properties related to the environment."	See Table 7-1.	None	None	Exposure is emphasized and assess in some detail as a "problematic property" of a chemical. With two categories: persistent, bioaccumulative and toxic and very persistent and very bioaccumulative.	Two categories of bioconcentration factors >2000 and >5000 (bioacummulatable vs very bioacummulatable).	Half-life criteria on the same order but not exactly the same as GHS criteria. Parsed by freshwater/estuarine vs marine (no explanation given).	Explicit evaluations that include release potential (solubility, vapor pressure) and sub criteria for short- and long- range transport and transport as a "dusty" chemical.	Use of resources; greenhouse gas potential as mass of carbon dioxide per kg of substance produced. "Origin of raw materials," including some value-laden criteria such as "supplier doesn't care about environmental protection" and social responsibility; Numerous "use of resources criteria" - such as renewability, energy consumption, water consumption, and waste production.
Interstate Chemicals Clearing-house (IC2) Alternatives Assessment Guidance	A detailed method for assessing chemicals through a long series of questions. The document specifies other tools that can be used to answer these questions (GreenScreen, GreenScreen® Plus), but does not offer its own methods for assessment.	Based on GreenScreen® and GreenScreen® Plus	See Table 7-1.	Lists wildlife development growth, reproductive, and survival impairment. Phytotoxicity recommended as end points but provides no method for assessment.	Lists eutrophication but offers no evaluative scheme.	Recognizes six levels of exposure from exposure assessment evaluation to full exposure assessment as found in risk assessment guidance.	See Table 5-2.	Uses GreenScreen® Criteria	Not explicitly addressed.	Not explicitly addressed.

TABLE B-1 (Continued)

	Level of Specificity	Risk Assessment Elements								Environmental Assessment Elements
		Hazard Assessment	*Hazard End Points*			*Exposure Assessment*	*Exposure Criteria*			
California Safer Consumer Products Regulation	A listing and brief description of various hazard assessment and exposure assessment tools with no specific recommendations.	Refers the reader to other specific evaluation methods.	Refers the reader to other specific evaluation methods.	Refers the reader to other specific evaluation methods.	Refers the reader to other specific evaluation methods.	Refers the reader to other specific evaluation methods.	Refers the reader to other specific evaluation methods.	Refers the reader to other specific evaluation methods.	Refers the reader to other specific evaluation methods.	Refers the reader to other specific evaluation methods.
Lowell Center Alternatives Assessment Framework	A high-level, general framework in flowchart format that includes three "core elements": Alternatives Assessment Foundation; Alternatives Assessment Process; and Evaluation modules. One of these modules is "Human Health and the Environment."	Refers the reader to other specific evaluation methods.	Refers to various other methods.	Refers to various other methods.	Refers to various other methods.	Refers to various other methods.	One of the decision- making rules in this alternatives framework is to "Avoid alternatives that are the direct source of persistent, bioaccumulative toxics (PBTs) across their lifecycle."	One of the decision- making rules in this alternatives framework is to "Avoid alternatives that are the direct source of persistent, bioaccumulative toxics (PBTs) across their lifecycle."	Refers to various other methods.	Refers to various other methods.
REACH Authorisation Analysis of Alternatives			Requires the use of a predicted no effect concentration.	Requires the use of a predicted no effect concentration.	Requires the use of a predicted no effect concentration, such as food chain effects.	If a substance is shown to be PBT or vPvB, then an exposure assessment and risk characterization is required.	Uses the concept of vPvB with described substances that are characterized by high persistence and high bioavailability but not necessarily by proven toxicity.	Uses the concept of vPvB with described substances that are characterized by high persistence and high bioavailability but not necessarily by proven toxicity.	None	None

TURI Alternatives Assessment Process Guidance P2OASys	TURI provides a review of methods for alternatives assessment, including the P2OSys, which it developed.	Uses a numerical scoring system (2,4,6,8) to characterize acute and chronic aquatic toxicity.	See Table 7-1.	None	None	Scores exposure potential environmental and worker impacts.	See Table 5-2.	Scored based on hydrolysis half- life ranging from 4 to 500 days	None
UCLA MCDA	Compendium of approaches that refer the reader to various assessment methods or software for evaluating environmental effects.	Provides summary of various methods.	Provides summary of various methods.	Provides summary of various methods.	Provides summary of various methods.	Provides summary of various methods.	Provides summary of various methods.	Provides summary of various methods.	Provides summary of various methods.
UNEP Persistent Organic Pollutants Review Committee General Guidance on Alternatives	High-level framework that specifies general steps to be taken in the environmental assessment of risk from persistent organic pollutants.	Recom-mended but not specified.	None	None	None	Recommends assessment of release to environment, especially for those chemicals that may be used in "dispersive" products, such as paint or possibly dispersed products such as lubricating oil.	Not considered	Not considered	Consideration of environ-mental exposures for POPs used in dispersive or possibly dispersive products.

BizNGO

Relative to ecological risk and environmental assessment, BizNGO presents a broad, step-by-step protocol to compare the safety of chemical alternatives. Although it is labeled as a protocol, it does not provide the level of methodological detail that allows the user to conduct a comparative evaluation. Rather, it offers seven broad steps to be taken in series when conducting a chemical alternatives assessment. The protocol explicitly emphasizes hazard assessment over exposure assessment and requires hazard assessment to occur in advance of technical or economic analysis of the chemicals that are being compared. The following discussion points show how the BizNGO steps are relevant to ecological risk or environmental assessment.

Step 1: Identify chemicals of concern. BizNGO generally relies on specific lists to complete this step. This approach does not conform to methods that ecologists usually use to identify chemicals of concern for purposes of environmental assessment or risk assessment. Ecologists generally rely on functional attributes of chemicals that characterize its potential for persistence, bioaccumulation, or toxicity.

Steps 2 and 3: Characterize end uses and function and identify alternatives. There is no ecological assessment involved at these steps. BizNGO defers human and ecological assessment to other resources.

Step 4: Assess chemical hazards. BizNGO directs the reader to other methods (for example, EPA DfE or GreenScreen® "benchmarking") and depends on GreenScreen® to assess and classify human and environmental health based on 17 end points into one of four benchmarks. GreenScreen® includes assessment of breakdown products. BizNGO also references other screening methods, such as Washington State DEP Quick Chemical Assessment Tool (WA Department of Ecology 2014) and Massachusetts Toxics Use Reduction Institutes Five Chemicals Alternative Assessment Study (TURI 2006b).

Step 5: Technical and economic performance. No ecological aspects included.

Step 6: Apply Life Cycle Thinking. This step suggests the use of Life Cycle Thinking (an undefined term) to assess "other human health and environmental impacts such as global warming, end-of-life management, and worker exposure."

DESIGN FOR THE ENVIRONMENT (DfE)

EPA's DfE assessment framework is a hazard-based assessment protocol that incorporates six general requirements into the alternatives assessment:

1. Data for all relevant exposure routes are evaluated.
2. The review of toxicological data uses the U.N. Globally Harmonized System of Classification and Labelling of Chemicals (GHS) criteria and EPA risk-assessment guidance to identify no observed adverse effect level (NOAEL) or no observed adverse effect concentration (NOAEC) and lowest observed adverse effect level (LOAEL) or lowest observable adverse effect concentration (LOEAC) data where possible.
3. EPA High Production Volume (HPV) Challenge Program and OECD HPV Programme data guidelines are used.
4. Peer-reviewed studies, government reports, and confidential sources of information are incorporated into the characterization of toxicity.
5. The sensitivity of test species is considered in the evaluation of data.
6. The hazard assessment considers degradation or metabolism of a chemical into a by-product that might itself be hazardous.

The hazard assessment parses end points into four hazard designations (very high, high, moderate, and low level of concern) on the basis of certain criteria (see Table 7-1 and Tables 5-2 to 5-5). Relevant environmental end points include acute aquatic toxicity (in water) based on LC_{50} or EC_{50} data; chronic aquatic toxicity (in water) based on NOEC or LOEC data; avian acute toxicity based on an acute oral dose or concentration in the diet; acute bee toxicity; persistence in water, soil, or sediment based on half-life; persistence in air based on a qualitative assessment of data; and bioaccumulation based on BAF or BCF or K_{ow}. The framework recognizes that other end points might be applicable if data are available. They include domestic animal toxicity, epigenetic toxicity, mobility in environmental media, ozone formation, eutrophication, global warming potential, lactational or transplacental transfer, loss of genetic diversity or biodiversity, non-target phytotoxicity, specific target organ toxicity from a single exposure, wildlife developmental impairment, wildlife growth impairment, wildlife survival impairment, wildlife reproductive impairment, and immunotoxicity.

GERMAN GUIDE ON SUSTAINABLE CHEMICALS

The Guide on Sustainable Chemicals (Reihlen et al. 2011) is for selecting sustainable chemicals on the basis of lists, dangerous chemical properties, human health toxicity, "problematic properties related to the environment, mobility," origin of raw materials, greenhouse-gas emissions, and resource consumption. It uses a color-coded system (green, yellow, red) and white (for insufficient information). This guidance emphasizes exposure to a greater degree than most other frameworks and incorporates mobility in terms of release potential, criteria for short- and long-range transport, and Aeolian transport as a dusty chemical.

INTERSTATE CHEMICALS CLEARINGHOUSE ALTERNATIVES ASSESSMENT GUIDANCE

IC2 (2013) is a detailed method for assessing chemicals through a long series of questions posed within two general types of modules: scoping modules and assessment modules. The assessment modules include performance evaluation, hazard, cost and availability, exposure assessment, materials management, social impact, and Life Cycle Thinking. The hazard module uses GreenScreen® and GreenScreen® Plus to assess hazard but does not offer its own methods for assessment. The method categorizes end points as low, moderate, or high based on the ranges shown in Table 7-1 and Tables 5-2 to 5-5. This framework recognizes the potential importance of terrestrial ecological hazards and eutrophication but offers no specific evaluative methods.

CALIFORNIA SAFER CONSUMER PRODUCTS REGULATION

California regulation (Safer Consumer Products, Regulations, R-20011-02) specifies that the California Department of Toxics Substances Control (the Department) shall provide on its website guidance materials for conducting alternatives assessment and that the assessment shall evaluate "environmental fate" and "adverse environmental impacts," among other topics. Subsequently, the Department (CA DTSC 2012) published a list and brief descriptions of the following:

- Hazard assessment methods that include GreenScreen®, Globally Harmonized System, EPA Source Ranking Database, EPA Cluster Scoring System, and OECD Screening Information Data Set.
- Exposure assessment methods that include EPA PBT profiler, EPA ChemSTEER, EPA E-FAST, and EPA EPI Suite (with the caution that the programs in this suite provide screening values and should not be used when direct property measurements are available); EPA PIRAT, EPA ReachScan, EPA ECOSAR (which the document recognizes as a predictor of toxicity but lists as an exposure assessment method), NIOSH Control Banding (human health only), UK COSHH (human health only), CleanGredients, UC Berkley PLUM, SUBSPORT Portal, P2OASys, and Pharos (human health only).

Neither the regulations nor the published descriptions make specific recommendations regarding the use of the hazard assessment or exposure assessment tools noted.

LOWELL CENTER ALTERNATIVES ASSESSMENT FRAMEWORK

The Lowell Center Alternatives Assessment (LCSP) Framework (Rossi et al. 2006) is in flowchart format and includes three core elements: alternatives assessment foundation, alternatives assessment process, and evaluation modules. One of the evaluation modules is "Human Health and the Environment." The LCSP framework prefers methods that present disaggregated data in their actual values for comparison across evaluation categories or hazards (as opposed to creating a single number to compare across options). That approach is used to increase transparency and the ability to identify trade-offs among categories. The framework promotes "creating summary tables from the evaluation modules to support the selection process."

The Human Health and the Environment module does not provide a framework-specific method for evaluating environmental effects. Rather, it directs the reader to various assessment methods or software that can serve that purpose. They include:

- The "Evaluation Matrix" developed for the German Federal Environmental Agency;
- "Quick Scan" developed by The Netherlands;

- "PRIO" developed by the Swedish Chemicals Inspectorate;
- "The Column Model" developed by the German Institute for Occupational Safety;
- The "Pollution Prevention Options Analysis System" (P2OASys) developed by the Massachusetts Toxics Use Reduction Institute;
- The "Cradle to Cradle Design Protocol" developed by McDonough Braungart Design Chemistry;
- The "Chemicals Assessment and Ranking System" designed by the Zero Waste Alliance;
- The "P2 Framework Models" developed by EPA;
- EPA DfE Program;
- EPA's chemical alternatives assessment developed in Furniture Flame Retardancy Partnership; and
- The "GreenList" process developed by the SC Johnson Company.

REACH AUTHORISATION ANALYSIS OF ALTERNATIVES

The REACH framework is published by the European Chemicals Agency and is used to conduct chemical safety assessments. It follows the familiar risk paradigm, incorporating hazard assessment and exposure assessment into a risk characterization. If a substance is shown to be a persistent, bioaccumulative, and toxic (PBT) chemical or a very persistent and very bioaccumulative (vPvB) chemical, an exposure assessment and risk characterization are required. The exposure assessment addresses operational conditions, such as duration and frequency of use, amount used, concentration in the product, and process temperature and local measures, such as ventilation, air filtering, wastewater treatment, and personal protection equipment.

TURI ALTERNATIVES ASSESSMENT PROCESS GUIDANCE

The Massachusetts Toxics Use Reduction Institute report (Edwards et al. 2005) is a survey of methods and tools used in alternatives assessment. It provides an appendix that summarizes more than 100 various methods and tools available for use in chemical assessment. The report reviews nine methods for alternatives assessment of chemicals and divides them into hazard display methods (several of which aggregate data to create a risk index for comparing substances) and screening methods that evaluate a range of hazards and recommend elimination of those that "are deemed to be a high risk." The hazard display methods include:

- Pollution Prevention Options Analysis System developed by the Institute;
- The Column model;
- Five-Step Evaluation Matrix created by the German Federal Environmental Agency; and
- Chemicals Assessment and Ranking System designed by the Zero Waste Alliance, a private consulting organization based in Oregon.

The screening methods contain built-in decision rules to determine priorities for eliminating a chemical on the basis of inherent hazard. The screening methods include Quick Scan, PRIO, Norwegian Guidelines, and C2C protocol and the PBT profiler. Among those methods, P2OASys is the software tool developed by TURI to determine the potential environmental, worker, and public-health impacts of alternative technologies. Table B-1 shows the categorization system used in P2OASys.

UNITED NATIONS ENVIRONMENT PROGRAM PERSISTENT ORGANIC POLLUTANTS REVIEW COMMITTEE GENERAL GUIDANCE ON ALTERNATIVES

UNEP (2009) provides a general description of the issues to be considered in identifying and evaluating alternatives to listed persistent organic pollutants and candidate chemicals. In assessing risks, it considers the hazardous properties of persistent organic pollutants. UNEP requires the collection of information on the release of a chemical into the environment if it is to be used in dispersive products (such as paints) and that some release to the environment should be considered for non-dispersive products (such as lubricants). The guidance requires at least a simple risk assessment, taking into account the weight of available evidence. This high-level guidance does not provide specific recommendations, categorization protocols, or end-point ranges to characterize ecological toxicity, environmental impact, or exposure.

UNIVERSITY OF CALIFORNIA LOS ANGELES MULTI-CRITERIA DECISION ANALYSIS

The MCDA method provides a comparative alternatives assessment based on a wide range of criteria, including physicochemical hazards; human health, ecological, and environmental impacts, as well as technical and economic feasibility. The method recognizes two categories of ecological impacts: adverse effects (aquatic animal or plant species, aquatic and terrestrial ecosystems, endangered or threatened species, and environmentally sensitive habitats) and exposure (volume in manufacturing, volume in consumer use, extent of dispersive use, persistence, and bioaccumulation). Environmental impacts include three broadly populated categories: adverse air quality effects, adverse water quality effects, and adverse soil quality effects.

REFERENCES

CA DTSC (California Department of Toxic Substances Control). 2012. Alternative Analysis Workshop Tools, Methodologies and Frameworks [online]. Available: www.dtsc.ca.gov/PollutionPrevention/upload/WorkshopTables.pdf [accessed July 22, 2014].

CA DTSC (California Department of Toxic Substances Control). 2013. Safer Consumer Products [online]. Available: https://dtsc.ca.gov/LawsRegsPolicies/Regs/SCPA.cfm [accessed June 26, 2014].

ECHA (European Chemical Agency). 2011. Guidance on the Preparation of an Application for Authorization. European Chemical Agency, Helsinki, Finland. January 2011 [online]. Available: http://echa.europa.eu/documents/10162/13643/authorisation_application_en.pdf [accessed June 26, 2014].

Edwards, S., M. Rossi, and P. Civie. 2005. Alternatives Assessment for Toxic Use Reduction: A Survey of Methods and Tools. Toxic Use Reduction Institute [online]. Available: http://www.turi.org/TURI_Publications/TURI_Methods_Policy_Reports/Alternatives_Assessment_for_Toxics_Use_Reduction_A_Survey_of_Methods_and_Tools._2005 [accessed July 22, 2014].

EPA (U.S. Environmental Protection Agency). 2011. Design for the Environment Program Alternatives Assessment Criteria for Hazard Evaluation, Version 2.0. Office of pollution Prevention and Toxics, U.S. Environmental Protection Agency [online]. Available: http://www.epa.gov/dfe/alternatives_assessment_criteria_for_hazard_eval.pdf [accessed July 22, 2013].

IC2 (Interstate Chemicals Clearinghouse). 2013. Interstate Chemicals Clearinghouse Alternatives Assessment Guide, Version 1.0. November 2013 [online]. Available: http://www.newmoa.org/prevention/ic2/IC2_AA_Guide-Version_1.pdf [accessed June 26, 2014].

Malloy, T.F., P.J. Sinsheimer, A. Blake, and I. Linkov. 2011. Developing Regulatory Alternatives Analysis Methodologies for the California Green Chemistry Initiative. Sustainable Technology and Policy Program, University of California, Los Angeles, CA. October 20, 2011[online]. Available: http://www.stpp.ucla.edu/sites/default/files/Final%20AA%20Report.final%20rev.pdf [accessed June 26, 2014].

Reihlen, A., D. Bunke, R. Gross, and D. Jepsen, eds. 2011. Guide on Sustainable Chemicals: A Decision Tool for Substance Manufacturers, Formulators and End Users of Chemicals. Federal Environmental Agency, Dessau-Rosslau, Germany. April 2011 [online]. Available: http://www.umweltbundesamt.de/sites/default/files/medien/publikation/long/4169.pdf [accessed June 26, 2014].

Rossi, M., J. Tickner, and K. Geiser. 2006. Alternatives Assessment Framework of the Lowell Center for Sustainable Production. Version 1.0. July 2006 [online]. Available: http://www.chemicalspolicy.org/downloads/FinalAltsAssess06.pdf [accessed June 26, 2014].

Rossi, M., C. Peele, and B. Thorpe. 2012. BizNGO Chemical Alternatives Assessment Protocol: How to Select Safer Alternatives to Chemicals of Concern to Human Health or to the Environment. The Business-NGO Working Group, April 12, 2012 [online]. Available: http://www.bizngo.org/static/ee_images/uploads/resources/BizNGOChemicalAltsAssessmentProtocol_V1.1_04_12_12-1.pdf [accessed June 25, 2014].

TURI (Toxics Use Reduction Institute), 2006a. Five Chemicals Study: Alternatives Assessment Process Guidance. March 2006 [online]. Available:

http://www.turi.org/Our_Work/Research/Alternatives_Assessment/TURI_Alternatives_Assessment_Method_-_5_Chemicals_Study [accessed June 25, 2014].

TURI (Toxics Use Reduction Institute), 2006b. Five Chemicals Alternatives Assessment Study. University Massachusetts Lowell. June 2006 [online]. Available: http://www.turi.org/TURI_Publications/TURI_Methods_Policy_Reports/Five_Chemicals_Alternatives_Assessment_Study._2006/Full_Report [accessed June 26, 2014].

UNEP (United Nations Environment Programme). 2009. General Guidance on Considerations Related to Alternatives and Substitutes for Listed Persistent Organic Pollutants and Candidate Chemicals. UNEP/POPS/POPRC.5/10/Add.1. Persistent Organic Pollutants Review Committee, Fifth Meeting, October 12-16, Geneva [online]. Available: http://www.subsport.eu/images/stories/pdf_archive/substitution_tools/7_guidance_alt_pop_en.pdf [accessed June 26, 2014].

WA Department of Ecology. 2014. The Quick Chemical Assessment Tool [online]. Available: http://www.ecy.wa.gov/programs/hwtr/chemalternatives/QCAT.html [accessed July 23, 2014].

Appendix C
Toxicological Priority Index (ToxPi)

The Toxicological Priority Index (ToxPi) was discussed in Chapters 7, 8, 9 and 12 as a tool for transparent integration and visualization of data across disparate information domains. While the committee's charge did not call for making specific recommendations about computational approaches, a tool that clarifies and documents the judgment and trade-offs entering into an assessment is a big step toward transparency. ToxPi is a tool familiar to committee members and was implemented with an EPA grant to the Carolina Center for Computational Toxicology. A graphic user interface is at http://comptox.unc.edu/toxpi.php.

At some point in an assessment, decisions that prioritize/rank chemicals against each other will need to be made, as explained in Chapter 9. To this effect, ToxPi software (Reif et al. 2013) is an example of a prioritization support tool for integrating evidence across end points and visualizing the relative prioritized ranks of the compounds under consideration. ToxPi was proposed by Reif and colleagues in 2010 as a dimensionless index score that enables multiple sources of evidence on exposure and safety to be integrated and transformed into visual rankings that are transparent and facilitate decision making. Different data are translated into ToxPi scores for all compounds, as explained below, in the publications describing the approach (Reif et al. 2010), and the associated software package (Reif et al. 2013). ToxPi takes the entire realm of information that goes into a decision and reduces it to one number, which can be used to prioritize or rank. While reducing the various types of information into one number could obfuscate the underlying information that goes into producing the number, ToxPi provides a transparent visualization of both the rankings and the individual compound's ToxPi score components.

CALCULATION OF TOXPI SCORES: CONGLOMERATION OF INFORMATION

ToxPi software calculates a unitless number that may be used for rank ordering chemicals being compared. First, the user needs to assemble the data that are intended to be integrated and analyzed by ToxPi into one data matrix, where columns are individual information types (e.g., numerical values of a chemical's potency in a particular assay) and rows are the compounds to be compared. All data need to be in a numeric format, and qualitative scores must be converted to numerical values. For example, a summary description of a panel of bacterial mutagenicity assays that may be characterized with qualitative descriptors, "clearly mutagenic," "ambiguous results" and "likely non-mutagenic" may be converted to the numerical values of 1, 0.5, and 0, respectively. Missing values (or alphanumeric entries) may remain in the matrix, and ToxPi visualization will report the percentage of the compounds with missing values for each data type integrated into one ToxPi "slice."

Second, the user needs to define how the information will be integrated. In other words what "data domains" or categories best describe the available database in the context of the decision to be made? Figure C-1 shows the quantitative information types available for alternatives analysis. They may consist of several broad categories, such as chemical properties, in vitro assay data, exposure data, in vivo study results, and/or biological pathways perturbation. Within one or more of these broad categories, additional sub-categories may be defined, such as different types of nuclear receptors probed with several assays. In Figure C-2, the example shows how the estrogen receptor (ER) slice within a broad category, "in vitro assay data," integrates data from six independent assay types.

Third, the data selected to be integrated into one slice are transformed into a slice score for each compound (Figure C-2).The values for each compound across all data columns to be integrated into one slice are summed up. The summed values are normalized to the interval [0,1] by dividing each compound's result by the slice maximum. If the data being integrated into one slice represent relative potency (e.g., in vitro or other assays assays), values closer to the unit score (equal to 1) translate to higher potency. Conversely, values closer to the origin (equal to 0) translate to lower potency within

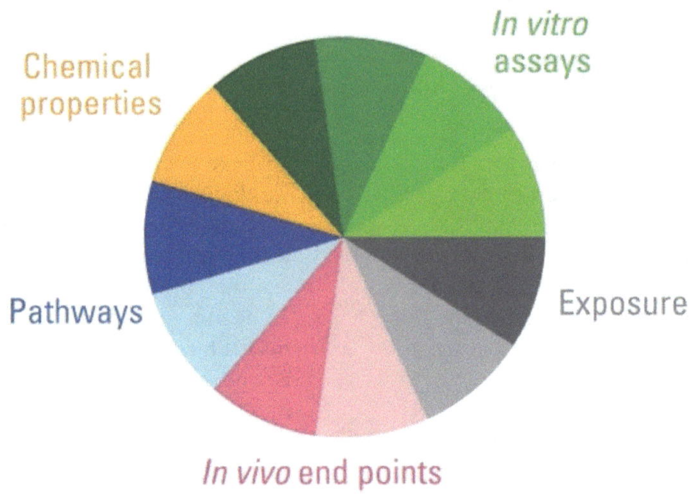

FIGURE C-1 Information types available for alternatives assessment. Reproduced with permission from Environmental Health Perspectives (Reif et al. 2010).

the corresponding data domain. Slices that do not extend at all from the origin represent "inactive/no activity."

The ToxPi analysis also allows, but doesn't require, data integration into a single slice. In the example shown in Figure C-2, the concentration-response curves for each of the six assays being integrated into one ER slice are shown for three example chemicals. On each concentration-response curve showed in Step 1, the red asterisk represents the AC_{50} (active concentration or "potency") for each chemical's activity in these assays, and flat blue lines indicate assays in which that chemical exhibited no activity. For the slices where information other than quantitative "potency" is to be integrated, the same procedure may be followed, with particular chemical property values, pathway scores, or other categorical values used to provide a notion of the "activity" of each compound relative to other compounds in this particular analysis. In addition, each data type may serve as its own slice; in that case, the normalization is performed on the actual values in the data column, rather than a sum of values across multiple columns.

Once each slice has been assembled and slice scores calculated, the scores for each slice are summed up to derive a final ToxPi score for each compound. These scores are then used to plot the relative rank of the compounds being compared (Figure C-3). The X-axis is the ToxPi score and the Y-axis is the relative rank. It is important to note that each ToxPi analysis is a relative comparison that yields ToxPi scores meaningful only when used in the context of both the compendium of the compounds included and the data with which they were analyzed.

In addition, the ToxPi charts for individual compounds (Figure C-3 left panel, insets overlaid onto the dot plot) can be easily visualized and downloaded either individually, or as a matrix (Figure C-3, bottom right panel). In Figure C-2, it is evident that HPTE is more potent in ER assays than is 2,4-D; this is shown visually by the HPTE slice that is extending farther from the origin than the slice for 2,4-D. The rankings can be used to compare chemical toxicity or assess the similarity of predicted compound activity. The value of ToxPi is that it provides explicit documentation of judgments and decisions made during the data integration step, thus providing the necessary documentation and clarity that is needed for transparency while still reducing the information to the point where it can be used for decision making. In essence, the explicit capturing and documentation at the stage of data integration enables scientific debate about either the science or the value judgments.

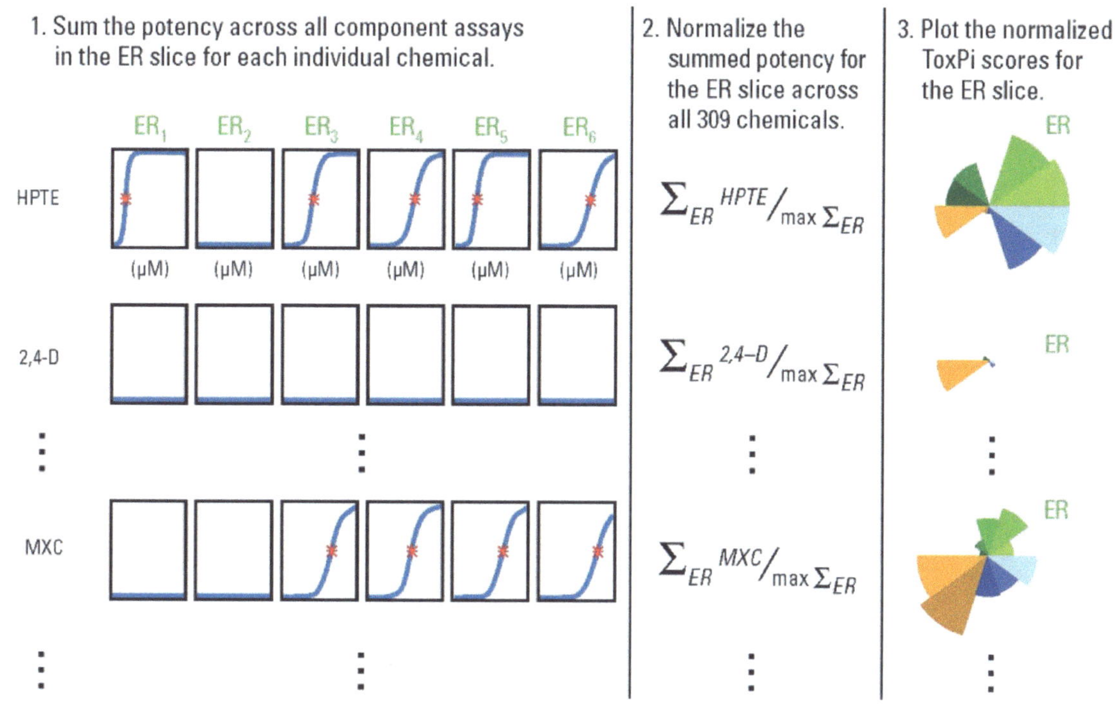

FIGURE C-2 Reproduced with permission from Environmental Health Perspectives (Reif et al. 2010).

USE OF TOXPI TO PERFORM AND TRANSPARENTLY COMMUNICATE DIFFERENTIAL WEIGHTING OF THE INFORMATION

Weighting, which is ultimately a value judgment, as discussed in Chapter 9 and elsewhere in this report, is frequently favored by the stakeholders performing or evaluating the alternatives assessments.

Examples of weighting factors frequently applied are:

- Weighting of end points *within* a domain (e.g., carcinogenicity weighted heavier than respiratory effects within the human health domain);
- Weighting *between* domains (e.g., human health effects weighted heavier than aquatic toxicity or human health effects weighted higher than resource use); and
- Weighting of different types of data (e.g., providing greater weight to the data from human studies, as compared to animal toxicity studies or in vitro assays).

The easiest way to analyze and visualize data with ToxPi is to not weight individual slices (e.g., data domains) differently. For example, in Figure C-3, each slice is of equal weight, which is evident from equal division of the ToxPi into slices that have the same width (in radians), indicating no preferential weighting of any slice in the overall ToxPi calculation. In the implementation presented in Figures C-2 and C-3, each of the 10 slices is weighted equally, so each is 36 degrees in width.

Alternative weighting schemes that differentially emphasize the information represented by the individual ToxPi slice(s) may also be performed. ToxPi software (v 1.3 or higher) allows for applying weight factors to each slice, which introduces an additional coefficient to the calculation of the overall ToxPi score for each compound. This information is easy to convey through ToxPi graphs, which will display different widths (in radians) of the slices. Two examples of the effect that value judgment-based differential weighting of the information may have on the outcome of data integration are illustrated in Chapter 12 (see Figures 12-9 and 12-10).

USE OF TOXPI TO CAPTURE UNCERTAINTIES IN THE RELATIVE RANKINGS OF COMPOUNDS

Two types of confidence limits are calculated by ToxPi software: (i) the uncertainty in the relative size (from 0 to 1) of each individual slice if multiple data types were integrated into one slice; and 2) the

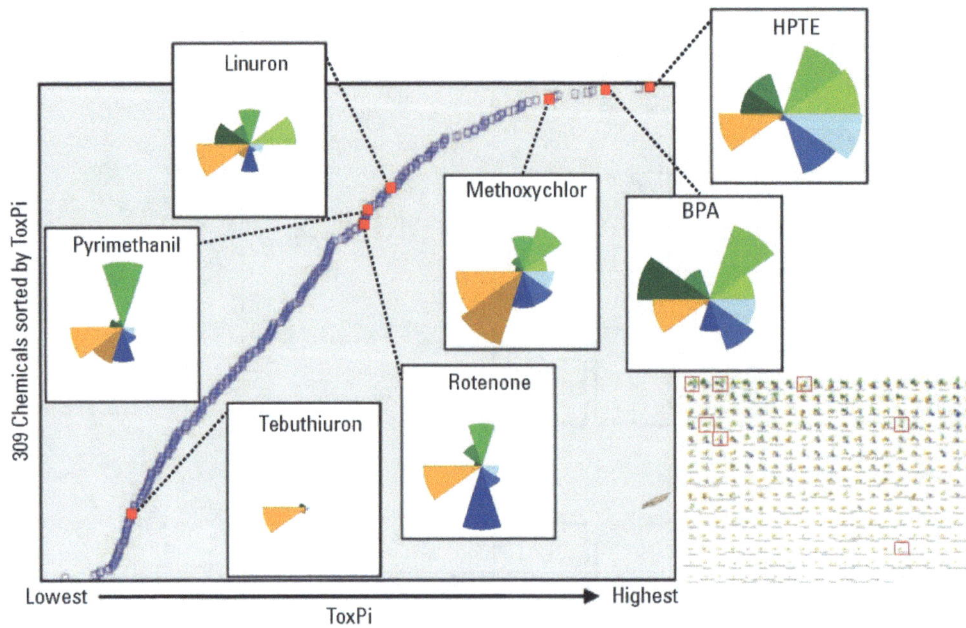

FIGURE C-3 ToxPi scores are used to plot relative rank of compounds (Y axis) and ToxPi score (X axis). Reproduced with permission from Environmental Health Perspectives (Reif et al. 2010).

confidence intervals on the overall ToxPi score and the relative rank of each compound in a given comparison. The former are displayed on the ToxPi chart for the individual agents (Figure C-4A) and the latter are displayed on the dot plot of the ranked compounds (Figure C-4B).

ToxPi software addresses the uncertainty around the data by calculating confidence intervals and scaling by bootstrap sampling[57] of the component values (source data) within each slice. If multiple data columns were integrated into a slice, the numerical values within each slice are sampled with replacement, and these resampled values are used to calculate a bootstrap ToxPi statistic. The bootstrap ToxPi statistic is calculated exactly as the original statistic, but on the resampled data. This process is repeated 1,000 times, and these 1,000 or more bootstrap statistics are used to assess the stability of the estimated ToxPi score for the chemical. In particular, a 95% confidence interval for the ToxPi score is generated in the standard way: the lower bound is given by the 2.5 percentile in the bootstrap statistics and the upper bound is given by the 97.5 percentile of the bootstrap statistics. These bounds are visualized as dashed lines within each slice where bootstrapping was possible (Figure C-4A). Intuitively, the width of the confidence interval for a chemical depends on the amount of variability within each slice. The ToxPi score will have a narrow confidence interval if the assay values within a slice are very similar and a wide confidence interval if the assay values within a slice are very different. The bootstrapping approach does not make any assumptions regarding the distribution of the data values, and should give appropriate confidence intervals in most contexts. However, there is a possibility that confidence intervals generated using the approach described here may be unreliable in the following situations: if the measurements within each slice are on dramatically different scales (while different scales can be combined within ToxPi, individual slices are best used to represent similar/related data); if there are just a small number of assays within each slice; and if there are extreme values (outliers) in the data.

The uncertainty of the ToxPi score and relative rank are calculated in a way identical to that for the confidence interval for each slice. The confidence intervals are visualized (Figure C-4B) as horizontal and vertical bars. Along the X-axis, the bootstrapped confidence intervals in an overall ToxPi score for a particular chemical are displayed. Along the Y-axis, the bootstrapped confidence intervals in the relative rank of a specific chemical are displayed.

[57] Bootstrapping is a statistical approach for assigning measures of accuracy to sample estimates, using resampling methods.

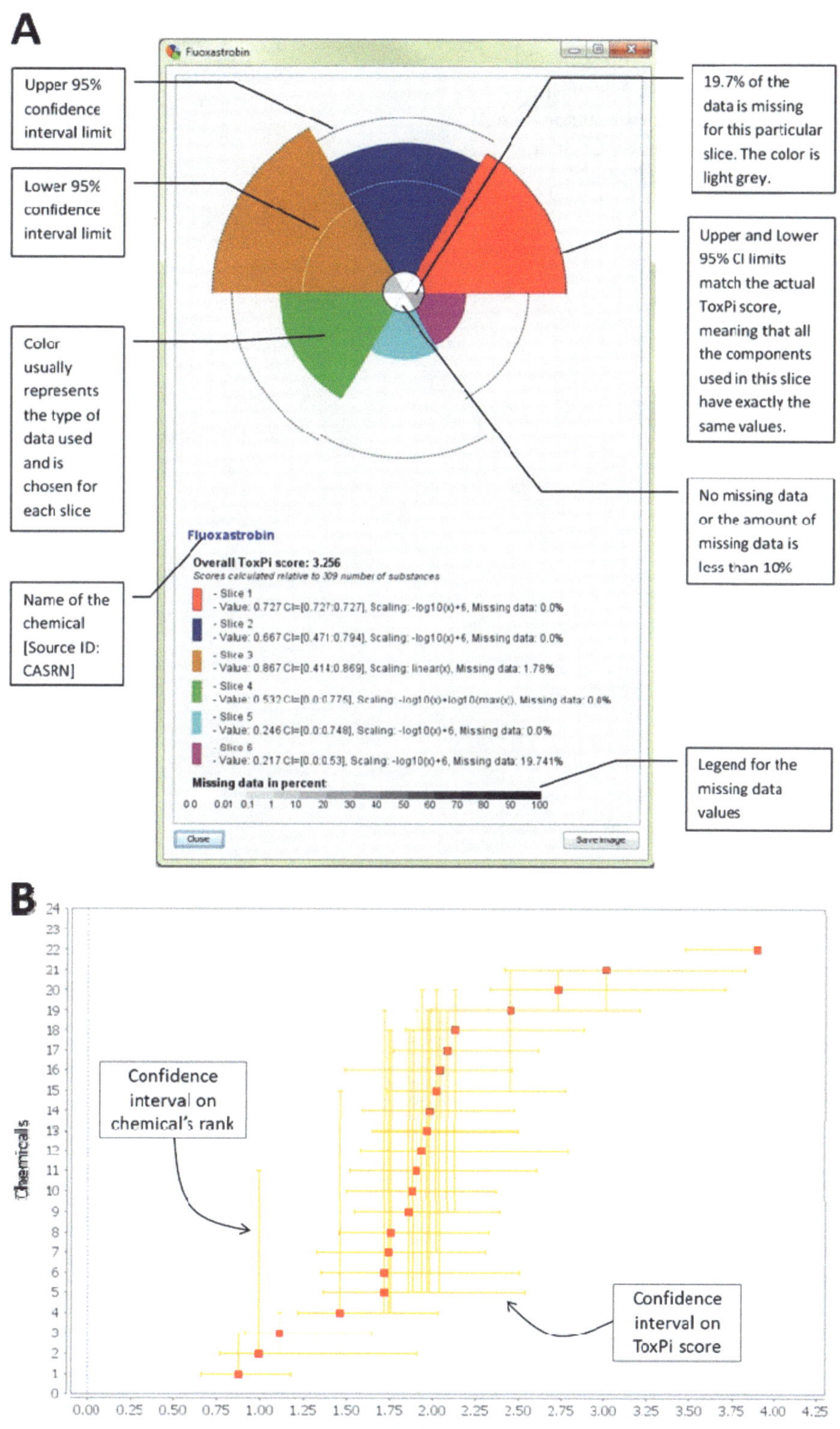

FIGURE C-4 ToxPi confidence intervals. A: Uncertainty in size of each slice. B: Confidence intervals in overall ToxPi score and relative rank of each compound.

REFERENCES

Reif, D.M., M.T. Martin, S.W. Tan, K.A. Houck, R.S. Judson, A.M. Richard, T.B. Knudsen, D.J. Dix, and R.J. Kavlock. 2010. Endocrine profiling and prioritization of environmental chemicals using ToxCast data. Environ. Health Perspect. 118(12):1714-1720.

Reif, D.M., M. Sypa, E.F. Lock, F.A. Wright, A. Wilson, T. Cathey, R.R. Judson, and I. Rusyn. 2013. ToxPi GUI: An interactive visualization tool for transparent integration of data from diverse sources of evidence. Bioinformatics 29(3):402-403.

Appendix D
Overview of the GHS Classification Scheme in Hazard Classification

The following provides a summary of the GHS system as it relates to classification of health hazards. Examples of how this classification system is used in DfE and the GreenScreen® tool are also provided. Although the committee's discussion primarily focuses on GHS classification schemes, the GreenScreen® tool, and the DfE framework, the committee also describes selected situations when slightly different approaches have been used to inform other alternatives assessment frameworks (e.g., TURI, REACH). This appendix also describes how authoritative lists are used to classify human health hazards and briefly describes approaches used to address end points not included in the GHS classification scheme.

USE OF THE GHS CLASSIFICATION SCHEME TO ASSESS HEALTH HAZARDS

Acute Mammalian Toxicity

The GHS defines acute "toxicity as adverse effects occurring following oral or dermal administration of a single dose of a substance, or multiple doses given within 24 hours, or an inhalation exposure of 4 hours" (UNECE 2013). Chemicals can be classified into five hazard categories based on animal LD_{50} (oral, dermal) or LC_{50} (inhalation) values (Table D-1). The criteria consist of hazard levels assigned to the five GHS categories (any exposure route). The hazard levels described by DFE range from *Very High Hazard* = (Category 1 or 2) to *Low* (Category 5, or adequate data available and negative studies, no structural alerts, and GHS not classified).

Carcinogenicity

The GHS classification criteria are based on strength of evidence of a chemical posing a carcinogenic hazard. The GHS guidance points out that classification is based on the inherent properties of a chemical and does not provide information on the level of the human cancer risk. The GHS classification criteria (Table D-2) are largely consistent with those of the International Agency for Research on Cancer (IARC) and the National Toxicology Program (NTP).

DfE and GreenScreen® have adopted similar criteria for assessing chemicals for carcinogenicity. According to DfE, their criteria mirror IARC's classification approach. Although the DfE and GreenScreen® systems incorporate the GHS carcinogen categories, they assign hazard designations differently. The impact, if any, of the differences on the outcomes of alternatives assessments, is unclear.

- GHS Categories 1A and 1B:
  - DfE: *Very High Hazard*/Green Screen: *High Hazard*
- GHS Category 2:
  - DfE: *High Hazard* /Green Screen: *Moderate Hazard*
- No GHS Category
  - DfE: *Moderate Hazard* = limited or marginal evidence of carcinogenicity in animals (and inadequate evidence in humans)
  - DfE: *Low Hazard* = negative studies or robust mechanism-based structure-activity-relationships as described in the DfE guidance document (EPA 2011a).
  - GreenScreen®: *Low Hazard* = adequate data available, and negative studies, no structural alerts, and GHS not classified.

DfE and GreenScreen® also assign hazard designations to authorized carcinogen lists. If significant difference in authoritative classification of a chemical occurs, then the GreenScreen® uses the most conservative health classification.

TABLE D-1 Acute Toxicity Hazard Categories and Acute Toxicity Estimate (ATE) Values Defining the Respective Categories

<table>
<tr><th>Exposure Route</th><th>Category 1</th><th>Category 2</th><th>Category 3</th><th>Category 4</th><th>Category 5</th></tr>
<tr><td>Oral (mg/kg bodyweight)</td><td>5</td><td>50</td><td>300</td><td>2000</td><td rowspan="2">5000</td></tr>
<tr><td>Dermal (mg/kg bodyweight)</td><td>50</td><td>200</td><td>1000</td><td>2000</td></tr>
<tr><td>Gases (ppmV)</td><td>100</td><td>500</td><td>2500</td><td>20000</td><td></td></tr>
<tr><td>Vapors (mg/l)</td><td>0.5</td><td>2.0</td><td>10</td><td>20</td><td></td></tr>
<tr><td>Dusts and Mists (mg/l)</td><td>0.05</td><td>0.5</td><td>1.0</td><td>5</td><td></td></tr>
</table>

SOURCE: Adapted from UNECE 2011.

TABLE D-2 GHS Criteria to Categorize the Carcinogenicity of a Single Substance

Category 1 Known or Presumed Carcinogen		Category 2 Suspected Carcinogen
Subcategory 1A Known Human Carcinogen Based on human evidence	Subcategory 1B Presumed Human Carcinogen Based on demonstrated animal carcinogenicity	Limited evidence of human or animal carcinogenicity

Mutagenicity/Genotoxicity

The GHS criteria used to assess chemicals for mutagenicity/genotoxicity health end points are adapted from criteria developed for the GHS health hazard "Germ Cell Mutagenicity." This hazard class is primarily concerned with chemicals that may cause human germ cell mutations. Also considered in classifying substances of this hazard class are mutagenicity/genotoxicity tests in vitro and mammalian somatic cells in vivo (UNECE 2013) (Table D-3).

DfE developed criteria for assessing chemicals as hazards for the mutagenicity /genotoxicity health end point. DfE supplemented the GHS criteria for germ cell mutagenicity with considerations for mutagenicity and genotoxicity in cells other than germ cells:

- GHS Categories 1A and 1B:
    - DfE: *Very High Hazard* for germ cell mutagenicity
- GHS Category 2:
    - DfE: *High Hazard* for germ cell mutagenicity and mutagenicity and genotoxicity in somatic cells. This DfE classification is also applied when there is in vitro evidence of mutagenicity plus in vivo evidence of mutagenicity in somatic cells or germ cells of humans or animals (EPA 2011a).
- No GHS Category
    - DfE: *Moderate Hazard* for germ cell mutagenicity and mutagenicity and genotoxicity in somatic cells = evidence of mutagenicity supported by positive results in in vitro or in vivo somatic cells of humans and animals.
    - DfE: *Low Hazard* for germ cell mutagenicity and mutagenicity and genotoxicity in somatic cells = Negative results for chromosomal aberrations and gene mutations, or no structural alerts.

Many frameworks, including DfE, use authoritative lists to assess chemicals for mutagenicity/genotoxicity end points.

Reproductive Toxicity

The GHS includes developmental toxicity in the definition of reproductive toxicity, but subdivides reproductive toxicity and developmental toxicity in the classification system. For classification purposes, reproductive toxicity is defined by GHS as adverse effects on sexual function and fertility in adult males and females, including, but not limited to, effects on sexual behavior, fertility, parturition, pregnancy outcomes. Adverse effects on or via lactation are also included in reproductive toxicity. The GHS

TABLE D-3 GHS Criteria to Categorize the Germ Cell Mutagenicity of a Single Substance

Category 1: Known/Presumed Known to produce heritable mutations in human germ cells		Category 2 Suspected/Possible
Subcategory 1A Positive evidence from epidemiological studies	Subcategory 1B Positive results in: In vivo heritable germ cell tests in mammals; human germ cell tests; in vivo somatic mutagenicity tests, combined with some evidence of germ cell mutagenicity	May include heritable mutations in human germ cells. Positive evidence from tests in mammals and somatic cell tests. In vivo somatic genotoxicity supported by in vitro mutagenicity data.

SOURCE: Adapted from UNECE 2013.

TABLE D-4 GHS Criteria to Categorize the Reproductive Toxicity of a Single Substance

Category 1 Known or presumed to cause effects on human reproduction or on development		**Category 2** Suspected	**Additional Category**
Category 1A Known: Based on human evidence	**Category 1B Presumed:** Based on experimental animals	Human or animal evidence possibly with other information	Effects on or via lactation

treats lactation effects separately, however, so that a specific hazard warning can be provided to lactating mothers. The GHS criteria for reproductive toxicity consist of placing substances into one of two categories based on the strength of the evidence (Table D-4).

With the exception of DfE, all of the frameworks reviewed by the committee use GHS criteria to establish evidence of reproductive toxicity in chemical hazard assessments (see Table 8.1). GreenScreen® developed the following reproductive toxicity hazard designations based on the GHS criteria:

- *High Hazard* = GHS Category 1A (Known) and 1B (Presumed)
- *Moderate* = GHS Category 2 (Suspected) or limited or marginal evidence of reproductive toxicity in animals (see Guidance)
- *Low* = Adequate data available, and negative, no structural alerts, and GHS not classified.

In contrast to the other frameworks, DfE combines reproductive and developmental toxicity into a single health end point in chemical hazard assessments, and does not use GHS criteria to establish evidence of reproductive toxicity. DfE uses criteria derived from EPA's Office of Pollution Prevention and Toxics (OPPT) criteria for HPV chemical categorization (EPA 2009) and the EU REACH criteria for Annex IV (EC 2007). The criteria consist of hazard levels assigned to dose/concentration ranges (oral, dermal, and respiratory routes) obtained from experimental animal tests. For inhalation exposure (vapor/gas), for example, the hazard designations in mg/L/day are: *High Hazard* = < 1; *Moderate Hazard* = 1-2.5; *Low Hazard* = > 2.5-20; *Very Low Hazard*= > 20. Parental (reproductive) and offspring (developmental) exposure to chemicals are evaluated using the criteria and in general, NOAELs and LOAELs as the metric.

DfE-assigned hazard designations from two authoritative lists also provide evidence of reproductive and developmental toxicity. *High Hazard* = H362 (May cause harm to breast-fed children) and *High or Moderate Hazard* = CA Proposition 65 List (chemicals known to the state to cause reproductive and developmental toxicity).

The basis or rationale for assigning a higher hazard level to H362 compared to the Prop 65 List is unclear, given the much larger number and broader range of reproductive/developmental toxicants on the Prop 65 List and the transparent and rigorous review and approval process required for listing chemicals. The impact of using different criteria and authoritative lists to identify reproductive and developmental toxicants on the outcome of alternatives assessments is unclear, but should be considered.

Repeated Exposure Human Health End Points

These health end points include neurotoxicity (repeated exposure); repeated dose toxicity; and systemic toxicity/organ effects (repeated exposure). The frameworks use the GHS criteria for the hazard class "specific target organ toxicity (repeated exposure) to provide evidence of the health end points. The GreenScreen® tool lists GHS criteria for "systemic toxicity/organ effects" as the "information source" for the neurotoxicity end point (Clean Production 2013). However, since "systemic toxicity/organ effects" is not a GHS health hazard class (UNECE 2013), the committee interpreted this as meaning "specific target organ toxicity."

As described by the GHS, repeated exposures to chemicals in the specific target organ (repeated exposure) hazard class produce significant toxic effects on specific target organs, including effects that impair function, are both reversible and irreversible, and are immediate or delayed. Classifying chemicals as specific target organ toxicants (repeated exposure) based on the GHS criteria requires using expert judgment to conduct weight-of-evidence evaluations of all available evidence. The GHS specifies that all existing data include peer-reviewed published studies and additional data acceptable to regulatory agencies. The information comes either from repeated exposure in humans or animal studies. The GHS states: "it is recognized that human data will be the primary source of evidence for this hazard class" (UNECE 2013).

Chemicals are placed in one of two categories based on the nature and severity of the observed effects. "*Category 1*: Chemicals that have produced significant toxicity in humans, or that on the basis of evidence from studies in animals can be presumed to have the potential to produce significant toxicity in humans (emphasis in original) following repeated exposure. Placing a chemical in Category 1 is based on: (a) reliable and good quality evidence from human cases or epidemiological studies; or (b) observations from appropriate studies in experimental animals in which significant and/or severe toxic effects, of relevance to human health, were produced at generally low exposure concentrations. *Category 2*: Chemicals that, on the basis of evidence from studies in experimental animals can be presumed to have the potential to be harmful to human health (emphasis in original) following repeated exposure. Placing a chemical in Category 2 is based on observations from animal studies in which significant toxic effects, of relevance to human health, were produced at generally moderate exposure concentrations" (UNECE 2013).

GHS guidance values (dose/concentration) for various exposure routes based on standard repeated exposure studies (e.g., 90-day sub-chronic) that provide information on specific target organ toxicity can be used in weight-of-evidence evaluations to assist in classifying chemicals as Category 1 and Category 2. Examples of guidance values (mg/l/6h/d) in a 90-day inhalation (vapor) toxicity study in rats: *Category 1* = ≤ 0.2; *Category 2* = 0.2 < C ≤ 1.0.

Neurotoxicity (Repeated Exposure)

GreenScreen® does not list the GHS guidance values as an information source for this neurotoxicity. GreenScreen® assigns the following hazard designations to the GHS categories: *High Hazard* = Category 1; *Moderate Hazard* = Category 2; *Low Hazard* = Adequate data available and negative studies, no structural alerts, and GHS not classified. GreenScreen® uses U.S. EPA Risk Assessment Guidance to define applicable neurotoxic effects.

In contrast to the IC2 and BizNGO frameworks, the DfE framework does not classify chemicals as Category 1 and Category 2 neurotoxicants. The DfE framework uses the GHS guidance values independently of conducting weight-of-evidence evaluations of human and animal studies. The GHS guidance values with DfE- assigned hazard designations are used as criteria to provide evidence of neurotoxicity (repeated exposure). For example, in a 90-day rat inhalation (vapor) study, a chemical for which target organ toxicity is observed at a given exposure concentration (mg/L/6h/day) is designated as a hazard according to the following criteria: *High Hazard* = < 0.2; *Moderate Hazard* = 0.2—1.0; *Low Hazard* = > 1.0.

Repeated Dose Toxicity

Repeated dose toxicity is identified as a health end point in the DfE frameworks as shown in Table 8.1. The DfE framework uses the results of repeated dose toxicity studies to evaluate chronic exposure (EPA 2011a). The DfE framework criteria for repeated dose toxicity are the same as the DfE criteria for neurotoxicty (repeated dose). As described above, the criteria consist of GHS guidance values for specific target organ toxicity (repeated exposure) with DfE-assigned hazard designations. DfE points out that the criteria mirror the EPA's OPPT criteria for HPV chemical categorization (EPA 2009). In addition, the following DfE suggested hazard designations for authoritative lists can be used to supplement the criteria: *High Hazard* = EU R48 (23/24/25)—Danger of serious damage to health by prolonged exposure (repeated exposure); EU H372—Causes damage to organs.

Moderate Hazard = H373—May cause damage to organs; *High or Moderate* Hazard = EU R48 (20/21/22)—Danger of serious damage to health by prolonged exposure.

Systemic Toxicity/Organ Effects (Repeated Exposure)

The IC2 and BizNGO frameworks criteria for specific target organ toxicity (repeated exposure) are GHS guidance values with DfE-assigned hazard designations. The criteria are the same as the DfE criteria for the repeated dose toxicity end point discussed earlier. The frameworks use the same authoritative lists with the DfE-suggested hazard designations that are described in the *Repeated Dose Toxicity* section to supplement the criteria.

Single Exposure Human Health End Points

These end points include: neurotoxicity (single exposure) and systemic toxicity/organ effects (single exposure). The frameworks use GHS criteria for the hazard class, "specific target organ toxicity (single exposure)" to provide evidence for these health end points. A single exposure to chemicals in this hazard class causes specific, non-lethal target organ toxicity (UNECE 2013). This toxicity includes all significant health effects, including both immediate and delayed, reversible and irreversible, that can impair function, but are not covered by other GHS health hazard classes (UNECE 2013). According to the GHS, human data will be the primary source of evidence for this hazard class.

Chemicals are classified using expert judgment based on the weight of all available evidence, including the use of recommended guidance values, and are placed into three categories based on the severity and nature of the observed effect(s). *Category 1 designation for*: chemicals that have produced significant toxicity in humans and chemicals which have the potential to produce significant toxicity in humans following a single exposure, based on animal study evidence (UNECE 2013). Chemicals are placed into the category based on: (a) reliable and good quality evidence from human cases or epidemiological studies, or (b) evidence of significant and/or severe toxic effects in experimental animals that are of relevance to humans and occurred with low exposures. *Category 2 describes*: chemicals that, based on animal studies, "can be presumed to have the potential to be harmful to human health following single exposure" (UNECE 2013). Chemicals are placed into the category based on observations in studies of experimental animals of significant toxic effects of relevance to human health that are produced at generally moderate exposure concentrations. *Category 3*: Transient target organ effects for which a chemical may not meet the criteria to be classified in Categories 1 and 2. The effects adversely alter human function for a short duration after exposure and recovery occurs in a reasonable period without significantly altering structure and function. This category only includes narcotic effects and respiratory irritation.

GHS guidance values (dose/concentration) for various exposure routes relevant to humans are used as part of weight of evidence evaluations to assist in classifying chemicals into Categories 1 and 2. Category 3 does not include guidance values because this classification is primarily based on human data (29CFR 1910. 1200). The guidance value ranges are proposed single-dose exposure concentrations that have been shown to produce significant non-lethal toxic effects in experimental animal studies. For example, in a rat inhalation study, guidance value ranges (single exposure) for vapor (mg/L/4h) for *Category 1* = $C \leq 10$; *Category 2* = $20 \geq C > 10$.

Neurotoxicity (Single Exposure)

Some frameworks use hazard designations assigned to neurotoxicant categories as criteria to provide evidence of neurotoxicity (single exposure). GreenScreen® does not list GHS guidance values as a part of the criteria, so it is unclear whether they are used to assist in classifying chemicals into categories. Criteria: *Very High Hazard* = Category 1; *High Hazard* = Category 2; *Moderate Hazard* = Category 3; *Low Hazard* = adequate data available and negative studies; no structural alerts, and GHS not classified. Screening lists are used in addition to the criteria to provide evidence of the health end point.

Systemic Toxicity/Organ Effects (Single Exposure)

Several frameworks use GHS guidance values to which GreenScreen® has assigned hazard designations as criteria to provide evidence of the systemic toxicity/organ effects health end point. For example, Inhalation-Gas or Vapor (mg/L/4h): *Very High Hazard* = ≤ 10 (GHS Category 1, Single Exposure, any route); *High Hazard* = > 10-20 (GHS Category 2, Single Exposure, any route). The frameworks use authoritative lists to supplement the criteria. For example: *Very High Hazard* = H370 (causes damage to organs); *High Hazard* = H371 (may cause damage to organs); *Moderate Hazard* = H335 (may cause respiratory irritation)

Respiratory Sensitization

GHS criteria provide evidence of the respiratory sensitization health end point. In the context of the GHS, a respiratory sensitizer is a chemical that will lead to hypersensitivity following inhalation exposure. Respiratory hypersensitivity usually means asthma, although other hypersensitivity reactions (rhinitis/conjunctivitis and alveolitis) are considered.

Respiratory sensitizers are classified into GHS Hazard Category 1 if: "(a) there is evidence in humans that the chemical can lead to specific hypersensitivity; and/or (b) if there are positive results from an appropriate animal test" (UNECE 2013). If required, and if there are sufficient data, chemicals can be further categorized in sub-category 1A and sub-category 1B. Category 1A chemicals "show a high frequency of occurrence in humans; or a probability of occurrence of a high sensitization rate in humans based on animal or other tests" (UNECE 2013). Category 1B chemicals "show a low to moderate frequency of occurrence in humans; or a probability of occurrence of a low to moderate sensitization rate in humans based on animal or other tests" (UNECE 2013). Reaction severity may also impact classification into Categories 1A and 1B.

Evidence that a chemical can lead to specific hypersensitivity (asthma) is based on human experience. Currently, there are no validated animal models for testing for respiratory hypersensitivity. The human evidence can include: (a) clinical history (medical and occupational) and data from appropriate lung function tests related to exposure to the chemical, confirmed by other supportive evidence (e.g. skin prick test) and (b) data from bronchial challenge tests (29CFR 1910.1200).

The DfE framework and GreenScreen® tool use the GHS criteria differently (and use different assigned hazard designations) to provide evidence of the respiratory sensitization end point. GreenScreen® classifies respiratory sensitizers as: *High* = GHS category 1A (high frequency of occurrence) and *Moderate* = GHS category 1B (low to moderate frequency of occurrence). In the DfE framework: *High* = GHS categories 1A and 1B (occurrence in humans or evidence of sensitization in humans based on animal or other tests) and *Moderate* = Limited evidence, including the presence of structural alerts.

GreenScreen® and the DfE framework also use the following authoritative lists with assigned hazard designations to supplement the GHS criteria: the EU hazard statement, H334, the Association of Occupational and Environmental Clinics (AOEC) Exposure Code List (asthmagens), and the MAK (Germany occupational exposure limits with "Sa" and "Sah" notations (DFG 2013). It is unclear why the frameworks identify chemicals in the AOEC database that cause reactive air dysfunction syndrome (RADS) as providing evidence of the respiratory sensitization endpoint. RADS results from single, high exposures to irritant chemicals. RADS does not fit the two phase-sensitization mechanism that defines respiratory sensitizers under the GHS. It also is unclear why DfE assigns the same hazard designation to chemicals identified as "generally accepted" asthmagens and "sensitizer-induced" asthmagens in the AOEC database. In contrast to "generally accepted" asthmagens, which are identified based on expert opinion, "sensitizer-induced" asthmagens are identified based on established AOEC criteria (AOEC 2009). The MAK designation does not appear to be consistent with the description in GreenScreen® of "high hazard" as "frequency of occurrence of sensitization." Sufficient evidence of a MAK respiratory sensitizer requires documentation in only two patients tested at two independent facilities (DFG 2013).

The frameworks do not provide a rationale for listing the MAK and AOEC as authoritative sources for identifying respiratory sensitizers (Quint et al. 2008). As a result, it is not clear why other government agencies (e.g., the UK HSE and NIOSH) and non-government organizations (e.g., ACGIH), which identify occupational respiratory sensitizers that conform to the GHS criteria, are not included as authoritative lists.

Skin Sensitization

The frameworks primarily use GHS criteria to provide evidence of the skin sensitization health end point. A skin sensitizer is defined in the GHS as a chemical that will lead to an allergic response following contact. Sensitization includes an induction phase in which the immune system learns to react. This is followed by an elicitation phase in which clinical symptoms arise upon subsequent exposure to the chemical, usually at a lower concentration.

Skin sensitizers are classified as GHS Category 1. They can be further classified into sub-categories 1A and 1B if required, or if there are sufficient data. A substance is classified as Category 1 if: "(a) there is evidence in humans that the substance can lead to sensitization by skin contact in a substantial number of persons or (b) there are positive results from an appropriate animal test. Chemicals in sub-category 1A show a high frequency of occurrence in humans and/or a high potency in animals and can be presumed to potentially produce significant sensitization in humans. Sub-category 1B chemicals show a low to moderate frequency of occurrence in humans and/or a low to moderate potency in animals and can be presumed to potentially produce significant sensitization in humans" (29CFR 1910.1200). Severity of reaction can be considered in sub-categories 1A and 1B.

"Effects seen in either humans or animals will normally justify classification in a weight of evidence approach" (UNECE 2013). The GHS specifies that evidence should include any or all of six types of data/information, including: (a) positive data from patch testing, usually obtained in more than one dermatology clinic; (b) positive data from appropriate animal studies; (c) well-documented episodes of allergic contact dermatitis, normally obtained in more than one dermatology clinic.

The DfE framework uses a similar hazard designations assigned to the GHS hazard categories to provide evidence of the skin sensitization health endpoint. The DfE framework assigns the following hazard designations: *High* = GHS Category 1A (high frequency of sensitization in humans and/or high potency in animals); *Moderate* = GHS Category 1B (low to moderate frequency of sensitization in humans and/or low to moderate potency in animals); *Low* = adequate data available and not GHS Category 1A or 1B. Sub-category 1A Animal Test Results = *High*; sub-category 1B Animal Test Results = *Moderate*.

The frameworks use authoritative lists with assigned hazard designations to establish evidence of skin sensitization in addition to the GHS criteria: H317 (may cause sensitization by skin contact) = *High or Moderate Hazard* (DfE, IC2, BizNGO). MAK (Germany occupational exposure limits denoted with "Sh" and "Sah"(DFG 2013)[x] = *High* (IC2 and BizNGO). The criterion for sufficient evidence of a MAK skin sensitizer, "case reports of clinically relevant sensitization (association of symptoms and exposure) for more than one patient from at least two independent centres" does not appear to meet the "high hazard" description (above) in certain frameworks.

The rationale for identifying the authoritative lists is not provided, so it is not clear why other similar agencies and organization that identify skin sensitizers are not included. The addition of NIOSH and the ACGIH as authoritative lists, for example, would increase the number of identified skin sensitizers that meet the GHS criteria. The TURI framework uses information from the HSDB, Sax (Sax's Dangerous Properties of Industrial Chemicals textbook), and MSDSs/SDSs. The up-to-date, peer-reviewed animal and human studies on chemicals in the HSDB enable TURI to use weight-of-evidence evaluations to classify skin sensitizers as specified in the GHS criteria. The Sax reference also provides information on animal and human studies. The 2012 Hazard Communication Standard requires health effects information in SDSs to be aligned with GHS criteria. However, compared to ACGIH and NIOSH, these information sources have limitations (see Information Sources Used by Existing Frameworks in Chapter 8).

Skin and Eye Irritation and Corrosion

The GHS addresses effects on the skin and eye as two separate hazard classes—skin corrosion/irritation and serious eye damage/eye irritation. The classification approaches for the hazards, however, are the same. The GHS specifies a "tiered approach with emphasis placed upon existing human data, followed by existing animal data, followed by *in vitro* data and then other sources of information" (UNECE 2013). All available information related to the health hazards of the skin or eye is considered together in a total weight of evidence approach. The available information includes the "results of appropriate validated *in vitro* tests, relevant animal data, and human data such as epidemiological and clinical studies and well-documented case reports and observations" (UNECE 2013). The GHS "tiered approach provides guidance on how to organize existing information on a chemical and to make a weight of evidence decision about hazard assessment and hazard classification" (UNECE 2013).

Skin Corrosion/Irritation

Chemicals classified based on standard animal test data can be placed in one of three categories: (a) *Category 1 (skin corrosion)* is comprised of chemicals that cause "destruction of skin tissue, namely, visible necrosis through the epidermis and into the dermis, in at least one tested animal after exposure for ≤ 4h" (UNECE 2013). Category 1 can be subdivided into three sub-categories (1A, 1B, 1C) if more than one corrosion designation is required. Corrosive responses are noted and observed at specified time periods. For example, in Category 1A, corrosive responses are noted following exposures greater than 3 minutes and up to 1 hour observation. (b) Category 2 (skin irritation) chemicals produce reversible damage to the skin following application for up to 4 hours. Criteria include: mean scores for erythema/eschar or for edema in at least 2 of 3 tested animals at specified time periods after patch removal; or inflammation that persists to the end of the observation period (normally 14 days) in at least 2 animals. Category 3 (mild skin irritation) is used by authorities (e.g., pesticides) that want to have more than one skin irritation category.

The IC2 and BizNGO frameworks use GHS categories with hazard designations assigned by GreenScreen® to provide evidence of the skin corrosion/irritation end point: *Very High Hazard* = Category 1 (corrosive); *High Hazard* = Category 2 (irritation); *Moderate Hazard* = Category 3 (mild irritation); *Low Hazard* = "Not classified." Authoritative lists with assigned hazard designations supplement the criteria: *Very High Hazard* = H314 (causes severe skin burns and eye damage); *High Hazard* = H315 (causes skin irritation).

The DfE framework uses criteria derived from the Office of Pesticide Programs Acute Toxicity Categories to provide evidence of skin irritation/corrosivity: *Very High Hazard* = Corrosive; *High Hazard* = Severe irritation at 72 hours; *Moderate Hazard* = Moderate irritation at 72 hours; *Low Hazard* = Not irritating. The DfE guidance document did not provide a rationale or reason for the use of the Office of Pesticides Programs criteria in the framework instead of the GHS criteria. One possible disadvantage is the inability to link the criteria to the EU hazard statements, H314 and H315.

In the TURI framework, information obtained from HSDB, NIOSH, and MSDSs provide evidence of skin irritation/corrosion. The information above under "Skin Sensitization" regarding the use of HSDB and MSDSs also applies to the use of these resources in providing evidence for skin irritation/corrosion. Information from NIOSH provides a source of existing human and animal toxicity studies on chemicals that have undergone a weight of evidence evaluation, consistent with GHS criteria. In addition, a search of the NIOSH Pocket Guide to Chemical Hazards (available online) using the key phrases "irrit skin" and "skin burns" under "SY" (Symptoms) identifies chemicals that cause skin irritation and skin corrosion, respectively (NIOSH 2005).

Eye Corrosion/Irritation

Based on the results of animal tests, GHS-classified chemicals are placed into one of two categories. Chemicals in Category 1 (serious eye damage/irreversible effects on the eye) produce: "(a) in at least one animal effects on the cornea, iris or conjunctiva that are not expected to reverse or have not fully reversed within an observation period (normally 21 days); and/or (b) in at least 2 of 3 tested animals, a positive response of corneal opacity and/or iritis for up to 72 hours after instillation of the test material" (UNECE 2013). Chemicals in *Category 2/2A* (eye irritation/reversible effects on the eye) "produce in at least 2 or 3 tested animals a positive response of: (a) corneal opacity; and/or (b) iritis; and/or (c) conjunctival redness; and/or (d) conjunctival oedma (chemosis)" for up to 72 hours after instillation of the chemical, which fully reverses within an observation period (usually 21 days) (UNECE 2013). Category 2B (mildly irritating) is comprised of Category 2A chemicals for which the effects are fully reversible within 7 days of observation.

IC2 and BizNGO use the following GHS categories with hazard designations assigned by GreenScreen® as evidence that a chemical causes eye corrosion or irritation: *Very High Hazard* = Category 1 (irreversible damage); *High* = Category 2 (irritating). Hazard designations assigned to EU statements supplement the criteria: *Very High Hazard* = H318 (causes severe eye damage); *High Hazard* = H319 (causes serious irritation). The ECHA database (harmonized classifications) lists 543 H318 substances and 431 H319 substances as of 2/14/2014, the most recent update of the database.

In the DfE framework, hazard designations (assigned by DfE) to the EPA OPPT categories (EPA 2011b) provide evidence of eye irritation and corrosivity. *Very High* = Irritation persists for > 21 days or corrosive; *High* = Clearing in 8-21 days, severely irritating; *Moderate* = Clearing in 7 days or less, moderately irritating; irritating; *Low* = Clearing in less than 24 hours, mildly irritating; *Very Low* = Not irritating. Although the DfE and GHS criteria for identifying eye irritants and corrosives as "Very High Hazard" and "High Hazard" do not appear to be substantially different, the DfE framework does not suggest using the H318 and H319 as authoritative lists to classify chemicals. It is not clear whether this means that the DfE considers the GHS and DfE criteria to be significantly different, and that chemicals classified as H318 and H319 do not meet the DfE criteria for the eye corrosion/irritation.

Respiratory Irritation

The REACH and the TURI frameworks identify respiratory irritation as a health endpoint. The REACH framework uses the GHS criteria for the hazard class specific target organ toxicity (single exposure) to provide evidence of the end point. Category 3 of the criteria addresses transient target organ effects that "adversely alter human function for a short duration after exposure and from which humans may recover in a reasonable period without leaving significant alteration of structure or function"(29CFR1910.1200 [2012]).

The specific GHS criteria for respiratory tract irritation as Category 3 are: (a) Respiratory irritant effects "include effects that impair function with symptoms such as cough, pain, choking, and breathing difficulties. The evaluation is based primarily on human data; (b) Subjective human observations can be supported by objective measurements of clear respiratory tract irritation (e.g., electrophysiological responses, biomarkers of inflammation in nasal or bronchoalveolar lavage fluids); (c) The symptoms observed in humans should also be typical of those that would be produced in the exposed population rather than being an isolated idiosyncratic reaction of response triggered only in individuals with hypersensitive airways" (29CFR 1910.1200 [2012]); (d) There are currently no validated animal tests that deal specifically with respiratory tract irritation; however, single and repeated inhalation toxicity may provide useful information.

The REACH framework provides guidance regarding potential sources of existing information that can be used as evidence of respiratory irritation (ECHA 2013). The guidance indicates that on a case-by-case basis, information where symptoms have been described associated with occupational exposures can be used. Information from acute and repeated dose inhalation toxicity studies may also be considered sufficient to show that a substance causes respiratory irritation at a specific concentration level or range. The EU hazard statement H335 (may cause respiratory irritation) provides evidence of the end point and supplements the GHS criteria.

The TURI framework's use of the HSDB, NIOSH, and MSDSs as information sources for providing evidence of the respiratory irritation end point is consistent with the GHS criteria and the REACH framework's approach to classifying respiratory irritants. The NIOSH Pocket Guide to Chemical Hazards identifies chemicals that cause respiratory irritation with the phrase "irrit resp sys" under "SY" (Symptoms), which can provide evidence of the health end point (NIOSH 2005).

Chemicals that provide evidence of respiratory irritation also can be identified from ATSDR Minimal Risk Levels (MRLs) where the MRL is based on an inhalation study and the respiratory system is listed as the health end point (ATSDR 2013). The Cal/EPA OEHHA acute and chronic inhalation Reference Exposure Levels for which the respiratory system is the target organ (OEHHA 2014) also can identify chemicals that provide evidence of the respiratory irritation end point.

End Point of Concern that is Not Identified as a GHS Health Hazard

Endocrine Activity

Endocrine activity is assessed in several existing frameworks. However, it is not yet identified as a health hazard in the GHS classification system. The criteria the frameworks use to provide evidence of endocrine-related health effects depend on how they define or describe the health end point. The DfE, IC2, and BizNGO frameworks identify the health end point as "endocrine activity." The German Guide and TURI frameworks describe the end point as "endocrine disruption." The CA SCP framework uses the term "endocrine toxicity," which includes endocrine disruption and metabolic syndrome. Endocrine toxicity is characterized by toxicological end points that include adverse effects on endocrine organs and adverse perturbations of the synthesis, secretion, transport, binding, action, or elimination on natural hormones or their receptors (OEHHA 2012). In the REACH framework, based on existing legislation, endocrine disruption per se is not identified as a health end point. Adverse endocrine-related effects on reproduction or disease states like cancer, however, are addressed. Endocrine disrupters can be identified as SVHCs under REACH on the basis that they cause probable serious human health effects that are equivalent to the level of concern for carcinogens, mutagens, and reproductive toxicants (ECHA 2014).

GreenScreen®'s definitions of endocrine activity and endocrine disruption point out differences in the descriptors. Endocrine active substances are defined as "having the inherent ability to interact or interfere with one or more components of the endocrine system resulting in a biological effect, but need not necessarily cause adverse effects" (EFSA 2013a). An endocrine disruptor is "an exogenous substance or mixture that alters function(s) of the endocrine system and consequently causes adverse health effects in an intact organism, or its progeny, or (sub) populations" (Clean Production Action 2013). The European Food and Safety Authority's definitions are similar: "endocrine active substances are chemicals that interact or interfere with normal hormonal activity; when this leads to adverse effects they are called endocrine disruptors" (EFSA 2013b).

The frameworks use all available data and authoritative lists to provide evidence of endocrine activity and/or disruption. The IC2 and BizNGO frameworks and the DfE framework use different approaches to provide evidence of the endocrine activity health end point. The DfE framework evaluates endocrine activity of chemicals, but does not characterize hazard in terms of endocrine disruption. Based on criteria developed by GreenScreen®, the IC2 and BizNGO frameworks evaluate chemicals for endocrine activity and assign hazard values based on adverse endocrine-related health effects (Clean Production Action 2013).

Classification of Endocrine Activity in the DfE Framework

In assessing endocrine activity, the DfE framework uses data resources that include: "(a) *in vitro* data such as hormone binding assays or *ex vivo* assays; (b) *in vivo* data from studies of intact animals or wildlife (including aquatic organisms); (c) ethically conducted human studies; (d) in vivo short term exposures or altered (e.g., ovariectomized) animal models; (e) structural similarity to known endocrine active substances using SAR tools such as AIM, QSAR, etc.; and (f) additional information from studies that indicate a chemical's endocrine system interactions, such as changes in hormone profiles or reproductive organ weights" (EPA 2011a).

Using the following criteria, DfE evaluates available data for each chemical for the presence of endocrine activity, noting caveats and limitations: (a) *No Data (ND)* = No data available to evaluate end point; endocrine activity is unknown, untested; (b) *Potentially Endocrine Active* = Data show evidence of endocrine activity; (c) *No Evidence of Endocrine Activity* = Data show no evidence of endocrine activity (no binding, perturbation, or evidence of endocrine-related adverse effects). "In consultation with EPA toxicologists and risk assessors, DfE provides a summary statement of the available data, including the presence of equivocal or conflicting data and any limitations to the available data. The level of confidence in the assessment is also noted" (EPA 2011a).

Classification of Endocrine Activity in IC2 and BizNGO Frameworks

Based on the GreenScreen® tool, the frameworks evaluate chemicals for endocrine activity and designate hazard levels using the following protocol: (a) assign a preliminary hazard level based on searching GreenScreen® specified lists and available data; (b) determine whether there is a plausibly related adverse health effect for chemicals identified as endocrine active; (c) identify the level of hazard associated with the plausibly related adverse effect(s); and (d) assign the final hazard level for endocrine activity using expert judgment and a weight of evidence approach.

IC2 and BizNGO classify chemicals as endocrine active using the following hazard levels: *High Hazard* = chemical on EU SVHC authorization list for endocrine activity; *Moderate / Moderate or High Hazard* = (1) indication of endocrine activity in scientific literature; (2) initial assignment of all chemicals with data suggesting endocrine activity associated with adverse effects; (3) listed for endocrine activity on Specified Lists, except EU SVHC list. Further review using scientific literature is required to confirm the list-based classifications (except EU SVHC list). *Low Hazard* = requires data for multiple endocrine pathways (e.g., androgenicity, anti-androgenicity, thyroid effects, estrogenicity, and anti-estrogenicity).

The frameworks modify the hazard level for endocrine activity from *Moderate* to *High* where there is a *High* (or *very High*) plausibly-related adverse effect for carcinogenicity, reproductive toxicity, developmental toxicity and/or systemic toxicity (repeated dose, typically thyroid). The endocrine activity level is not modified where an adverse health effect is not plausibly related.

The DfE and GreenScreen® guidance documents do not indicate whether the frameworks have developed (or use existing) guidance or criteria related specifically to identifying endocrine active chemicals to help ensure that the process is consistent and transparent. The use of hazard identification guidance or criteria, developed a priori and modified as appropriate, may be particularly important for this end point, given the lack of validated tests and the developing nature of the science related to endocrine-related effects of chemicals and their potential adverse health impacts. For example, it is not clear how the various data resources are weighted regarding strength of what data, test results, or combination of test results provide sufficient evidence of endocrine activity.

REFERENCES

AOEC (Association of Occupational and Environmental Clinics). 2009. AOEC Exposure Codes. Description of the AOEC Exposure Code System [online]. Available: http://aoecdata.org/ [accessed April 27, 2014].ATSDR (Agency for Toxic Substances and Disease Registry). 2013. Minimal Risk Levels (MRLs) [online]. Available: http://www.atsdr.cdc.gov/mrls/index.asp [accessed April 28, 2014].

Clean Production Action. 2013. GreenScreen® for Safer Chemicals Chemical Hazard Assessment Procedure V 1.2 Final [online]. Available: http://tcodevelopment.com/files/2014/04/GreenScreenv1-2_Guidance_Assessment_Procedure_FINAL_2013_9_18.pdf [accessed April 27, 2014].

DFG (Deutsche Forschungsgemeinschaft). 2013. List of MAK and BAT Values 2013: Maximum Concentrations and Biological Tolerance Values at the Workplace. Report No. 49. Weinheim, Germany: Wiley-VCH [online]. Available: http://onlinelibrary.wiley.com/book/10.1002/9783527675128[accessed April 27, 2014].

EC (European Commission). 2007. Corrigendum to Regulation (EC) No 1907/2006 of the European Parliament and of the Council of 18 December 2006 concerning the Registration, Evaluation, Authorisation and Restriction of Chemicals (REACH), establishing a European Chemicals Agency. OJ. EU I.136 [online]. Available: http://www.reach-compliance.eu/english/legislation/docs/launchers/launch-2006-1907-EC.html [accessed March 6, 2014].

ECHA (European Chemicals Agency). 2013. Guidance on Information Requirements and Chemical Safety Assessment. Chapter R.7a: Endpoint Specific Guidance, Version 2.3. European Chemicals Agency, Helsinki, Finland.

ECHA (European Chemicals Agency). 2014. Endocrine Disruptor Expert Group [online]. Available: http://echa.europa.eu/addressing-chemicals-of-concern/substances-of-potential-concern/substance-specific-groups/endocrine-disruptor-expert-group [accessed April 28, 2014].

EFSA (European Food Safety Authority). 2013a. Scientific opinion on the hazard assessment of endocrine disruptors: Scientific criteria for identification of endocrine disruptors and appropriateness of existing test methods for assessing effects mediated by these substances on human health and the environment . EFSA J. 11(3):3132 [online]. Available: http://www.efsa.europa.eu/en/efsajournal/doc/3132.pdf [accessed July 23, 2014].

EFSA (European Food Safety Authority). 2013b. EFSA Supports European Commission in Defining Scientific Criteria for Endocrine Disruptors. Press Release: March 20, 2013 [online]. Available: http://www.efsa.europa.eu/en/press/news/130320.htm?wtrl=01[accessed April 28, 2014].

EPA (U.S. Environmental Protection Agency). 2009. Methodology for Risk-Based Prioritization under ChAMP. Office of Pollution Prevention and Toxics, U.S. Environmental Protection Agency [online]. Available: http://www.epa.gov/champ/pubs/hpv/RBPMethodology_Web_April%202009.pdf [accessed July 9, 2014].

EPA (U.S. Environmental Protection Agency). 2011a. Design for the Environment Program Alternatives Assessment Criteria for Hazard Evaluation, Version 2.0. Office of Pollution Prevention and Toxics, U.S. Environmental Protection Agency [online]. Available: http://www.epa.gov/dfe/alternatives_assessment_criteria_for_hazard_eval.pdf [accessed July 9, 2014].

EPA (U.S. Environmental Protection Agency). 2011b. Label Review Manual [online]. Available: http://www.epa.gov/oppfead1/labeling/lrm/label-review-manual.pdf[accessed April 27, 2014].

NIOSH (National Institute for Occupational Safety and Health). 2005. Pocket Guide to Chemical Hazards. DHHS (NIOSH) Publication No. 92-100 [online]. Available: http://www.cdc.gov/niosh/npg/default.html[accessed April 27, 2014].

OEHHA (Office of Environmental Health Hazard Assessment). 2012. Division 4.5, Title 22, California Code of Regulations: Chapter 54. Green Chemistry Hazard Traits for California's Toxics Information Clearinghouse. Office of Environmental Health Hazard Assessment, State of California [online]. Available: http://www.oehha.ca.gov/multimedia/green/gc01 1912.html[accessed April 28, 2014].

OEHHA (Office of Environmental Health Hazard Assessment). 2014. OEHHA Acute, 8-hour and Chronic Reference Exposure Levels as of January 2014. Office of Environmental Health Hazard Assessment, State of California [online]. Available: http://www.oehha.ca.gov/air/allrels.html [accessed April 28, 2014].

OSHA(Occupational Safety and Health Information) 2010. A Guide to The Globally Harmonized System of Classification and Labelling of Chemicals (GHS) [online]. Available: https://www.osha.gov/dsg/hazcom/ghs.html [accessed July 9, 2014].

Quint, J., W.S. Beckett, S.L. Campleman, P. Sutton, J. Prudhomme, J. Flattery, R. Harrison, B. Cowan, and R. Kreutzer. 2008. Primary prevention of occupational asthma: Identifying and controlling exposures to asthma-causing agents. Am. J. Ind. Med. 51(7):477-491.

UNECE (United Nations Economic Commission for Europe). 2011. Globally Harmonized System of Classification and Labelling of Chemicals (GHS), 4th Rev. Ed. New York: United Nations [online]. Available: http://www.unece.org/fileadmin/DAM/trans/dang er/publi/ghs/ghs_rev04/English/ST-SG-AC10-30-Rev4e.pdf [accessed March 29, 2014].

UNECE (United Nations Economic Commission for Europe). 2013. Globally Harmonized System for the Classification and Labeling of Chemicals (GHS). Part 3. Health Hazards. Geneva: United Nations [online]. Available: http://www.unece.org/fileadmin/DAM/trans/dang er/publi/ghs/ghs_rev05/English/03e_part3.pdf [accessed July 9, 2014].